SERVICE DESIGN FOR SIX SIGMA

SERVICE DESIGN FOR SIX SIGMA

A Road Map for Excellence

BASEM EL-HAIK
DAVID M. ROY

A WILEY-INTERSCIENCE PUBLICATION

Published by John Wiley & Sons, Inc., Hoboken, New Jersey
Published simultaneously in Canada.

Library of Congress Cataloging-in-Publication Data is available.

ISBN-13 978-0-471-68291-2
ISBN-10 0-471-68291-8

Printed in the United States of America.

10 9 8 7 6 5 4 3 2 1

To our parents, families and friends
for their continuous support

CONTENTS

PREFACE

Today's service design solutions of current development practices in many industries are generally suffering from deficiencies or vulnerabilities such as modest quality levels, ignorance of customer wants and desires, and too much complexity. These are caused by a lack of a systematic design methodology to address these issues. Such vulnerabilities are common and generate hidden and unnecessary developmental effort, in terms of non-value-added elements, and, later, operational costs as experienced by the customer. Design vulnerabilities manifest themselves in a high degree of customer dissatisfaction, low market share, a rigid organizational structure, and high complexity of operations. Complexity in design creates operational bureaucracies that can be attributed to the lack of adherence to sound design processes. This root cause is coupled with several sources of variation in the service delivery processes, inducing variability in customer attributes, which are commonly known as critical-to-satisfaction characteristics (CTSs).

The success of Six Sigma deployments in many industries has generated enormous interest in the business world. In demonstrating such successes, Six Sigma combines the power of teams and process. The power of teams implies organizational support and trained teams tackling objectives. The power of process means effective Six Sigma methodology deployment, risk mitigation, project management, and an array of statistical and system-based methods. Six Sigma focuses on the *whole quality* of a business. Whole quality includes product or service quality to external customers and also the *operation quality* of all internal processes, such as accounting, billing and so on. A whole-quality business with whole-quality perspectives will not only provide high-quality products or services, but will also operate at lower cost and higher efficiency because all of its business processes are optimized.

Compared with the defect-correction Six Sigma methodology that is characterized by "DMAIC" processes (define, measure, analyze, improve, control), service design for six sigma (identify, characterize, optimize, verify) is proactive. The DMAIC Six Sigma objective is to improve a process without redesigning it.

Design for Six Sigma focuses on design by doing things right the first time—a proactive approach. The ultimate goal of service DFSS is whole quality, that is, do the right things, and do things right all the time. This means achieving absolute excellence in design, whether it is a service process facing the customer or an internal business process facing the employee. Superior service design will deliver superior functions to generate high customer satisfaction. A design for Six Sigma (DFSS) entity will generate a process that delivers the service in a most efficient, economic, and flexible manner. Superior service process design will generate a service process that exceeds customer wants, and delivers with quality and low cost. Superior business process design will generate the most efficient, effective, economical, and flexible business process. This is what we mean by whole quality. That is, not only should we provide superior service, the service design and supporting processes should always deliver what they were intended to do and at Six Sigma quality levels. A company will not survive if it develops some very superior service products, but also develops some poor products as well, leading to inconsistent performance. It is difficult to establish a service business based on a highly defective product.

Service design for Six Sigma (DFSS), as described in this book, proactively produces highly consistent processes with extremely low variation in service performance. The term "Six Sigma" indicates low variation; it means no greater than 3.4 defective (parts) per million opportunities (DPMO[1]) as defined by the distance between the specification limits and the mean, in standard deviation units. We care about variation because customers notice inconsistency and variation, not the averages. Can you recall the last time you experienced the average wait time? Nowadays, high consistency is not only necessary for a sound reputation, it is also a matter of survival. For example, the dispute between Ford and Firestone over tires only involved an extremely small fraction of tires, but the negative publicity and litigation impacted a giant company like Ford significantly.

Going beyond Six Sigma DMAIC, this book will introduce many new methods that add to the effectiveness of service DFSS. For example, key methodologies for managing innovativeness and complexity in design will be introduced. Axiomatic design, design for X, theory of inventive problem solving (TRIZ), transfer function, and scorecards are really powerful methods to create superior service designs; that is, to do the right things within our whole quality perspective.

This book also adds another powerful methodology, the Taguchi method (robust design), to its toolbox. A fundamental objective of the Taguchi method is to create a superior entity that can perform consistently in light of many external disturbances and uncertainties called *noise factors,* thus performing robustly all the time.

Because of the sophistication of DFSS tools, DFSS operative training (Black Belts, Green Belts, and the like) is quite involved. However, this incremental investment is rewarded by dramatically improved results. A main objective of this book is to provide a complete picture of service DFSS to readers, with a focus on supply chain applications.

[1]See Chapter 2 for more details on this terminology.

OBJECTIVES OF THIS BOOK

This book aims to

1. Provide in-depth and clear coverage of philosophical, organizational, and technical aspects of service DFSS to readers.
2. Illustrate very clearly all the service DFSS deployment and execution processes—the DFSS road map.
3. Present the know-how of all the key methods used in service DFSS, discussing the theory and background of each method clearly. Examples are provided, with detailed step-by-step implementation processes for each method.
4. Assist in developing the readers' practical skills in applying DFSS in service environments.

BACKGROUND REQUIRED

The background required to read this book includes some familiarity with simple statistics, such as normal distribution, mean, variance, and simple data analysis techniques.

SUMMARY OF CHAPTER CONTENTS

In Chapter 1, we introduce service design. We highlight how customers experience service, the process through which the service is delivered, and the roles that people and other resources play in this context. We discuss the relationship between different quality tasks and tools and at various stages of service development. This chapter presents the Six Sigma quality concept, the *whole quality,* business excellence, quality assurance, and service life cycle. It provides a detailed chronology of the evolution of quality, the key pioneers in the field, and supply chain applications.

In Chapter 2, we explain what Six Sigma is and how it has evolved over time. We explain that it is a process-based methodology and introduce the reader to process modeling, with a high-level overview of process mapping, value stream mapping, and value analysis, as well as the business process management system (BPMS). The criticality and application of measurement systems analysis (MSA) is introduced. The DMAIC methodology and how it incorporates these concepts into a road map method is also explained, and a design for Six Sigma (DFSS) briefing is presented.

Chapter 3 offers a high-level DFSS process. The DFSS approach as introduced helps design teams frame their project with financial, cultural, and strategic implications to the business. In this chapter, we formed and integrated several strategic, tactical, and synergistic methodologies to enhance service DFSS capabilities and to deliver a broad set of optimized solutions. It highlights and presents the service

DFSS phases: *i*dentify, *c*haracterize, *o*ptimize, and *v*erify, or ICOV for short. In this book, the ICOV and DFSS acronyms will be used interchangeably.

In Chapter 4, we discuss the deployment of a service DFSS initiative starting from a white paper. We present the deployment plan, roles, and responsibilities of deployment operatives, project sources, and other aspects of sound deployment strategy in three phases: predeployment, initial deployment, and steady-state deployment. We also discuss certain desirable characteristics of design teams and offer several perspectives on cultural transformation and initiative sustainability.

In Chapter 5, we present the service design for a Six Sigma project road map. The road map highlights at a high-level the *i*dentify, *c*harcaterize, *o*ptimize, and *v*alidate phases over the seven development stages (idea creation, voice of the customer and business, concept development, preliminary design, design optimization, verification, launch readiness). In this chapter, the concept of the tollgate is introduced. We also highlight the most appropriate DFSS tools and methods for each DFSS phase, indicating where it is most appropriate to start tool usage. The methods are presented in the subsequent chapters.

In Chapter 6, the transfer function and design scorecard tools are introduced. The use of these DFSS tools parallels the design mappings. A transfer function is a mathematical relationship relating a design response to design elements. A design scorecard is used to document the transfer function as well the performance.

In Chapter 7, quality function deployment (QFD) is presented. It is used to translate customer needs and wants into focused design actions, and parallels design mapping as well. QFD is key to preventing problems from occurring once the design is operational. The linkage to the DFSS road map allows for rapid design cycles and effective utilization of resources while achieving Six Sigma levels of performance.

Design mapping is a design activity that is presented in Chapter 8. The service DFSS project road map recognizes two different mappings: the functional mapping and the process mapping. In this chapter, we present the functional mapping as a logical model, depicting the logical and cause–effect relationships between design elements through techniques such as axiomatic design and value engineering. A process map is a visual aid for picturing work processes; it shows how inputs, outputs, and tasks are linked. In this chapter, we feature the business process management system (BPMS), an effective tool for improving overall business performance within the design context. The Pugh concept selection method is used after design mapping to select a winning concept for further DFSS road map processing.

The use of creativity methods such as the theory of problem solving (TIPS, also known as TRIZ) in service DFSS is presented in Chapter 9. TRIZ, based on the discovery that there are only 40 unique innovative principles, provides design teams a priceless toolbox for innovation so they can focus on the true design opportunity and provide principles to resolve, improve, and optimize concepts. TRIZ is a useful innovative problem solving method that, when applied successfully, replaces the trial-and-error method in the search for vulnerability-free concepts. It is the ultimate library of lessons learned. TRIZ-based thinking for management tasks helps to identify the technology tools that come into play, such as innovation principles for business and management, separation principles for resolving organizational contradictions and conflicts, operators for revealing and utilizing system resources, and patterns of evolution of technical systems to support conceptual optimization.

In Chapter 10, we introduce the concept of design for X (DFX) as it relates to service transactions and builds from the work performed for product design. In this context, we show that DFX for service requires that the process content be evaluated, much the same as in assembly processes, to minimize complexity and maximize commonality. The end result will be a robust design that meets the customer's needs profitably, through implementation of methods such as design for serviceability, processability, and inspectability.

Chapter 11 discusses failure mode and effect analysis (FMEA). FMEA is a very important design review method to remove potential failures in the various design stages. We discuss all aspects of FMEA, as well as the difference between design FMEA and process FMEA and the linkages to service DFSS road maps.

In Chapter 12, we present the service DFSS approach to the design of experiments (DOE), a prime optimization tool, with many service-related examples. DOE is a structured method for determining the transfer function relationship between factors affecting a process and the output of that process. DOE refers to experimental methods used to quantify indeterminate measurements of factors and interactions between factors statistically through observance of forced changes made methodically, as directed by systematic tables called design arrays. The main DOE data analysis tools include analysis of variance (ANOVA), empirical transfer function model building, and main effects and interaction charts.

Chapter 13 presents the employment of robust design methodology in service design environments. Thinking about robustness helps the DFSS team classify design parameters and process variables mapped into the design as controlled and uncontrolled. The objective is to desensitize the design to the uncontrolled disturbance factors, also called noise factors, thus producing a consistently performing, on-target design with minimal variation.

The Discrete event simulation (DES) technique presented in Chapter 14 is a powerful method for business process simulation of a transactional nature within DFSS and Six Sigma projects. A DES provides modeling of service entity flows with capabilities that allow the design team to see how flow objects are routed through the process. DES leads to growing capabilities, software tools, and a wide spectrum of real-world applications in DFSS.

In Chapter 15, we present validation as a critical step in the DFSS road map and discuss the need for it to be addressed well in advance of production of a new design. The best validation occurs when it is done as near to production configuration and operation as possible. Service design validation often requires prototypes that need to be near "final design" but are often subject to trade-offs in scope and completeness due to cost or availability. Once prototypes are available, a comprehensive test plan should be followed in order to capture any special event and to populate the design scorecard, and should be based on statistically significant criteria. This chapter concludes the DFSS deployment and core methods that were presented in Chapters 1 through 15. The last two chapters present the supply chain thread through a design case study.

Chapter 16 discusses the supply chain process that covers the life cycle of understanding customer needs to producing, distributing, and servicing the value chains from customers to suppliers. We describe how supply chains apply to all contexts of acquiring resources to be transformed into value for customers. Because of its broad applicability to all aspects of consumption and fulfillment, the supply chain is the ultimate "service" for design consideration. A case study is presented in Chapter 17.

In Chapter 17, we apply the DFSS road map to accelerate the introduction of new processes and align the benefits for customers and stakeholders. In this supply chain case study, we describe how service DFSS tools and tollgates allow for risk management, creativity, and a logical documented flow that is superior to the "launch and learn" mode that many new organizations, processes, or services are deployed with. Not all projects will use all of the complete toolkit of DFSS tools and methodologies, and some will use some to a greater extent than others. In the supply chain case study, the design scorecard, quality function deployment, and axiomatic design applications are discussed, among others.

WHAT DISTINGUISH THIS BOOK FROM OTHERS IN THE AREA?

This book is the first to address service design for Six Sigma and to present an approach to applications via a supply chain design case study. Its main distinguishing feature is its completeness and comprehensiveness, starting from a high-level overview of deployment aspects and the service design toolbox. Most of the important topics in DFSS are discussed clearly and in depth. The organizational, implementation, theoretical, and practical aspects of both the DFSS road map and DFSS toolbox methods are covered very carefully and in complete detail. Many of the books in this subject area give only superficial descriptions of DFSS without any details. This is the only book that discusses all service DFSS perspectives, such as transfer functions, axiomatic design,[2] and TRIZ and Taguchi methods in great detail. The book can be used either as a complete reference book on DFSS, or as a complete training manual for DFSS teams. We remind readers that not every project requires full use of every tool.

With each copy of this book, purchasers can access a copy of Acclaro DFSS Light® by downloading it from the Wiley ftp site. This is a training version of the Acclaro DFSS software toolkit from Axiomatic Design Solutions, Inc. (ADSI[3]) of Brighton, MA. Under license from MIT, ADSI is the only company dedicated to supporting axiomatic design methods with services and software solutions. Acclaro

[2]Axiomatic design is a methodology used by individuals as well as Fortune 100 design organizations. The axiomatic design process aids design and development organizations in diverse industries including automotive, aerospace, semiconductor, medical, government, and consumer products.

[3]Browse their site at http://www.axiomaticdesign.com/default.asp.

software, a Microsoft Windows-based solution implementing DFSS quality frameworks around axiomatic design processes, won *Industry Week*'s Technology of the Year award. Acclaro DFSS Light® is a JAVA-based software package that implements axiomatic design processes as presented in Chapter 8.

John Wiley & Sons maintains an ftp site at: ftp://ftp.wiley.com/public/sci_tech_med/six_sigma

ACKNOWLEDGMENTS

In preparing this book, we received advice and encouragement from several people. We are thankful to Peter Pereira, Sheila Bernhard, Sherly Vogt, Eric Richardson, Jeff Graham, and Mike Considine. We are also thankful to Dr. Raid Al-Aomar of Jordan University of Science and Technology (JUST) for his contribution to Chapter 14. The authors are appreciative of the help of many individuals, including George Telecki and Rachel Witmer of John Wiley & Sons, Inc. We are very thankful to Invention Machine Inc. for their permission to use TechOptimizer™ software and to Generator.com for many excellent examples in Chapter 9.

CONTACTING THE AUTHORS

Your comments and suggestions about this book are greatly appreciated. We will give serious considerations to your suggestions for future editions. We also conduct public and in-house Six Sigma and DFSS workshops and provide consulting services. Dr. Basem El-Haik can be reached via e-mail at basemhaik@hotmail.com. Dave Roy can be reached via e-mail at gundroy@cox.net.

software, a Microsoft Windows-based [illegible] implementing DESS quality [illegible] works around [illegible] design processes [illegible]. The [illegible] of the [illegible] DESS [illegible]-based software package that implements [illegible] design [illegible] in Chapter 8.

John Wiley & Sons maintains [illegible]

ACKNOWLEDGMENTS

[illegible]

1

SERVICE DESIGN

1.1 INTRODUCTION

Throughout the evolution of quality control processes, the focus has mainly been on manufacturing (parts). In recent years, there has been greater focus on process in general; however, the application of a full suite of tools to service design is rare and still considered risky or challenging. Only companies that have mature Six Sigma deployment programs see the application of design for Six Sigma (DFSS) to processes as an investment rather than a needless expense. Even those companies that embark on DFSS for processes seem to struggle with confusion over the DFSS "process" and the process being designed.

There are multiple business processes that can benefit from DFSS. A sample of these are listed in Table 1.1.

If properly measured, we would find that few if any of these processes perform at Six Sigma performance levels. The cost per transaction, timeliness, or quality (accuracy, completeness) are never where they should be and hardly world class.

We could have chosen any one of the processes listed in Table 1.1 for the common threaded service DFSS book example but we have selected a full supply chain process because it either supports each of the other processes or is analogous in its construct. Parts, services, people, or customers are all in need of being sourced at some time.

A service is typically something that we create to serve a paying customer. Customers may be internal or external; if external, the term consumer (or end user) will be used for clarification purposes. Some services, for example dry cleaning, consist of a single process, whereas many services consist of several processes linked together. At each process, transactions occur. A transaction is the simplest process step and typically consists of an *input, procedures, resources* and a resulting *output*. The resources can be people or machines and the procedures can be written, learned or even digitized in software code. It is important to understand that some services are enablers to other services, whereas some provide their output to the end customer.

Table 1.1 Examples of organizational functions

Marketing	Sales	Human Resources	Design
• Brand Management • Prospect	• Discovery • Account Management	• Staffing • Training	• Change Control • New Product
Production Control	**Sourcing**	**Information Technology**	**Finance**
• Inventory Control • Scheduling	• Commodity • Purchasing	• Help Desk • Training	• Accounts Payable • Accounts Receivable

For example, the transaction centered around the principal activities of an order-entry environment include transactions such as entering and delivering orders, recording payments, checking the status of orders, and monitoring the stock levels at the warehouse. Processes may involve a mix of concurrent transactions of different types and complexity either executed on-line or queued for deferred execution.

Services span the range from ad hoc to designed. Our experience indicates that the vast majority of services are ad hoc and have no metrics associated with them, and many consist solely of a person with a goal and objectives. These services have large variation in their perceived quality and are very difficult to improve; doing so is akin to building a house on a poor foundation.

Services affect almost every aspect of our lives. They include restaurants, health care, financial, transportation, entertainment, and hospitality, and they all have the same elements in common.

1.2 WHAT IS QUALITY?

We all use services and interact with processes each day. When was the last time you remember feeling really good about a service? What about the last poor service you received? It is usually easier for us to remember the painful and dissatisfying experiences than it is to remember the good ones. One of the authors recalls sending a first-class registered letter and after eight business days he still could not see that the letter was received, so he called the postal service provider's toll-free number and had a very professional and caring experience. It is a shame they could not provide the same level of service when delivering a simple letter. It turns out that the letter was delivered but their system failed to track it. So how do we measure quality for services?

In a traditional manufacturing environment, conformance to specification and delivery are the common quality items that are measured and tracked. Often times, lots are rejected because they do not have the correct documentation supporting them. Quality in manufacturing, then, is a conforming product, delivered on time and having all of the supporting documentation. In services, quality is measured as conformance to expectations, availability, and experience of the process and people interacting with the service delivery.

If we look at Figure 1.1, we can see that customers experience service in three ways:

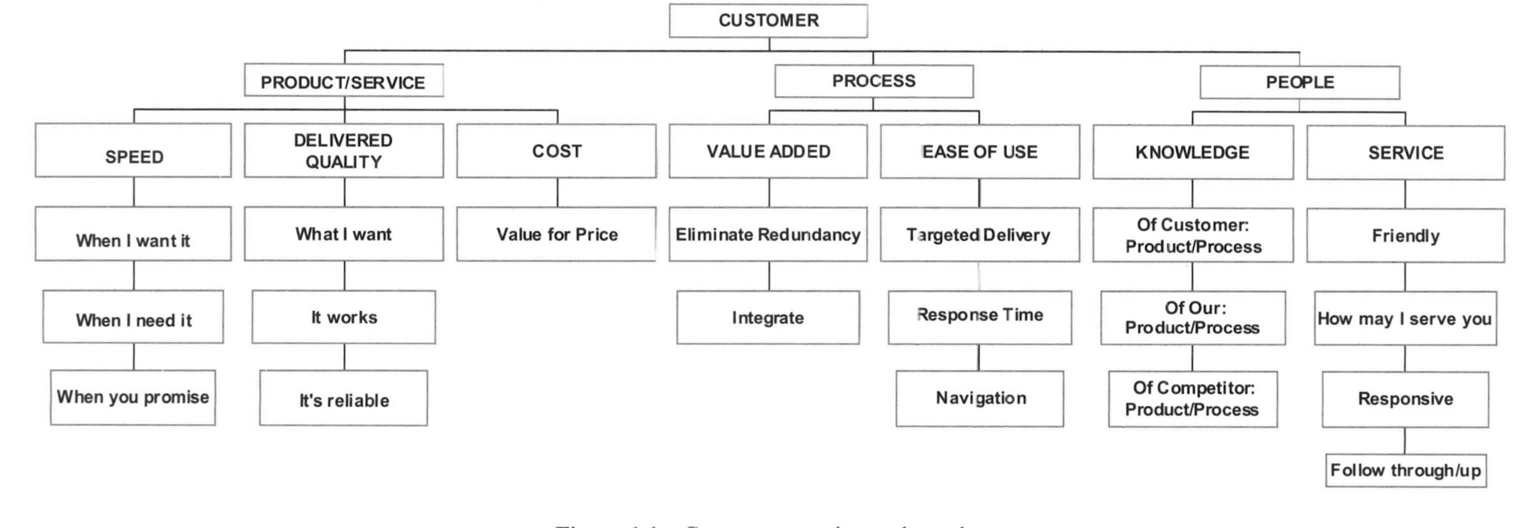

Figure 1.1 Customer experience channels.

1. The specific service or product has attributes such as availability, it's what I wanted, and it works.
2. The process through which the service is delivered can be ease of use or value added.
3. The people (or system) should be knowledgeable and friendly.

To fulfill these needs, there is a service life cycle to which we apply a quality operating system.

1.3 QUALITY OPERATING SYSTEM AND SERVICE LIFE CYCLE

To deliver a high-performing service, we need a system of methods and activities that can provide the overarching structure to successfully plan and develop the service. Such a system is called a quality operating system. The quality operating system includes all the planned and systematic activities performed within the system that can demonstrate with confidence that the service will fulfill the requirements for quality. Figure 1.2 depicts a graphical flow of the service life cycle that encompasses the service life cycle from ideation through phase out/retirement. Below, we enumerate the service life cycle stages.

1.3.1 Stage 1: Idea Creation

The need for a new process or service can come from benchmarking, technology road maps, or multigenerational plans (see Chapter 4). Many times, new processes come about because of "revolution," not "evolution." For example, when a new management team is brought in and they staff the organization with knowledgeable people to execute the new strategies and methods, the switching costs are often huge and it takes time for the new process to start delivering benefit. The change brought on by the new team is a revolution compared to the case in which the legacy team is able to evolve slowly.

It is the premise of this book that, based on performance metrics and benchmarking, natural evolution via DFSS deployment can provide process redesign that is manageable and controllable.

1.3.2 Stage 2: Voice of the Customer and Business

Customer and business requirements must be studied and analyzed in this second stage, even in a redesign environment. We need to understand the key functional requirements (in a solution-free environment) that will fulfill stated or implied needs of both external and internal customers (the business). We also need to understand the relationships between the voice of the customer and the voice of the business. The quality function deployment (QFD) house of quality is an ideal method for this purpose.

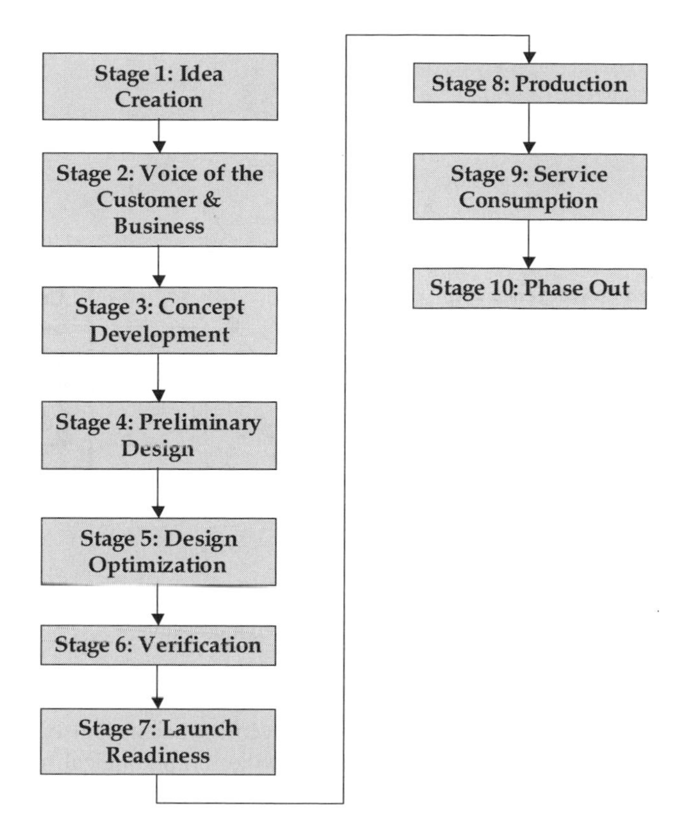

Figure 1.2 Service life cycle. (See Yang and El-Haik (2003) for the product life cycle equivalent.)

1.3.3 Stage 3: Concept Development

Concept development is the third stage in the service life cycle. In this stage, concepts are developed that fulfill the functional requirements obtained from the previous stage. This stage of the life cycle process is still at a high level and remains solution free; that is, the design team is able to specify "*what* needs to be accomplished to satisfy the customer wants" and not "*how* to accomplish these wants." The strategy of the design team is to create several innovative concepts and use selection methodologies to narrow down the choices. At this stage, we can highlight the Pugh concept selection method (Pugh, 1996) as summarized in Figure 1.3.

The method of "controlled convergence" was developed by Dr. Stuart Pugh (Pugh, 1991) as part of his solution selection process. Controlled convergence is a solution-iterative selection process that allows alternate convergent (analytic) and divergent (synthetic) thinking to be experienced by the service design team. The method alternates between generation and convergence selection activities.

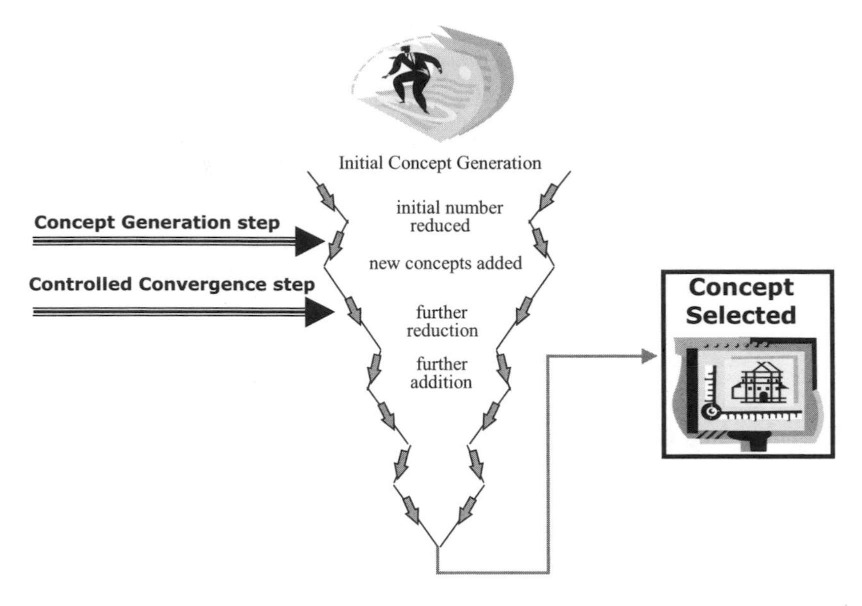

Figure 1.3 Pugh phased innovation.

Tools such as TRIZ (theory of Russian inventive science, also known as TIPS for theory of inventive problem solving) and the morphological matrix are better suited to the creative process, whereas the Pugh selection matrix helps with critical selection. TRIZ will be explored in Chapter 9.

1.3.4 Stage 4: Preliminary Design

In Stage 4, the prioritized functional requirements must be translated into design parameters with detail specifications. Appropriate tools for this purpose are QFD or axiomatic design[1] (see Chapter 8).

A preliminary design, which could consist of a design structure (architecture or hierarchy) with subsystem requirements flow-down, should be developed in this phase. QFD (see Chapter 7) and process modeling (see Chapter 8) are very beneficial at this stage. Design targets for reliability, quality, processability, and ease of use are established in Stage 4. When one potential design solution has been selected, the team can begin to focus on the specific failure modes of that design using a services design failure modes and effects analysis (DFMEA). Concurrently, from all of these design requirements the first elements for inclusion in the design scorecard (see Chapter 6) can be identified and recorded.

[1]Download the axiomatic design software "Acclaro DFSS Light®" from (AUTHOR- PROVIDE URL). See Section 8.5. Acclaro software is protected under copyright and patents pending. Acclaro is a registered trademark of Axiomatic Design Solutions, Inc.

1.3.5 Stage 5: Design Optimization

In stage 5, the design team will make sure the final design matches the customer requirements that were originally identified in Stage 2 (capability flow-up to meet the voice of the customer). There are techniques [DFx (see Chapter 10), Poka Yoke (see Chapter 10), FMEA (see Chapter 11)] that can be used at this point to ensure that the design cannot be used in a way that was not intended, processed or maintained incorrectly, or that if there is a mistake it will be immediately obvious.

A final test plan is generated in order to assure that all requirements are met at the Six Sigma level by the pilot or prototype that is implemented or built in the next stage.

In Stage 5, detail designs are formulated and tested either physically or through simulation. Functional requirements are flowed down from the system level into subsystem design parameters using transfer functions (see Chapter 6), process modeling (see Chapter 8), and design of experiments (see Chapter 12). Designs are made robust to withstand the "noise" introduced by external uncontrollable factors (see Chapter 13). All of the activities in Stage 5 should result in a design that can be produced in a pilot or prototype form.

1.3.6 Stage 6: Verification

Test requirements and procedures are developed and the pilot is implemented and/or the prototype is built in this stage. The pilot is run in as realistic a setting as possible with multiple iterations and subjected to as much "noise" as possible in an environment that is as close to its final usage conditions as possible. The same philosophy applies to the testing of a prototype. The prototype should be tested at the extremes of its intended envelope and sometimes beyond. To the extent possible or allowed by regulation, simulation should replace as much testing as is feasible in order to reduce cost and risk.

The results of pilot or prototype testing allow the design team the opportunity to make final adjustments or changes in the design to ensure that product, service, or business process performance is optimized to match customer expectations.

In some cases, only real-life testing can be performed. In this situation, design of experiments is an efficient way to determine if the desired impact is created and confirmed.

1.3.7 Stage 7: Launch Readiness

Based on successful verification in a production environment, the team will assess the readiness of all of the process infrastructure and resources. For instance, have all standard operating procedures been documented and people been trained in the procedures? What is the plan for process switch-over or ramp-up? What contingencies are in place? What special measures will be in place to ensure rapid discovery? Careful planning and understanding the desired behavior is paramount to successful transition from the design world into the production environment.

1.3.8 Stage 8: Production

In this stage, if the team has not already begun implementation of the design solution in the production or service environment, the team should do so now. Validation of some services or business processes may involve approval by regulatory agencies (e.g., approval of quality systems) in addition to your own rigorous assessment of the design capability. Appropriate controls are implemented to ensure that the design continues to perform as expected and anomalies are addressed as soon as they are identified (update the FMEA) in order to eliminate waste, reduce variation, and further error-proof the design and any associated processes.

1.3.9 Stage 9: Service Consumption

Whether supporting the customers of the service with help lines or the production processes, which themselves need periodic maintenance (e.g., toner cartridges in a printer or money in an ATM), the support that will be required and how to provide it are critical in maintaining Six Sigma performance levels. Understanding the total life cycle of the service consumption and support are paramount to planning adequate infrastructure and procedures.

1.3.10 Stage 10: Phase-Out

Eventually, all products and services become obsolete and are either replaced by new technologies or new methods. Also, the dynamic and cyclical nature of customer attributes dictates continuous improvement to maintain adequate market share. Usually, it is difficult to turn off the switch, as many customers have different dependencies on services and processes. Just look at the bank teller and the ATM machine. One cannot just convert to a single new process. There must be a coordinated effort, and often change management is required to provide incentives for customers to shift to the new process. In terms of electronic invoicing, a discount may be offered for use of the electronic means or an extra charge imposed for the nonstandard method.

1.3.11 Service Life Cycle and Quality Operating System

The service life cycle is depicted in Figure 1.2. In addition, Table 1.2 highlights the key DFSS tools and objectives of service life cycle stages. The DFSS topics in this book span the first seven phases of the life cycle. Opportunities still exist for application in the remainder of the stages; however, the first seven have the highest potential life cycle impact.

1.4 DEVELOPMENTS OF QUALITY IN SERVICE

The earliest societies—Egyptians, Mayans, and Aztecs—provide archeological evidence of precision and accuracy nearly unmatched today. Following the example of

Table 1.2 Life cycle key DFSS objectives and methods*

Design life cycle stages	Stage objective	Key service DFSS methods/tools
1. Ideation creation	Ensure that new technology/ ideas are robust for downstream development	Market and customer research Technology road maps Growth strategy
2. Customer and business requirements study	Ensure new design concept to come up with right functional requirements that satisfy customer needs	QFD Kano analysis Customer engagement methods (survey, interview, etc.) Multigeneration plan
3. Concept development	Ensure that the new concept can lead to sound design and be free of design vulnerabilities Ensure the new concept to be robust for downstream development	QFD TRIZ Design mapping (process and functional mappings) Failure mode and effect analysis (FMEA) Design for serviceability Simulation/optimization Design scorecard
4. Preliminary design	Design parameters with detailed specification Design targets for reliability, quality, processability, and ease of use	Design mapping (process and functional mappings) Creativity tools: TRIZ and axiomatic design FMEA Pugh concept selection Robust design Design reviews Process managemnt
5. Design optimization	Capability flow-up to prove customer needs are met Final test plan	Transfer functions detailing (DOE, simulation, etc.) Taguchi Robust design Reliability Simulation Change managemnt Process capability Design scorecard Mistake proofing Hypothesis testing
6. Verification	Ensure designed product (design parameters) to deliver desired product functions over its useful life Ensure the process/service design to be robust for variations in production, consumption, and disposal stages	Piloting/prototyping DOE Simulation Process capability Confidence intervals Sample size Hypothesis testing

(*continued*)

Table 1.2 Life cycle key DFSS objectives and methods*

Design life cycle stages	Stage objective	Key service DFSS methods/tools
7. Launch readiness	Ensure the process to be able to deliver designed service consistently	Control plans Statistical proces control (SPC) Transition planning Mistake proofing Trouble shooting and diagnosis Training plans
8. Production	Produce designed service with high degree of consistency and free of defects	SPC Inspection Mistake proofing
9. Service consumption	Ensure that customer will have a satisfactory experience in consumption	Quality in after-sale service Service quality
10 Phase-out	Ensure that customer will be trouble free when disposing of used design	

*This table is replicated as a DFSS project road map in Figure 5.1.

these societies, the world entered into an extended period of "apprenticeship" in which we developed conformance to customer requirements with never more than one degree of separation between the producer and the customer. During the industrial revolution, societies began to separate the producers from the consumers, and this predicated the discovery and development of quality methodologies to improve the customer experience. These practices evolved around products based processes, and during the era of globalization and services growth, transference began to evolve into the services processes.

There are three components that drive the evolution of quality. These components are (1) knowledge, (2) technology, and (3) resources. The basic knowledge of quality philosophy, quality methods, and quality tools precedes the automation of these tools via technology and then is followed by their general awareness and adoption by practitioners.

In the early days of pioneer Walter A. Shewhart, slide rules were the prevalent technology and even the simplest calculations were tedious. The high level of effort required for calculations resulted in simplification of statistical process control (SPC) by the use of X-bar and R-charts and prevented rapid adoption of statistical process control. Today, we have mature knowledge with automated data capture systems and the ability to analyze large datasets with personal computers and statistical software. Today's resources have higher math skills then the average person in Shewhart's time and the penetration of quality methods has expanded into customer-centered support processes as well as product-based processes. The adoption of enabling processes such as human resources, supply chain, legal, and sales, although analogous to customer-centered processes, is weak due to a perceived cost benefit deficit and a lack of process-focused metrics in these processes.

Let us look at an abbreviated chronological review of some of the pioneers who added much to our knowledge of quality. Much of the evolution of quality has occurred in the following five disciplines:

1. Statistical analysis and control
2. Root cause analysis
3. Total quality management
4. Design quality
5. Process simplification

The earliest evolution began with statistical analysis and control, so we start our chronology there.

1.4.1 Statistical Analysis and Control

In 1924, Walter A. Shewhart introduced the first application of control charting to monitor and control important production variables in a manufacturing process. This charting method introduced the concepts of special cause and common cause variation. He developed his concepts and published *Economic Control of Quality of Manufactured Product* in 1931, which successfully brought together the disciplines of statistics, engineering, and economics, and with this book, Shewhart became known as the father of modern quality control. Shewhart also introduce the plan-do-study-act cycle (Shewhart cycle) later made popular by Deming as the PDCA cycle.

In 1925, Sir Ronald Fisher published the book, *Statistical Methods for Research Workers* (Fisher, 1925), and introduced the concepts of randomization and the *an*alysis *o*f *va*riance (ANOVA). Later in 1925 he wrote *Design of Experiments* (DOE). Frank Yates was an associate of Fisher and contributed the Yates standard order for ANOVA calculations. In 1950, Gertrude Cox and William Cochran coauthored *Experimental Design,* which became the standard of the time. In Japan, Dr. Genechi Taguchi introduced orthogonal arrays as an efficient method for conducting experimentation within the context of robust design. He followed this up in 1957 with his book, *Design of Experiments.* Taguchi robustness methods have been used in product development since the 1980s. In 1976, Dr. Douglas Montgomery published *Design and Analysis of Experiments.* This was followed by George Box, William Hunter, and Stuart Hunter's *Statistics for Experimenters* in 1978.

1.4.2 Root Cause Analysis

In 1937, Joseph Juran introduced the Pareto principle as a means of narrowing in on the vital few. In 1943, Kaoru Ishikawa developed the cause and effect diagram, also known as the fishbone diagram. In 1949, the use of multivari charts was promoted first by Len Seder of Gillette Razors and then service marked by Dorian Shainin, who added it to his Red X toolbox, which became the Shainin techniques from 1951 through 1975. Root cause analysis as known today relies on seven basic tools: the cause and effect diagram, check sheet, control chart (special cause verses common cause), flowchart, histogram, Pareto chart, and scatter diagram.

1.4.3 Total Quality Management/Control

The integrated philosophy and organizational alignment for pursuing the deployment of quality methodologies is often referred to as total quality management. The level of adoption has often been directly related to the tools and methodologies referenced by the thought leaders who created the method and tools as well as the perceived value of adopting these methods and tools. Dr. Armand V. Feigenbaum published *Total Quality Control* while still at MIT pursuing his doctorate in 1951. He later became the head of quality for General Electric and interacted with Hitachi and Toshiba. His pioneering effort was associated with the translation into Japanese of his 1951 book, *Quality Control: Principles, Practices and Administration,* and his articles on total quality control.

Joseph Juran followed closely in 1951 with the *Quality Control Handbook,* the most comprehensive "how-to" book on quality ever published. At this time, Dr W. Edwards Deming was gaining fame in Japan following his work for the U.S. government in the Census Bureau developing survey statistics, and published his most famous work, *Out of the Crisis,* in 1968. Dr. Deming was associated with Walter Shewhart and Sir Ronald Fisher and has become the most notable TQM proponent. Deming's basic quality philosophy is that productivity improves as variability decreases, and that statistical methods are needed to control quality. He advocated the use of statistics to measure performance in all areas, not just conformance to design specifications. Furthermore, he thought that it is not enough to meet specifications; one has to keep working to reduce the variations as well. Deming was extremely critical of the U.S. approach to business management and was an advocate of worker participation in decision making. Later, Kaoru Ishikawa gained notice for his development of "quality circles" in Japan and published the *Guide to Quality Control* in 1968. The last major pioneer is Philip Crosby, who published *Quality is Free* in 1979, in which he focused on the "absolutes" of quality, the basic elements of improvement, and the pursuit of "Zero Defects."

1.4.4 Design Quality

Design quality includes philosophy and methodology. The earliest contributor in this field was the Russian Genrich Altshuller, who provided us with the theory of inventive problem sSolving (TIPS or TRIZ) in 1950. TRIZ is based on inventive principles derived from the study of over 3.5 million of the world's most innovative patents and inventions. TRIZ provides a revolutionary new way of systematically solving problems based on science and technology. TRIZ helps organizations use the knowledge embodied in the world's inventions to quickly, efficiently, and creatively develop "elegant" solutions to their most difficult design and engineering problems. The next major development was the quality function deployment (QFD) concept, promoted in Japan by Dr. Yoji Akao and Shigeru Mizuno in 1966 but not westernized until the 1980s. Their purpose was to develop a quality assurance method that would design customer satisfaction into a product before it was manufactured. Prior quality control methods were primarily aimed at fixing a problem during or after manufacturing. QFD is a structured approach to defining customer needs or requirements and translating them into specific plans to produce products or services to meet those needs. The "voice of the customer" is the term used to describe these stated and unstated customer needs or requirements.

In the 1970s Dr. Taguchi promoted the concept of the quality loss function, which stated that any deviation from nominal was costly and that designing with the noise of the system the product would operate within, one could optimize designs. Taguchi packaged his concepts in the methods named after him, also called robust design or quality engineering.

The last major development in design quality was by Dr. Nam P. Suh, who formulated the approach of axiomatic design. Axiomatic design is a principle-based method that provides the designer with a structured approach to design tasks. In the axiomatic design approach, the design is modeled as mapping between different domains. For example, in the concept design stage, it could be a mapping between customer attribute domains to design the function domain; in the product design stage, it is a mapping from function domain to design parameter domain. There are many possible design solutions for the same design task. However, based on its two fundamental axioms, the axiomatic design method developed many design principles to evaluate and analyze design solutions and give designers directions to improve designs. The axiomatic design approach not only can be applied in engineering design, but also in other design tasks such as organization systems.

1.4.5 Process Simplification

Lately, "lean" is the topic of interest. The pursuit of the elimination of waste has led to several quality improvements. The earliest development in this area was "poka yoke" (mistake proofing), developed by Shigeo Shingo in Japan in 1961. The essential idea of poka-yoke is to design processes in a way that mistakes are impossible to make or at least easily detected and corrected. Poka-yoke devices fall into two major categories: prevention and detection. A prevention device affects the process in such a way that it is impossible to make a mistake. A detection device signals the user when a mistake has been made, so that the user can quickly correct the problem. Shingo later developed the single minute exchange of die (SMED) in 1970. This trend has also seen more of a system-wide process mapping and value analysis, which has evolved into value stream maps.

1.4.6 Six Sigma and Design For Six Sigma (DFSS)

The initiative known as Six Sigma follows in the footstep of all of the above. Six Sigma was conceptualized and introduced by Motorola in the early 1980s. It spread to Texas Instruments and Asea Brown Boveri before Allied Signal and then to GE in 1995. It was enabled by the emergence of the personal computer and statistical software packages such as Minitab™, SAS™, BMDP™, and SPSS™. It combines each of the elements of process management and design. The define-measure-analyze-improve-control (DMAIC) and design for six sigma (DFSS) processes will be discussed in detail in Chapter 2 and Chapter 3, respectively.

1.5 BUSINESS EXCELLENCE: A VALUE PROPOSITION?

At the highest level, business excellence is characterized by good profitability, business viability, and growth in sales and market share, based on quality (Peters, 1982). Achieving business excellence is the common goal for all business leaders and their employees. To achieve business excellence, design quality by itself is not sufficient; quality has to be replaced by "*whole quality,*" which includes quality in business operations such as those in Table 1.1. We will see this in our exploration of supply chain design throughout the book. To understand business excellence, we need to understand business operations and other metrics in business operations, which we cover in the next section.

1.5.1 Business Operation Model

Figure 1.4 shows a typical high-level business operation model for a manufacturing-based company. For companies that are service oriented, the business model

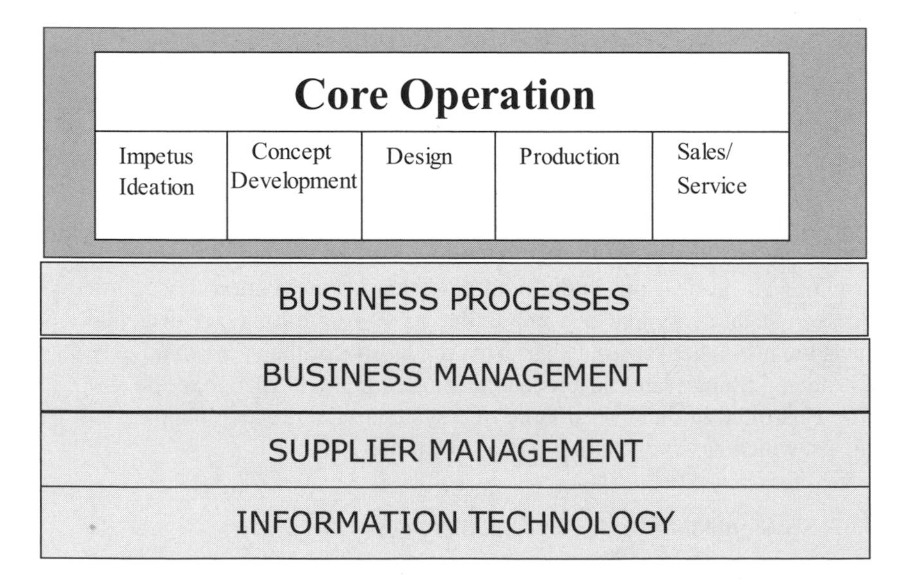

Figure 1.4 Typical business operation model.

could look somewhat different. However, for every company, there is always a "core operation" and a number of other enabling business elements. The core operation is the collection of all activities (processes and actions) used to provide service designs to customers. For example, the core operation of Federal Express is to deliver packages around the world, and the core operation of Starbucks is to provide coffee service all over the world. Core operations extend across all activities in the service design life cycle.

For a company to operate, the core operation alone is not enough. Figure 1.4 listed several other typical elements that are needed in order to make a company fully operational, such as business process and business management. The success of the company depends on the success of all aspects of the business operation. In addition to the structure depicted in Figure 1.4, each of these functions also has a "life cycle" of its own, as shown in Figure 1.5. Each of the blocks from Figure 1.4 can be dropped into the function chevron of Figure 1.5 and each of these functions requires the other chevrons of strategy, and planning, training and organizational development, and reporting to support their core function.

Before Six Sigma, quality was narrowly defined as the quality of the design that the company provided to external customers; therefore, it only related to the core operation. Clearly, from the point of view of a business leader, this "quality" is only part of the story, because other critical factors for business success, such as cost, profit, time to market, capital acquisition, and so on, are also related to other aspects of the business operation.

The key difference between Six Sigma and all other previously developed quality systems and methods, such as TQM, is that Six Sigma is a strategy for the *whole quality* (every quality dimension concurrently), which is a dramatic improvement for the *whole business operation.*

The following sections will show that improving whole quality will lead to business excellence, because improving whole quality means improving all major performance metrics of business excellence, such as profit, cost, and time to market.

1.5.2 Quality and Cost

Given that you have a viable product or service, low cost is directly related to high profitability. Cost can be roughly divided into two parts: life cycle costs related to all service designs offered by the company, and the cost of running the supporting functions within the company, such as various enabling operations in related de-

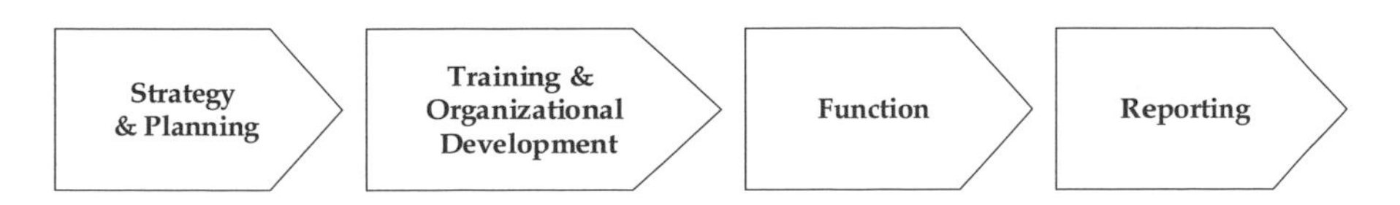

Figure 1.5 Business functional core operation and auxiliary requirements model.

partments. For a particular product or service, life cycle cost includes production/service cost plus the cost for design development.

The relationship between quality and cost is rather complex; the "quality" in this context is referred to as the design quality, not the whole quality. This relationship is very much dependent on what kind of quality strategy is adopted by a particular company. If a company adopted a quality strategy heavily focused on the downstream part of the design life cycle, that is, fire fighting, rework, and error corrections, then that "quality" is going to be very costly. If a company adopted a strategy emphasizing upstream improvement and problem prevention, then improving quality could actually reduce the life cycle cost because there will be less rework, less recall, less fire fighting, and, therefore, less design development cost. In the service-based company, it may also mean fewer complaints, higher throughput, and higher productivity. For more discussion on this topic, see Chapter 3 of Yang & El-Haik (2003).

If we define quality as the "whole quality," then higher whole quality will definitely mean lower total cost. Because whole quality means higher performance levels of all aspects of the business operation, it means high performance of all supporting functions, high performance of the production system, less waste, and higher efficiency. Therefore, it will definitely reduce business operation cost, production cost, and service cost without diminishing the service level to the customer.

1.5.3 Quality and Time to Market

Time to market is the time required to introduce new or improved products and services to the market. It is a very important measure for competitiveness in today's marketplace. If two companies provide similar designs with comparable functions and price, the company with the faster time to market will have a tremendous competitive position. The first company to reach the market benefits from a psychological effect that will be very difficult to be matched by latecomers.

There are many techniques to reduce time to market, such as:

- Concurrency: encouraging multitasking and parallel working
- Complexity reduction (see Suh (2001) and El-Haik (2005))
- Project management: tuned for design development and life cycle management

In the Six Sigma approach and whole quality concept, improving the quality of managing the design development cycle is a part of the strategy. Therefore, improving whole quality will certainly help to reduce time to market.

1.6 INTRODUCTION TO THE SUPPLY CHAIN

Supply chain can mean many things to many different businesses and may be called sourcing, full supply chain, or integrated supply chain. The function of the supply

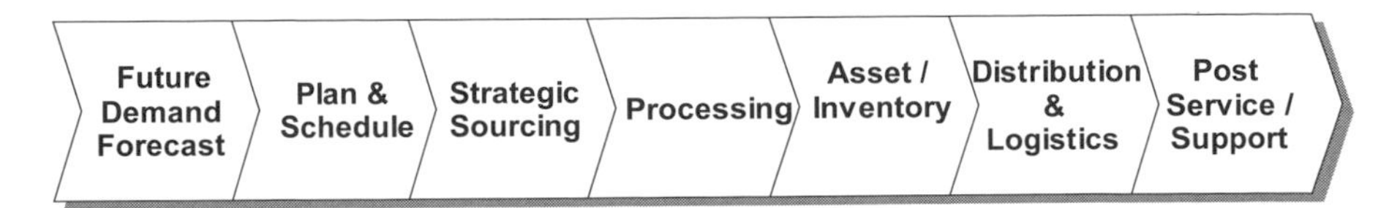

Figure 1.6 Supply chain process chevrons.

chain is integral to every organization and process, as resource planning and acquisition are pertinent whether applied to people, materials, information, or infrastructure. How often do we find ourselves asking, should I buy or make whatever we need or provide? Even in staffing and recruiting activities, we question whether to purchase the services of an outside recruiter or hire our own staff. What type of workload will there be? Where will we store the resumes and interview results? Do we source the applicant screening?

Common functions within the supply chain are contracts, sourcing, purchasing, production control, expediting, supplier selection, and supplier management. The common commodities are classified as direct and indirect; they may include production materials, leases and rentals, IT services, and professional services. The supply chain may involve globalization or logistics. An example of what the supply chain may look like is shown in Figure 1.6. Figure 1.7 shows that strategic sourcing can be expanded.

1.7 SUMMARY

1. Customers experience three aspects of service: the specific service or product has customer attributes such as availability, it is what they want, and it works. The process through which the service is delivered can be ease of use or value added. And the people (or system) should be knowledgeable and user friendly.

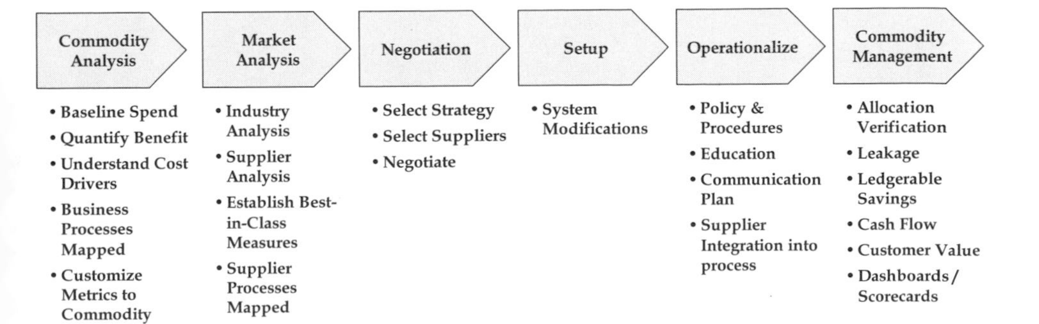

Figure 1.7 Strategic sourcing.

2. Quality assurance and service life cycle are not the result of happenstance, and the best way to ensure customer satisfaction is to adhere to a process that is enabled by tools and methodologies aimed at preventing defects rather than correcting them.
3. Quality has evolved over time from the early 1920s, when Shewhart introduced the first control chart, through the transformation of Japan in the 1950s, to the late 1980s when Six Sigma first came on the scene. Six Sigma and "lean" have now been merged and enabled by personal computers and statistical software to provide easy-to-use and high-value methodologies to eliminate waste and reduce variation, both on new designs as well as existing processes, in order to fulfill customer requirements.
4. The supply chain is applicable to every business function, so using it as an example for design for Six Sigma (as an objective) should allow every practitioner to leverage the application of design for Six Sigma (DFSS as a process).

We will next explore Six Sigma fundamentals in Chapter 2.

2

WHAT IS SIX SIGMA?

2.1 INTRODUCTION

In this chapter, we will provide an overview of Six Sigma and its development, as well as the traditional deployment for process/product improvement called DMA-IC, and three components of the methodology. We will also introduce the design application, which will be detailed in Chapter 3 and beyond.

2.2 WHAT IS SIX SIGMA?

Six Sigma is a philosophy, a measure, and a methodology that provides businesses with the perspective and tools to achieve new levels of performance both in services and products. In Six Sigma, the focus is on process improvement to increase capability and reduce variation. The vital few inputs are chosen from the entire system of controllable and noise variables and the focus of improvement is on controlling these vital few inputs.

Six Sigma as a philosophy helps companies believe that very low defects per million opportunities over long-term exposure is achievable. Six Sigma gives us a statistical scale to measure our progress and benchmark other companies, processes, or products. The defect per million opportunities measurement scale ranges from zero to one million, whereas the sigma scale ranges from 0 to 6. The methodologies used in Six Sigma, which will be discussed in more detail in the following chapters, build upon all of the tools that have evolved to date but puts them into a data-driven framework. This framework of tools allows companies to achieve the lowest defects per million opportunities possible.

Six Sigma evolved from the early TQM efforts mentioned in Chapter 1. Motorola initiated the movement and then it spread to Asea Brown Boveri, Texas Instruments Missile Division, and Allied Signal. It was at this juncture that Jack Welch learned from Larry Bossidy of the power of Six Sigma and with a convert's zeal committed GE to embracing the movement. It was GE that bridged the gap between just a manufacturing process and product focus and took it to what was first called *transactional* processes and later *commercial* processes. One of the reasons that Welch was so interested in this program was that an employee survey had just been completed and it revealed that the top-level managers of the company believed that GE had invented quality. After all, Armand Feigenbaum worked at GE. However, the vast majority of employees did not think GE could spell quality. Six Sigma has turned out to be the methodology to accomplish Crosby's goal of zero defects. Understanding what the key process input variables are and that variation and shift can occur, we can create controls that maintain Six Sigma, or 6σ for short, performance on any product or service and in any process. The Greek letter σ is used by statisticians to indicate standard deviation, a statistical parameter, of the population of interest. Before we can clearly understand the process inputs and outputs, we need to understand process modeling.

2.3 INTRODUCTION TO PROCESS MODELING

Six Sigma is a process-focused approach to achieving new levels of performance throughout any business or organization. We need to focus on a process as a system of inputs, activities, and output(s) in order to provide a holistic approach to all the factors and the way they interact together to create value or waste. Many products and services, when used in a productive manner, are also processes. An ATM machine takes your account information, personal identification number, energy, and money and processes a transaction that dispenses funds or an account rebalance. A computer can take keystroke inputs, energy, and software to process bits into a Word document. At the simplest level, the process model can be represented by a process diagram, often called an IPO diagram, for input–process–output (Figure 2.1).

If we take the IPO concept and extend the ends to include the suppliers of the inputs and the customers of the outputs, then we have the SIPOC, which stands for supplier-input–process–output–customer (Figure 2.2). This is a very effective tool for gathering information and modeling any process. A SIPOC tool can take the form of a table, with a column per each category.

2.3.1 Process Mapping

Whereas the SIPOC is a linear flow of steps, process mapping is a means of displaying the relationship between process steps, and allows for the display of various aspects of the process including delays, decisions, measurements, and rework and decision loops. Process mapping builds upon the SIPOC information by using standard symbols to depict varying aspects of the processes flow linked together with lines and arrows demonstrating the direction of flow.

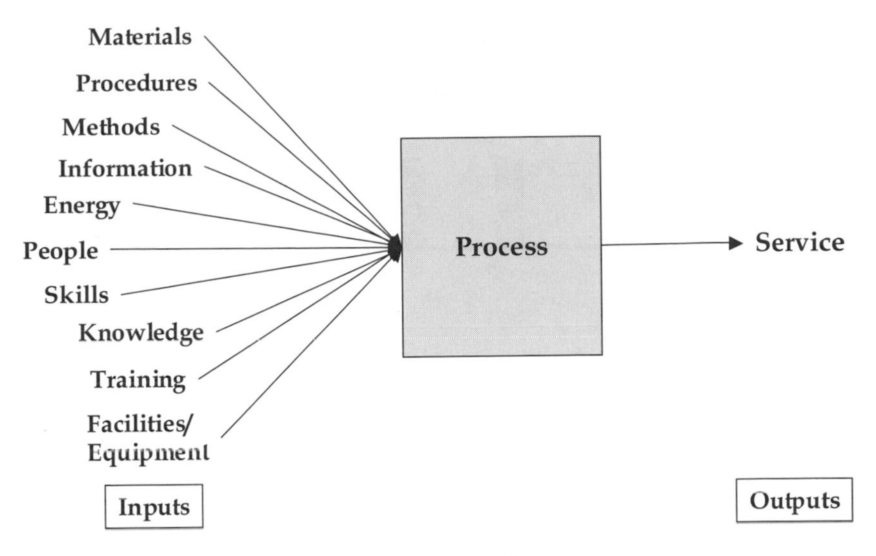

Figure 2.1 The IPO diagram.

2.3.2 Value Stream Mapping

Process mapping can be used to develop a value stream map to understand how well a process is performing in terms of value and flow. Value stream maps can be performed at two levels. They can be applied directly to the process map by evaluating each step of the process map as value added or non-value added (see Figure 2.3 and Figure 2.4). This type of analysis has been in existence since at least the ear-

Suppliers	Inputs	Inputs Characteristic	Process	Outputs	Output Characteristic	Customers
7. Who are the suppliers of the inputs?	6. What are the inputs of the process?	8. What are the characteristics of the inputs?	2a. What is the start of the process? 1. What is the process? 2b. What is the end of the process?	3. What are the outputs of the process?	5. What are the characteristics of the outputs?	4. Who are the customers of the outputs?

Figure 2.2 SIPOC table.

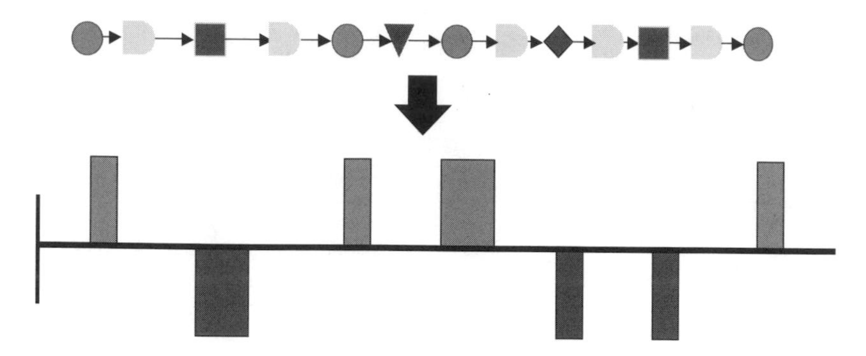

Figure 2.3 Process map transition to value stream map.

ly 1980s, and a good reference is the book *The Hunters and the Hunted: A Non-Linear Solution for Re-Engineering the Workplace* (Swartz, 1996). This is effective if the design team is operating at a local level. However, if the design team is at more of an enterprise level and needs to be concerned about the flow of information as well as the flow of products or services, then the higher-level value stream map is needed (see Figure 2.5). This methodology is best described in Rother and Shook's (2003), *Learning to See.*

2.4 INTRODUCTION TO BUSINESS PROCESS MANAGEMENT

Most processes are ad hoc or allow great flexibility to the individuals operating them. This, coupled with the lack of measurements of efficiency and effectiveness, result in the variation that we have all become accustomed to. In this case, we use the term efficiency for the within-process step performance (often called the "voice of the process"), whereas effectiveness is how all of the process steps interact to perform as a system (often called the "voice of the customer"). This variation we have become accustomed to is difficult to address due to the lack of measures that allow traceability to the root cause. Transactional businesses that have embarked on

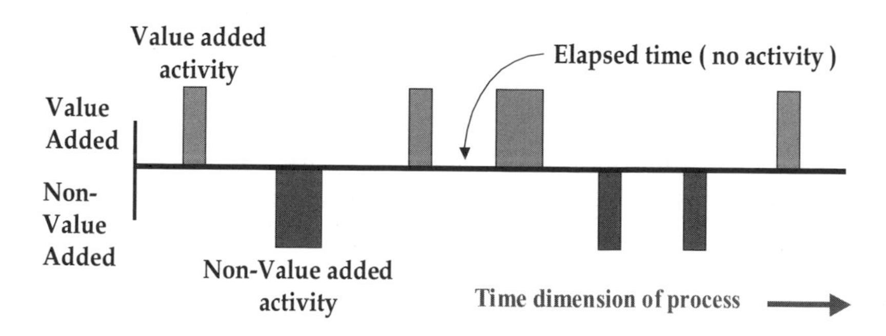

Figure 2.4 Value stream map definitions.

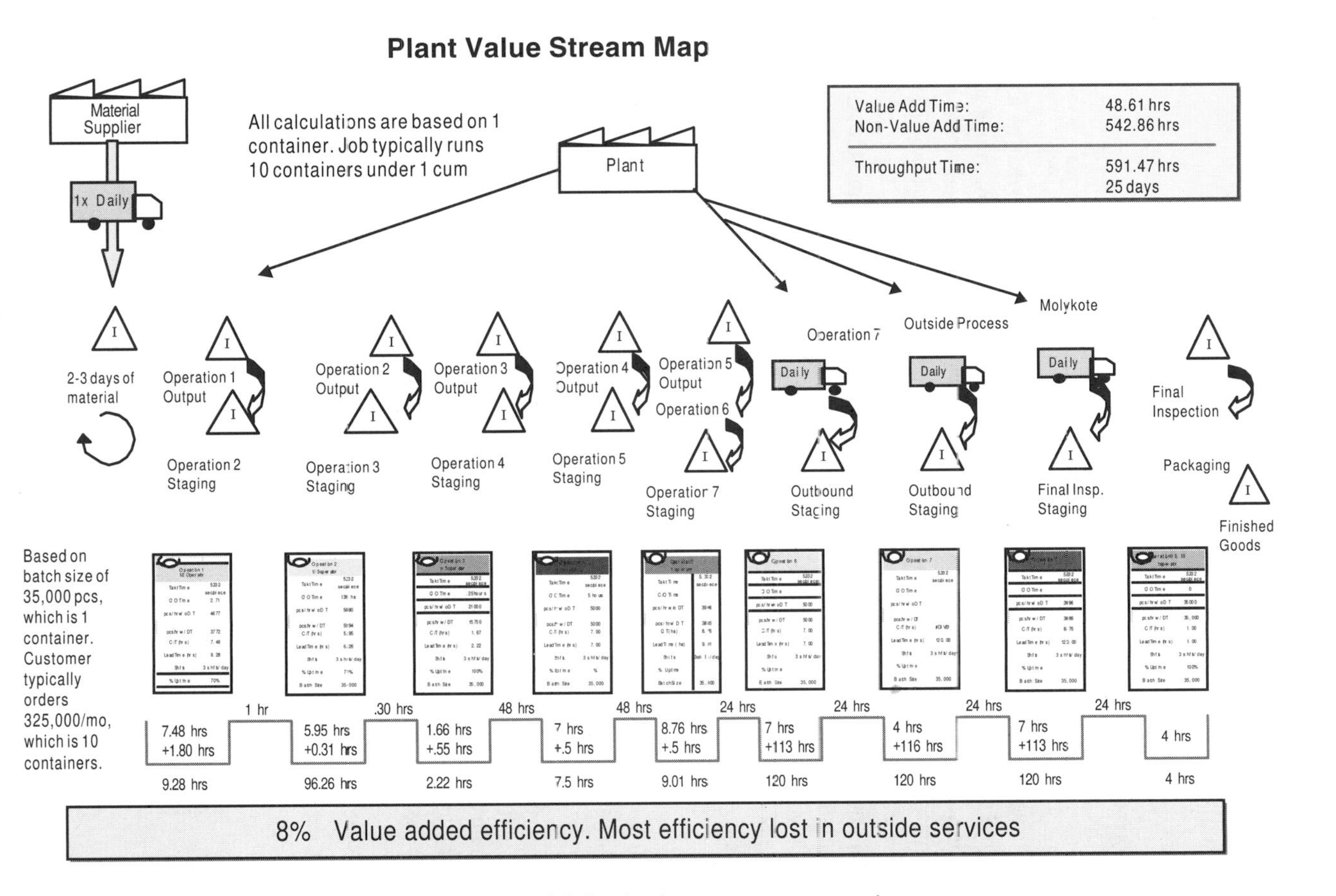

Figure 2.5 High-level value stream map example.

Six Sigma programs have learned that they have to develop process management systems and implement them in order to establish baselines from which to improve. The deployment of a business process management system (BPMS) often results in a marked improvement in performance as viewed by the customer and associates involved in the process. The benefits of implementing a BPMS are magnified in cross-functional processes.

2.5 MEASUREMENT SYSTEMS ANALYSIS

Once we have created some form of documented process—IPO, SIPOC, process map, value stream map, or BPMS—we can begin our analysis of what to fix and what to enhance. Before we can focus on what to improve and how much to improve it, we must be certain of our measurement system. Measurements can start at benchmarking and continue through to operationalization. We must know how accurate the measurement system is versus a known standard, how repeatable the measurement is, and how reproducible it is. Many service measures are the result of calculations. When performed manually, the reproducibility and repeatability can astonish you if you take the time to perform a measurement system analysis (MSA). In a supply chain, we might be interested in promises kept, on-time delivery, order completeness, deflation, lead time, and acquisition cost. Many of these measures require an operational definition in order to provide repeatable and reproducible measures.

Referring to Figure 2.6, is on-time delivery the same as on-time shipment? Many companies cannot know when a client takes delivery or processes a receipt transaction, so how do we measure these? Is it when the item arrives, when the paperwork is complete, or when the customer can actually use the item? We have seen a customer drop a supplier for a 0.5% lower cost component only to discover that the new multiyear contract that they signed did not include transportation and they ended up paying a 3.5% higher price for 3 years.

The majority of measures in a service or process will focus on:

- Speed
- Cost

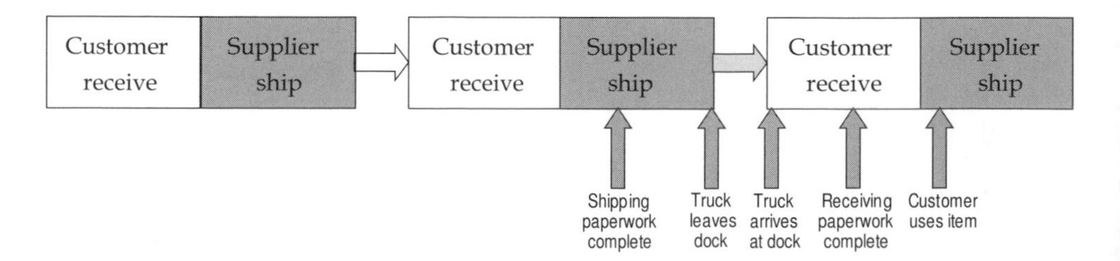

Figure 2.6 Supplier-to-customer cycle.

- Quality
- Efficiency as defined as first-pass yield of a process step
- Effectiveness as defined as the rolled throughput yield of all process steps

All of these can be made robust by creating operational definitions, defining the start and stop points, and determining sound methodologies for assessing. It should come as no surprise that "if you can't measure it, you can't improve it" is a statement worth remembering, and it is important to ensure that adequate measurement sytems are available throughout the project life cycle.

2.6 PROCESS CAPABILITY AND SIX SIGMA PROCESS PERFORMANCE

Process capability is revealed when we measure a process's performance and compare it to the customer's needs (specifications). Process performance may not be constant and usually exhibits some form of variability. For example, we may have an accounts payable (A/P) process that has measures of accuracy and timeliness. For the first two months of the quarter, the process has few errors and is timely, but at the quarter point, the demand goes up and the A/P process exhibits more delays and errors. If the process performance is measurable in real numbers (continous) rather than as pass or fail (discrete), then the process variability can be modeled with a normal distribution. The normal distribution is usually used due to its robustness in modeling many real-world-performance random variables. The normal distribution has two parameters quantifying the central tendency and variation. The center is the average performance and the degree of variation is expressed by the standard deviation. If the process cannot be measured in real numbers, then we convert the pass/fail, good/bad (discrete form) into a yield and then convert the yield into a sigma value. Several transformations from discrete to continuous distributions can be borrowed from mathematical statistics.

If the process follows a normal probability distribution, 99.73% of the values will fall between the $\pm 3\sigma$ limits, where σ is the standard deviation, and only 0.27% will be outside of the $\pm 3\sigma$ limits. Since the process limits extend from -3σ to $+3\sigma$, the total spread amounts to 6σ total variation. This total spread is the process spread and is used to measure the range of process variability.

For any process performance metrics there are usually some performance specification limits. These limits may be single sided or double sided. For the A/P process, the specification limit may be no less than 95% accuracy. For receipt of material into a plant, it may be 2 days early and 0 days late. For a call center, we may want the phone conversation to take between 2 to 4 minutes. Each of the last two double-sided specifications can also be stated as a target and a tolerance. The material receipt could be 1 day early $\pm$ 1 day and for the phone conversation it could be 3 minutes $\pm$ 1 minute.

If we compare the process spread with the specification spread, we can usually observe three conditions:

1. Condition I: Highly Capable Process (see Figure 2.7). The process spread is well within the specification spread:

$$6\sigma < (\text{USL} - \text{LSL})$$

The process is capable because it is extremely unlikely that it will yield unacceptable performance.

2. Condition II: Marginally Capable Process (see Figure 2.8). The process spread is approximately equal to the specification spread:

$$6\sigma = (\text{USL} - \text{LSL})$$

When a process spread is nearly equal to the specification spread, the process is capable of meeting the specifications. If we remember that the process center is likely to shift from one side to the other, then a significant amount of the output will fall outside of the specification limit and will yield unacceptable performance.

3. Condition III: Incapable Process (see Figure 2.9). The process spread is greater than the specification spread:

$$6\sigma > (\text{USL} - \text{LSL})$$

When a process spread is greater than the specification spread, the process is incapable of meeting the specifications and a significant amount of the output will fall outside of the specification limit and will yield unacceptable performance.

2.6.1 Motorola's Six Sigma Quality

In 1986, the Motorola Corporation won the Malcolm Baldrige National Quality Award. Motorola based its success in quality on its Six Sigma program. The goal of

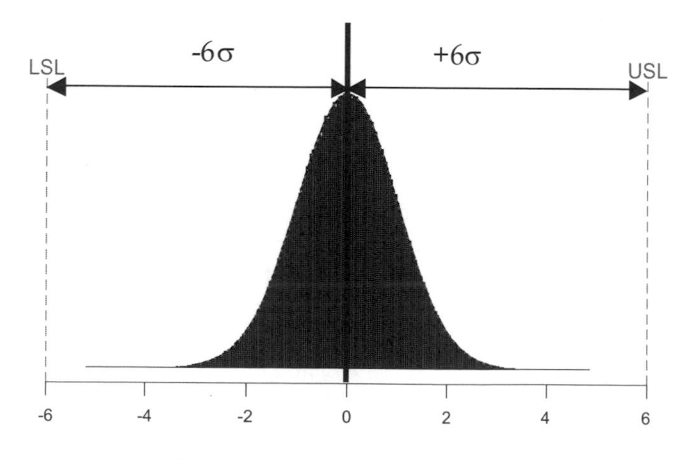

Figure 2.7 Highly capable process.

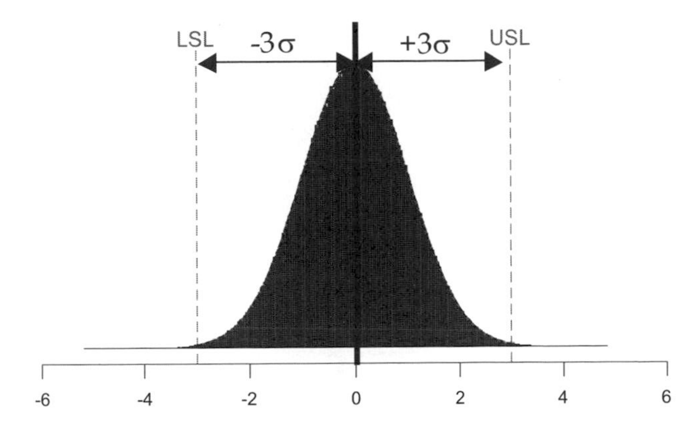

Figure 2.8 Marginally capable process.

the program was to reduce the variation in every process such that a spread of 12σ (6σ on each side of the average) fit within the process specification limits (see Figure 2.10).

Motorola accounted for the side-to-side shift of the process average over time. In this situation, one side shrinks to a 4.5σ gap and the other side grows to 7.5σ (see Figure 2.11). This shift accounts for 3.4 parts per million (ppm) on the small gap and a fraction of parts per billion on the large gap. So over the long term, a 6σ process will generate only 3.4 ppm of defect.

In order to achieve Six Sigma capability, it is desirable to have the process average centered within the specification window and to have the process spread over approximately one-half of the specification window.

There are two approaches to accomplishing Six Sigma levels of performance. When dealing with an existing process, the process improvement method, also known as DMAIC, may be used and if there is a need for a new process, the design

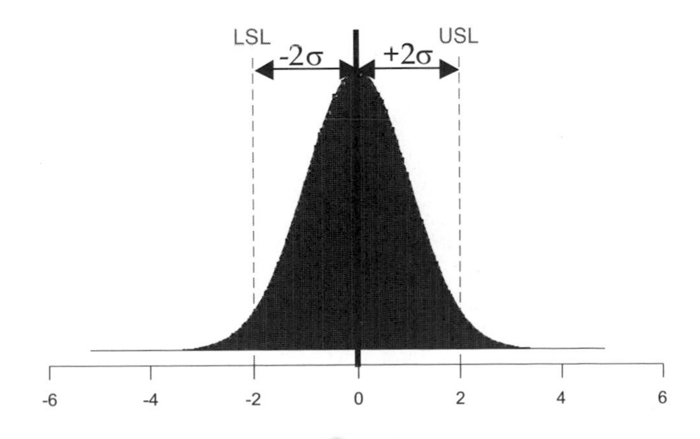

Figure 2.9 Incapable process.

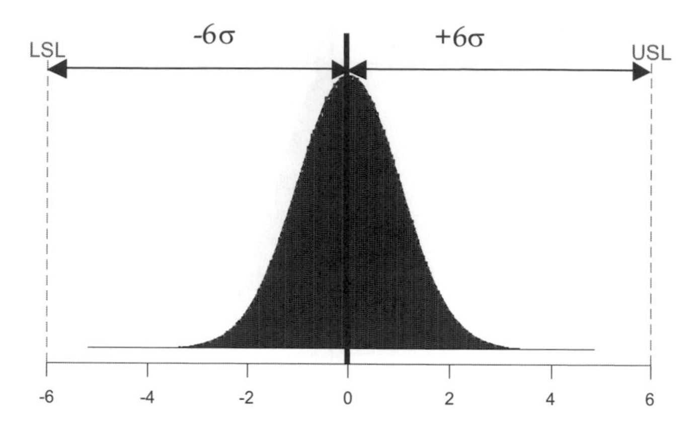

Figure 2.10 Six Sigma capable process (short-term).

for Six Sigma (DFSS) method may be used. Both of these will be discussed in the following sections.

2.7 OVERVIEW OF SIX SIGMA IMPROVEMENT (DMAIC)

Applying Six Sigma methodology to improve an existing process or product follows a five-phase process of:

1. *D*efine—define the opportunity and customer requirements
2. *M*easure—ensure adequate measures, process stability, and initial capability
3. *A*nalyze—analyze the data and discover the critical inputs and other factors

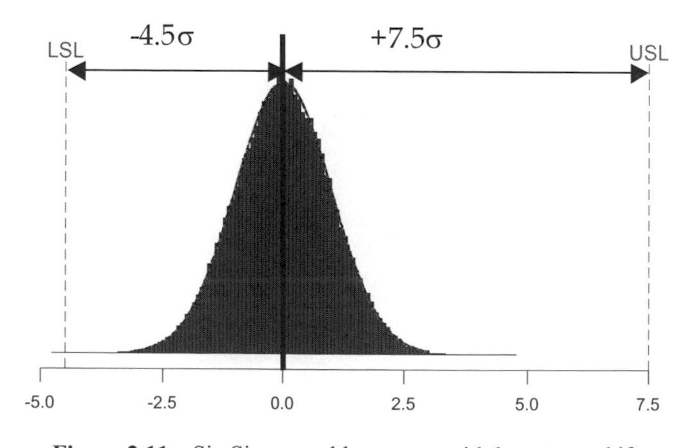

Figure 2.11 Six Sigma capble process with long-term shift.

4. *I*mprove—improve the process based on the new knowledge
5. *C*ontrol—implement adequate controls to sustain the gain

This five-phase process is often referred to as DMAIC, and each phase is briefly described below.

2.7.1 Phase 1: Define

First, we create the project definition that includes the problem/opportunity statement, the objective of the project, the expected benefits, what items are in scope and what items are out of scope, the team structure, and the project timeline. The scope will include details such as resources, boundaries, customer segments, and timing. The next step is to determine and define the customer requirements. Customers can be both external consumers or internal stakeholders. At the end of this step, you should have a clear operational definition of the project metrics (called big Ys or the outputs) and their linkage to critical business levers as well as the goal for improving the metrics. Business levers can consist of, for example, return on invested capital, profit, customer satisfaction, and responsiveness.

The last step in this phase is to define the process boundaries and high-level inputs and outputs using the SIPOC as a framework, and define the data collection plan.

2.7.2 Phase 2: Measure

The first step is to make sure that we have good measures of our outputs (*Y*s) through validation or measurement system analysis.

Next we verify that the metric is stable over time and then determine what our baseline process capability is using the method discussed earlier. If the metric is varying wildly over time, then we must first address the special causes creating the instability before attempting to improve the process. Many times, stabilizing the performance provides all of the improvement desired.

Lastly, in the measure phase we define all of the possible factors that affect the performance and use the qualitative methods of Pareto, cause and effect diagrams, cause and effect matrices, failure modes and their effects, and detailed process mapping to narrow down to the potential influential (significant) factors (the *x*s).

2.7.3 Phase 3: Analyze

In the analyze phase, we use graphical analysis to search out relationships between the input factors (*x*s) and the outputs (*Y*s).

Next, we follow this up with a suite of statistical analysis techniques, including various forms of hypothesis testing, confidence intervals, or screening design of experiments, to determine the statistical and practical significance of the factors on the project *Y*s. A factor may prove to be statistically significant; that is, with a certain confidence, the effect is true and there is only a small chance it could have been due to a mistake. The statistically significant factor is not always practical, in that it may only account for a small percentage of the effect on the Ys, in which case controlling this factor would not provide much improvement. The transfer function $Y = f(x)$ for every Y measure usually represents the regression of several influential factors on the project outputs. There may be more than one project metric (output), hence the Ys.

2.7.4 Phase 4: Improve

In the improve phase, we first identify potential solutions through team meetings and brainstorming or the use of TRIZ (covered in Chapter 9). It is important at this point to have completed a measurement system analysis on the key factors (*x*s) and possibly to have performed some confirmation design of experiments.

The next step is to validate the solution(s) identified through a pilot run or through optimization design of experiments. Following confirmation of the improvement, a detailed project plan and cost benefit analysis should be completed.

The last step in this phase is to implement the improvement. This is a point at which change management tools can prove to be beneficial.

2.7.5 Phase 5: Control

The control phase consists of four steps. In the first step we determine the control strategy based on the new process map, failure mode and effects, and a detailed control plan. The control plan should balance between the output metric and the critical few input variables.

The second step involves implementing the controls identified in the control plan. This is typically a blend of poka yoke's, control charts, clear roles and responsibilities, and operator instructions depicted in operational method sheets.

Third, we determine what the final capability of the process is with all of the improvements and controls in place.

The final step is to perform the ongoing monitoring of the process based on the frequency defined in the control plan. The DMAIC methodology has allowed businesses to achieve lasting breakthrough improvements that break the paradigm of reacting to the symptoms rather than the causes. This method allows design teams to make fact-based decisions using statistics as a compass and implement lasting improvements that satisfy the external and internal customers.

2.8 SIX SIGMA GOES UPSTREAM—DESIGN FOR SIX SIGMA

The DMAIC methodology is excellent when dealing with an existing process in which reaching the entitled level of performance will provide all of the benefit required. Entitlement is the best the process or product is capable of performing at adequate control. Reviewing historical data, it is often seen to be the best performance point. But what do we do if reaching entitlement is not enough or there is a need for an innovative solution never before deployment? We could continue with the typical build it and fix it process or we can utilize the most powerful tools and methods available for developing an optimized, robust, risk-free design. These tools and methods can be aligned with an existing new product/process development process or used in a stand-alone manner. The rest of this book is devoted to explaining and demonstrating the design for Six Sigma (DFSS) tools and methodology. Chapter 3 is the introductory chapter for DFSS, giving overviews for DFSS theory, DFSS gated processes, and DFSS applications. Chapter 4 gives a detailed description about how to deploy DFSS in a company, covering the training, organization support, financial management, and deployment strategy. Chapter 5 gives a very detailed "road map" of the whole DFSS project execution, which includes very detailed descriptions of the DFSS stages, task management, scorecards, and how to integrate all DFSS methods in developmental stages. Chapters 6 through 17 give detailed descriptions with examples of all of the major methods used in DFSS.

2.9 SUMMARY

In this chapter, we have explained what Six Sigma is and how it has evolved over time. We explained that it is a process-based methodology and introduced the reader to process modeling with a high-level overview of IPO, process mapping, value stream mapping, and value analysis, as well as business process management systems (BPMSs). We have explained the criticality of understanding the measurements of the process or system and how this is accomplished with masurement systems analysis (MSA). Once we understand the goodness of our measures, we can evaluate the capability of the process to meet customer requirements and demonstrate what Six Sigma capability is. Next, we moved to an explanation of the DMAIC methodology and how it incorporates these concepts into a road-map method. Finally, we covered how Six Sigma moves upstream to the design environment with the application of DFSS. In the next chapter, we will introduce the reader to the service design for Six Sigma (DFSS) process.

3

INTRODUCTION TO SERVICE DESIGN FOR SIX SIGMA (DFSS)

3.1 INTRODUCTION

The objective of this chapter is to introduce the service design for Six Sigma (DFSS) process theory and to lay the foundation for the next chapters of the book. DFSS combines design analysis (e.g., requirements cascading) with design synthesis (e.g., process engineering) within the framework of the deploying company's service- (product-) development systems. Emphasis is placed on CTSs (critical-to-satisfaction requirements, also known as big *Y*s), identification, optimization, and verification using the transfer function and scorecard vehicles. A transfer function in its simplest form is a mathematical relationship between the CTSs and/or their cascaded functional requirements (FRs) and the critical influential factors (called the *x*s). Scorecards help predict risks to the achievement of CTSs or FRs by monitoring and recording their mean shifts and variability performance.

DFSS is a disciplined and rigorous approach to service, process, and product design that ensures that new designs meet customer requirements at launch. It is a design approach that ensures complete understanding of process steps, capabilities, and performance measurements by using scorecards, transfer functions, and tollgate reviews to ensure the accountability of all the design team members (black belts, project champions, and deployment champions[1]) and the rest of the organization.

The service DFSS objective is to attack the design vulnerabilities in both the conceptual and operational phases by deriving and integrating tools and methods for their elimination and reduction. Unlike the DMAIC methodology, the phases or steps of DFSS are not universally defined, as evidenced by the many customized training curriculums available in the market. Many times, the deploying companies will implement the version of DFSS used by their vendor, thus assisting in the deployment. On the other hand, a company may implement DFSS to suit their business, industry, and culture by creating their own version. However, all approaches share common themes, objectives, and tools.

[1]We will explore the roles and responsibilities of these Six Sigma operatives and others in Chapter 4.

Service Design for Six Sigma. By Basem El-Haik and David M. Roy

DFSS is used to design or redesign a product or service. The expected process sigma level for a DFSS product or service is at least 4.5[2], but can be 6σ or higher depending on the designed entity. The achievement of such a low defect level from product or service launch means that customer expectations and needs must be completely understood before a design can be operationalized. That is quality as defined by the customer.

The material presented here is intended to give the reader a high-level understanding of service DFSS, its uses, and benefits. After reading this chapter, readers should be able to assess how it can be used in relation to their jobs and identify their needs for further learning.

DFSS as defined in this book has a two-track deployment and application. By deployment, we mean the strategy adopted by the deploying company to launch the Six Sigma initiative. It includes putting into play the deployment infrastructure, strategy, and plan for initiative execution (see Chapter 4). Six Sigma deployment will be covered in Chapter 4. In what follows, we are assuming that the deployment strategy is in place as a prerequisite for application and project execution. The DFSS tools are laid on top of four phases, as detailed in Chapter 5, in what we will call the service DFSS project road map.

There are two distinct tracks within the Six Sigma initiative, as discussed in the previous chapters: the retroactive Six Sigma DMAIC[3] approach that takes existing process improvement as an objective, and the proactive design for Six Sigma (DFSS) that targets redesign and new service introduction in both the development and production (process) arenas. DFSS is different from the Six Sigma DMAIC approach in that it is a proactive prevention approach to design.

The service DFSS approach can be broken down into *i*dentify, *c*haracterize, *o*ptimize, and *v*erify, or ICOV for short. These are defined below:

*I*dentify customer and design requirements. Prescribe the CTSs, design parameters, and corresponding process variables.

*C*haracterize the concepts, specifications, and technical and project risks.

*O*ptimize the design transfer functions and mitigate risks.

*V*erify that the optimized design meets the intent (customer, regulatory, and deploying service functions).

In this book, both ICOV and DFSS acronyms will be used interchangeably.

3.2 WHY USE SERVICE DESIGN FOR SIX SIGMA?

Generally, customer-oriented design is a development process of transforming customers' wants into design service solutions that are useful to the customer. This

[2]No more than approximately 1 defect per thousand opportunities.

[3]*D*efine project goals and customer deliverables; *M*easure the process and determine baseline; *A*nalyze—determine root causes; *I*mprove the process by optimization (i.e., eliminating/reducing defects); and *C*ontrol—sustain the optimized solution.

process is carried out over several development stages, starting at the conceptual stage. In this stage, conceiving, evaluating, and selecting good design solutions are difficult tasks with enormous consequences. It is usually the case that organizations operate in two modes: "proactive," that is, conceiving feasible and healthy conceptual entities; and "retroactive," that is, problem solving such that the design entity can live up to its committed potentials. Unfortunately, the latter mode consumes the largest portion of an organization's human and nonhuman resources. The design for Six Sigma approach highlighted in this book is designed to target both modes of operation.

DFSS is the premier approach to process design that is able to embrace and improve developed homegrown supportive processes (e.g., sales and marketing) within its development system. This advantage will enable the deploying company to build on current foundations while enabling them to reach unprecedented levels of achievement that exceed the set targets.

The linking of the Six Sigma initiative and DFSS to the company vision and annual objectives should be direct, clear, and crisp. DFSS has to be the crucial mechanism to develop and improve business performance and increase customer satisfaction and quality metrics. Significant improvements in all health metrics are the fundamental source of DMAIC and DFSS projects that will, in turn, transform the company's culture one project at a time. Achieving a Six Sigma culture is very essential for the future well-being of the deploying company and represents the biggest return on investment beyond the obvious financial benefits. Six Sigma initiatives must apply to all elements of the company's strategy in all areas of the business if massive impact is the objective.

The objective of this book is to present the service design for Six Sigma approach, concepts, and tools that eliminate or reduce both the conceptual and operational types of vulnerabilities of service entities and enables such entities to operate at Six Sigma quality levels in meeting all of their requirements.

Reducing operational vulnerability means adjustment of the critical-to-quality, critical-to-cost, and critical-to-delivery requirements (the CTSs). This objective has been the subject of many knowledge fields such as parameter design, DMAIC Six Sigma, and tolerance design/tolerancing techniques. On the contrary, the conceptual vulnerabilities are usually overlooked due to the lack of a compatible systemic approach to finding ideal solutions, ignorance of the designer, pressure of deadlines, and budget limitations. This can be partly attributed to the fact that traditional quality methods can be characterized as after-the-fact practices since they use lagging information for developmental activities, such as bench tests and field data. Unfortunately, this practice drives design toward endless cycles of design–test–fix–retest, creating what is broadly known as the "fire fighting" mode of the design process, that is, the creation of design-hidden factories. Companies that follow these practices usually suffer from high development costs, longer time to market, lower quality levels, and marginal competitive edge. In addition, corrective actions to improve their conceptual vulnerabilities via operational vulnerabilities improvement means are marginally effective if at all useful. Typically, these corrections are costly and hard to implement as the service project progresses in the development process. Therefore, implementing DFSS in the conceptual stage is a goal that can be achieved when systematic design methods are integrated with quality concepts and methods upfront. Specifically, on the technical side, we developed an approach to DFSS by borrowing from the following fundamental knowledge arenas: process engineering, quality engineering, TRIZ (Altshuller, 1969), axiomatic design (Suh, 1990), and theories of probability and statistics. At the same time, there are several venues in our DFSS approach that enable transformation to a data-driven and customer-centric culture such as concurrent design teams, deployment strategies, and plans.

In general, most of the current design methods are empirical in nature. They represent the best thinking of the design community that, unfortunately, lacks the design scientific base while relying on subjective judgment. When the company suffers because of poor customer satisfaction, judgment and experience may not be sufficient to obtain an optimal Six Sigma solution, providing another motivation to devise a DFSS method to address such needs.

With DFSS, the attention shifts from improving performance in the later stages of the service design life cycle to the front-end stages where design development takes place at higher levels of abstraction, that is, to prevention instead of problem solving. This shift is also motivated by the fact that the design decisions made during the early stages of the service design life cycle have the largest impact on total cost and quality of the system. It is often claimed that up to 80% of the total cost is committed in the concept development stage (Fredrikson, 1994). The research area of design is currently receiving increasing focus to address industry efforts to shorten lead times, cut development and manufacturing costs, lower total life cycle cost, and improve the quality of the design entities in the form of products, services, and/or processes. It is the experience of the authors that at least 80% of the design quality is also committed in the early stages, as depicted in Figure 3.1 (see Yang & El-Haik, 2003). The "potential" in the figure is defined as the difference between

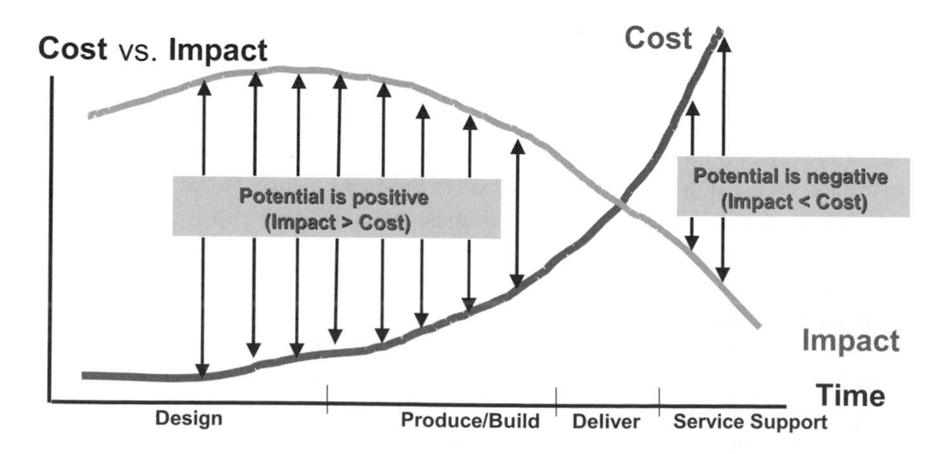

Figure 3.1 Effect of design stages on life cycle.

the impact (influence), of the design activity at a certain design stage and the total development cost up to that stage. The potential is positive but decreases as design progresses, implying reduced design freedom over time. As financial resources are committed (e.g., buying process equipment and facilities, hiring staff) the potential starts changing sign, going from positive to negative. For the consumer, the potential is negative and the cost overcomes the impact tremendously. At this stage, design changes for corrective actions can only be achieved at high cost, including customer dissatisfaction, warranty problems, and marketing promotions, in many cases under the scrutiny of the government (e.g., recall costs).

3.3 WHAT IS SERVICE DESIGN FOR SIX SIGMA?

Service DFSS is a structured, data-driven approach to design in all aspects of service functions (e.g., human resources, marketing, sales, IT) where deployment is launched, to eliminate the defects induced by the design process and to improve customer satisfaction, sales, and revenue. To deliver these benefits, DFSS applies design methods like axiomatic design,[4] creativity methods like TRIZ,[5] and statistical techniques to all levels of design decision making in every corner of the business, identifies and optimizes the critical design factors (the *x*s), and validates all design decisions in the use (or surrogate) environment of the end user.

DFSS is not an add-on but represents a cultural change within different functions and organizations in which deployment is launched. It provides the means to tackle weak or new processes, driving customer and employee satisfaction. DFSS and Six Sigma should be linked to the deploying company's annual objectives, vision, and mission statements. They should not be viewed as another short-lived initiative. They are vital, permanent components to achieve leadership in design, customer satisfaction, and cultural transformation. From marketing and sales to development, operations, and finance, each business function needs to be headed by a deployment leader or a deployment champion. This local deployment team will be responsible for delivering dramatic change, thereby removing a number of customer issues and internal problems and expediting growth. The deployment team can deliver on their objective through Six Sigma operatives called black belts and green belts, who will be executing scoped projects that are in alignment with the objectives of the company. Project champions are responsible for scoping projects from within their realm of control and handing project charters (contracts) over to the Six Sigma resources group. The project champion will select projects consistent with corporate goals and remove barriers. The Six Sigma resources will complete successful projects using Six Sigma methodology and will train and mentor the local organization on Six Sigma. The deployment leader, the highest initiative operative, sets meaningful

[4]A perspective design method that employs two design axioms: the independence axiom and the information axiom. See Chapter 8 for more details.

[5]TRIZ is the Russian acronym for theory of inventive problem solving, also known as TIPS. It is a systematic method to conceive creative, innovative, and predictable design solutions. See Chapter 9 for more details.

goals and objectives for the deployment in his or her function area and drives the implementation of Six Sigma publicly.

The Six Sigma black belt resources group consists of full-time Six Sigma operatives as opposed to green belts, who should be completing smaller projects of their own as well as assisting black belts. The black belts play a key role in raising the competency of the company as they drive the initiative into day-to-day operations.

Black belts are the driving force of service DFSS deployment. They are project leaders who are removed from day-to-day assignments for a period of time (usually two years) to focus exclusively on design and improvement projects. They have intensive training in Six Sigma tools, design techniques, problem solving, and team leadership. The black belts are trained by master black belts, who are initially hired if not available homegrown.

A black belt should have process and organizational knowledge, know some basic design theory, have statistical skills, and be eager to learn new tools. A black belt is a "change agent" who carries the initiative to his or her team, the staff, and the entire company. In doing so, communication and leadership skills are vital. Black belts need effective intervention skills. They must understand why some team members may resist the Six Sigma cultural transformation. Some soft training on leadership should be embedded in their training curriculum. Soft-skills training may target deployment maturity analysis, team development, business acumen, and individual leadership. In training, it is wise to share several maturity initiative indicators that are being tracked in the deployment scorecard, for example, alignment of the project to company objectives in its own scorecard (the big *Y*s), readiness of the project's mentoring structure, preliminary budget, team member identification, and scoped project charter.

DFSS black belt training is intended to be delivered in tandem with a training project for hands-on application. The training project should be well scoped with ample opportunity for tool application and should have cleared tollgate "0" prior to class. Usually, project presentations will be weaved into each training session. More details are given Chapter 4.

While handling projects, the role of the black belts spans several functions such as learning, mentoring, teaching, and coaching. As a mentor, the black belt cultivates a network of experts in the project on hand, working with the process operators; process owners, and all levels of management. To become self-sustained, the deployment team may need to task their black belts with providing formal training to green belts and team members.

Service design for Six Sigma (DFSS) is a disciplined methodology that applies the transfer function [CTSs $= f(x)$] to ensure that customer expectations are met and embedded into the design, design performance is predicted prior to the pilot, performance measurement systems (scorecards) are built into the design to ensure effective ongoing process management, and a common language for design is leveraged within the design tollgate process.

DFSS projects can be categorized as design or redesign of an entity, whether product, process, or service. "Creative design" is the term that we will be using to indicate new design, design from scratch, and "incremental design," to indicate the redesign case or design from a datum. In the later case, some data can be used to baseline current performance. The degree of deviation of the redesign from the datum is the key factor in deciding on the usefulness of relative existing data. Service DFSS projects can come from historical sources (e.g., service redesign due to customer issues) or from proactive sources like growth and innovation (new service introduction). In either case, the service DFSS project requires greater emphasis on:

- Voice of the customer collection scheme
- Addressing all (multiple) CTSs as cascaded by the customer
- Assessing and mitigating technical failure modes and project risks in their own environment as they are linked to the tollgate process reviews
- Project management with some communication plan to all affected parties and budget management
- Detailed project change management process

3.4 SERVICE DFSS: THE ICOV PROCESS

As mentioned in Section 3.1, design for Six Sigma has four phases spread over seven development stages. They are identify, characterize, optimize, and verify. The acronym ICOV is used to denote these four phases. The service life cycle is depicted in Figure 3.2. Notice the position of service ICOV phases of a design project.

Naturally, the process of service design begins when there is a need (an impetus). People create the need, whether it is a problem to be solved (e.g., if waiting time at a call center becomes very long, then the customer satisfaction deteriorates and the process needs to be redesigned) or a new invention. Design objective and scope are critical in the impetus stage. A design project charter should describe simply and clearly what is to be designed. It cannot be vague. Writing a clearly stated design charter is just one step. In stage 2, the design team must write down all the information they may need, in particular, the voice of the customer (VOC) and the voice of the business (VOB). With the help of the quality function deployment (QFD) process, such consideration will lead to the definition of the service design functional requirements that will be later grouped into processes (systems and subsystems) and subprocesses (components). A functional requirement must contribute to an innovation or to a solution of the objective described in the design charter. Another question that should be on the minds of the team members relates to how the end result will look. The simplicity, comprehensiveness, and interfaces should make the service attractive. What options are available to the team? And at what cost? Do they have the right physical properties, such as bandwidth, completeness, language, and reliability? Will the service be difficult to operate and maintain? Consider what methods you will need to employ to process, store, and deliver the service.

In stage 3, the design team should produce a number of solutions. It is very important that they write or draw every idea on paper as it occurs to them. This will help them remember and describe them more clearly. It is also easier to discuss them with other people if drawings are available. These first drawings do not have to be very detailed or accurate. Sketches will suffice and should be made quickly. The important

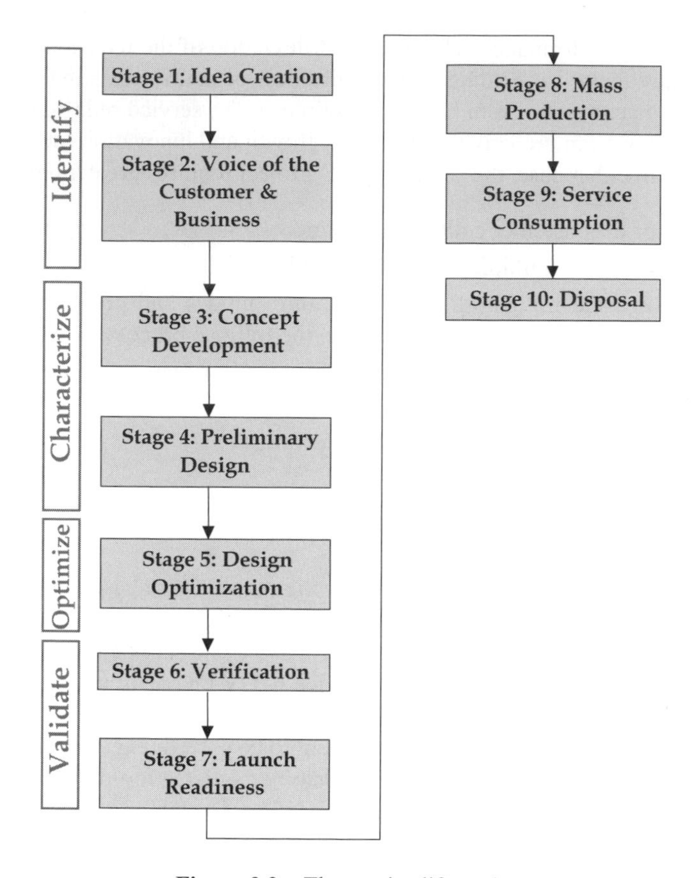

Figure 3.2 The service life cycle.

thing is to record all ideas and develop solutions in the preliminary design stage (stage 4). The design team may find that they like several of the solutions. Eventually, the design team must choose one. Usually, careful comparison with the original design charter will help them to select the best solution subject to the constraints of cost, technology, and skills available. Deciding among the several possible solutions is not always easy. It helps to summarize the design requirements and solutions and put the summary into a matrix called the morphological matrix.[6] An overall design alternative set is synthesized from this matrix, which contains conceptually high-potential and feasible solutions. Which solution should they choose? The Pugh matrix, a concept selection tool named after Stuart Pugh, can be used to reach a decision. The selected solution will be subjected to a thorough design-optimization stage (stage 5). This optimization could be deterministic and/or statistical in nature. On the statistical front, the design solution will be made insensitive to uncontrollable

[6]A morphological matrix is a way to show all functions and corresponding possible design parameters (solutions).

factors (called the noise factors) that may affect its performance. Factors like customer usage profile, environment, and production case-to-case variation should be considered as noise. To assist in this noise insensitivity task, we rely on the transfer function as an appropriate vehicle. In stage 5, the team needs to make detailed documentation of the optimized solution. This documentation must include all of the information needed to produce the service. Consideration for design documentation, process maps, operational instructions, software code, communication, marketing, and so on should be put in place. In stage 6, the team can make a model, assuming the availability of the transfer functions and, later, a prototype; or they can go directly to making a prototype or a pilot. A model is a full-size or small-scale simulation of a process or product. Architects, engineers, and most designers use models. Models are one more step in communicating the functionality of the solution. For most people, understanding is made clearer when the project end result is seen in three-dimensional form. A scale model is used when design scope is very large. A prototype is the first working version of the team's solution. It is generally full-size and often uses homegrown expertise. Design verification, stage 6, also includes testing and evaluation, which is basically an effort to answer these very basic questions:

- Does it work? (i.e., does it meet the design charter?)
- If failures are discovered, will modifications improve the solution?

These questions have to be answered. When satisfactory answers are obtained, the team can move to the next development and design stage.

In stage 7, the team needs to prepare the production facilities where the service will be produced for launch. At this stage, they should assure that the service is marketable and that there are no competitors to beat them to the market. The team together with the project stakeholders must decide how many to make. Similar to products, service may be mass produced in low volume or high volume. The task of making the service is divided into jobs. Each worker trains to do his or her assigned job. As workers complete their special jobs, the service/product takes shape. After stage 7, mass production saves time and other resources. Since workers train to do a certain job, each becomes skilled in that job.

3.5 SERVICE DFSS: THE ICOV PROCESS IN SERVICE DEVELOPMENT

Due to the fact that service DFSS integrates well with service life cycle systems, it is an event-driven process, in particular, the development (design) stage. In this stage, milestones occur when the entrance criteria (inputs) are satisfied. At these milestones, the stakeholders, including the project champion, process/design owner, and deployment champion (if necessary), conduct reviews called "tollgate" reviews. A development stage has some *thickness,* that is, entrance criteria and exit criteria for the bounding tollgates create a set of tasks that define the amount of effort that will be required. The ICOV DFSS phases as well as the seven stages of the development

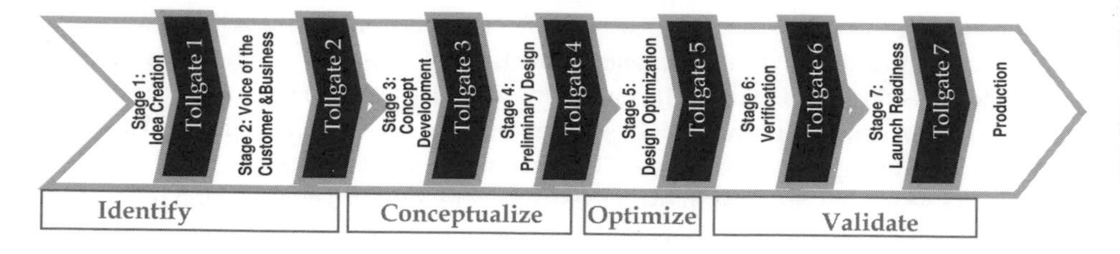

Figure 3.3 The ICOV DFSS process.

process are depicted in Figure 3.3. In these reviews, a decision should be made as to whether to proceed to the next phase of development, recycle back for further clarification on certain decisions, or cancel the project altogether. Cancellation of problematic projects, as early as possible, is a good thing. It stops nonconforming projects from progressing further while consuming resources and frustrating people. In any case, the black belt should quantify the size of the benefits of the design project in language that will have impact on upper management, identify major opportunities for improving customer dissatisfaction and associated threats to salability, and stimulate improvements through publication of the DFSS approach.

In tollgate reviews, work proceeds when the exit criteria (required decisions) are made. As a DFSS deployment side bonus, a standard measure of development progress across the deploying company, using a common development terminology, is achieved. Consistent exit criteria are obtained in each tollgate from the following: (1) DFSS ICOV process deliverables, (2) deploying business company-wide deliverables, and (3) a select subset of deliverables specific to certain business functions (for example, sales or human resources) within the deploying company. The detailed entrance and exit criteria by stage are presented in Chapter 5.

3.6 OTHER DFSS APPROACHES

Design for Six Sigma (DFSS) can be accomplished using any one of many other methodologies besides the one presented in this book. IDOV[7] is one popular methodology for designing products and services to meet six sigma standards. It is a four-phase process that consists of identify, design, optimize, and verify. These four phases parallel the four phases of the ICOV process presented in this book:

- Identify Phase. It begins the process by formally tying the design to the voice of the customer (VOC). This phase involves developing a team and team charter, gathering the VOC, performing competitive analysis, and developing CTSs.
- Design Phase. This phase emphasizes CTSs and consists of identifying functional requirements, developing alternative concepts, evaluating alternatives,

[7]See Dr. David Woodford's article at http://www.isixsigma.com/library/content/c020819a.asp.

selecting a best-fit concept, deploying CTSs, and predicting sigma capability.

- Optimize Phase. The optimize phase requires use of process capability information and a statistical approach to tolerancing. Developing detailed design elements, predicting performance, and optimizing design take place within this phase.
- Validate Phase. The Validate phase consists of testing and validating the design. As increased testing using formal tools occurs, feedback of requirements should be shared with production operations and sourcing, and future operations and design improvements should be noted.

Another popular design for Six Sigma methodology is called DMADV. It retains the same number of letters, number of phases, and general feel as the DMAIC acronym. The five phases of DMADV are:

- Define: Define the project goals and customer (internal and external) requirements.
- Measure: Measure and determine customer needs and specifications; benchmark competitors and industry.
- Analyze: Analyze the process options to meet the customer needs.
- Design: Design (in detail) the process to meet the customer needs.
- Verify: Verify the design performance and ability to meet customer needs.

Another flavor of the DMADV methodology is DMADOV—design, measure, analyze, design, optimize, and verify. Other modified versions include DCCDI and DMEDI. DCCDI is being promoted by Geoff Tennant and is defined as define, customer concept, design, and implement, a replica of the DMADV phases. DMEDI is being promoted by PriceWaterhouseCoopers and stands for define, measure, explore, develop, and implement. The fact is that all of these DFSS methodologies use almost the same tools (quality function deployment, failure modes and effects analysis, benchmarking, design of experiments, simulation, statistical optimization, error proofing, robust design, etc.), presenting little difficulty to their alternate use. In addition to these common elements, the ICOV offers a thread-through road map with overlaid tools that are based on nontraditional tools such as design mappings, design axioms, creativity tools such as TRIZ, as well as cultural treatments.

3.7 SUMMARY

Service DFSS offers a robust set of tools and processes that address many of today's complex business design problems. The DFSS approach helps design teams frame their project based on a process with financial, cultural, and strategic implications to the business. The service DFSS comprehensive tools and methods described in this book allow teams to quickly assess service issues and identify financial and operational improvements that reduce costs, optimize investments, and maximize returns. Service DFSS leverages a flexible and nimble organization and maintains low development costs, allowing deploying companies to pass these benefits on to their customers. Service DFSS employs a unique gated process that allows teams to build tailor-made approaches—not all the tools need to be used in each project. Therefore, designs can accommodate the specific needs of the project charter. Project by project, the competency level of the design teams will be enhanced, leading to deeper knowledge and broader experience.

In this book, we form and integrate several strategic and tactical methodologies that produce synergies to enhance service DFSS capabilities to deliver a broad set of optimized solutions. The methods presented in this book can have wide-spread application to help design teams and the "belt" population in different project areas (e.g., staffing and other human resources functions, finance, operations and supply chain functions, organizational development, financial services, training, technology and software tools and methods, etc.)

Service DFSS provides a unique commitment to the project customers by guaranteeing agreed upon financial and other results. Each project must have measurable outcomes and the design team must be responsible for defining and achieving those outcomes. The service DFSS approach ensures these outcomes through risk identification and mitigation plans, variable (DFSS tools that are used over many stages) and fixed (DFSS tools that are used once) tool structures, and advanced conceptual tools. The DFSS principles and structure should motivate design teams to provide businesses and customers with a substantial return on their design investment.

4

SERVICE DESIGN FOR SIX SIGMA DEPLOYMENT

4.1 INTRODUCTION

Service design for Six Sigma (DFSS) is a disciplined methodology that embeds customer expectations into the design, applies the transfer function approach to ensure that customer expectations are met, predicts design performance prior to the pilot, builds performance measurement systems (scorecards) into the design to ensure effective ongoing process management, leverages a common language for design, and uses tollgate reviews to ensure accountability

This chapter examines the service DFSS deployment team, which launches the Six Sigma program. A deployment team includes different levels of the deploying company's leadership, including initiative senior leaders, project champions, and other deployment sponsors. As such, the material of this chapter should be used as a deployment guideline with ample room for customization. It provides the considerations and general aspects required for a smooth and successful initial deployment experience.

The extent to which service DFSS produces the desired results is a function of the adopted deployment plan. Historically, we can observe that many sound initiatives became successful when commitment was secured from involved people at all levels. An initiative is successful when it is adopted as the new norm. Service Six Sigma and DFSS are no exception. A successful DFSS deployment depends on people, so almost every level, function, and division involved with the design process should participate in it, including the customer.

4.2 SERVICE SIX SIGMA DEPLOYMENT

The extent to which a service Six Sigma program produces results is directly affected by the plan with which it is deployed. This section presents a high-level perspective of a sound plan by outlining the critical elements of successful deployment. We must point out up front that a successful Six Sigma initiative is the result of key

Service Design for Six Sigma. By Basem El-Haik and David M. Roy

contributions from people at all levels and functions of the company. In short, successful Six Sigma initiatives require buy-in, commitment, and support from officers, executives, and management staff before and during execution of improvement projects by operational and process-level employees.

This top-down approach is critical to the success of a service Six Sigma program. Although black belts are the focal point for executing projects and generating cash from process improvements, their success is inextricably linked to the way leaders and managers establish the Six Sigma culture, create motivation, allocate goals, institute plans, set procedures, initialize systems, select projects, control resources, and provide recognition and rewards.

Several scales of deployment may be used; however, maximum benefit can only be achieved when all affected functions are engaged. A full-scale, company-wide deployment program requires that senior leadership install the proper culture of change before embarking on their support for training, logistics, and other resources required. People empowerment is key, as is leadership by example.

Benchmarking the DMAIC Six Sigma program in several successful deployments, we can conclude that a top-down deployment approach will work for service DFSS deployment as well. This conclusion reflects the critical importance of securing and cascading the buy-in from the top leadership level. The black belts and the green belts are the focused force of deployment under the guidance of the master black belts and champions. Success is measured by increase in revenue, customer satisfaction, and generated cash, one project at a time. Belted projects should be diligently scoped and aligned to company's objectives with some prioritization scheme. Six Sigma program benefits cannot be harvested without a sound strategy and the long-term vision of establishing the Six Sigma culture. In the short term, deployment success is dependent on motivation, management commitment, project selection and scoping, institutionalized reward and recognition systems, and optimized resources allocation. This chapter is organized into several sections containing the information for use by the deployment team.

4.3 SERVICE DFSS DEPLOYMENT PHASES

We categorize the deployment process, in term of evolution time, into three phases:

1. The predeployment phase to build the infrastructure
2. The deployment phase, where most activities will take place
3. The postdeployment phase, where effort needs to be sustained

4.3.1 Predeployment

Predeployment is a phase representing the period of time in which a leadership team lays the groundwork and prepares the company for service Six Sigma design implementation, ensures the alignment of their individual deployment plans, and creates synergy and heightened performance.

The first step in an effective service DFSS deployment starts with the top leadership of the deployment company. It is at this level that the team tasked with deployment works with the senior executives to develop a strategy and plan for deployment that is designed for success. The "marketing" and "selling" of the Six Sigma initiative culture should come from the top. Our observation is that senior leadership benchmark themselves across corporate America in terms of results, management style, and company aspirations. Six Sigma, and in particular DFSS, is no exception. The process usually starts with a senior leader or a pioneer who learns about Six Sigma and the benefits it brings to the culture and results. The pioneer starts the deployment one step at a time and begins challenging old paradigms. The old paradigms' guards come to its defense and try to block the deployment. Defense mechanisms fall one after the other in the face of the undisputable results achieved by several benchmarked deploying companies (GE, 3M, Motorola, Textron, Allied Signal, Bank of America, etc.). Momentum builds up and a team is formed to be tasked with deployment. As a first step, it is advisable that select senior leadership meet jointly as a team with the assigned deployment team in an off-site location (with limited distractions). This entails a balanced mix of strategic thinking, Six Sigma high-level education, interaction, and hands-on planning. On the education side, overviews of Six Sigma concepts, presentations of successful deployments, and demonstrations of Six Sigma statistical methods, improvement measures, and management controls will be very useful. Specifically, the following should be the minimum set of objectives of this launch meeting:

- Understand the philosophy and techniques of service DFSS and Six Sigma in general.
- Experience the application of some of the tools during the meeting.
- Brainstorm a deployment strategy and a corresponding deployment plan with high first-time capability.
- Understand the organizational infrastructure requirements for deployment.
- Set financial and cultural goals, targets, and limits for the initiative.
- Discuss the project pipeline and black belt resources in all phases of deployment.
- Put a mechanism in place to mitigate deployment risks and failure modes; for example, failure mode and effect analysis (FMEA) of a typical service DFSS deployment. Failure modes like the following are indicative of problematic strategies: training black belts before champions; deploying DFSS without multigenerational service and process plans and, if possible, technology road maps; and lack of valid data and measurement systems, leadership development, compensation plan, or change management process.
- Design a mechanism for tracking the progress of the initiative. Establish a robust "financial" management and reporting system for the initiative.

Once this initial joint meeting has been held, the deployment team could be replicated in other tiers of leadership whose buy-in is deemed necessary to *push* the initiative through the different functions of the company. A service Six Sigma *pull* system needs to be created and sustained in deployment and postdeployment phases. Sustainment indicates the establishment of bottom-up pulling power.

Service Six Sigma, including DFSS, has revolutionized many companies in the last 20 years. On the service side, companies in various industries can be found implementing service DFSS as a vehicle to plan growth, improve product/process quality and delivery performance, and reduce cost. Similarly, many deploying companies also find themselves reaping the benefits of increased employee satisfaction through the true empowerment Six Sigma provides. Factual studies of several successful deployments indicate that push and pull strategies need to be adopted based on needs and that they differ strategically by objective and phase of deployment. A push strategy is needed in the predeployment and deployment phases to jump start and operationalize deployment efforts. A pull system is needed in the postdeployment phase once sustainment is accomplished, to improve deployment process performance on a continuous basis. In any case, top and medium management should be on board at deployment, otherwise, the DFSS initiative will fade away eventually.

4.3.2 Predeployment Considerations

The impact of a DFSS initiative depends on the effectiveness of deployment, that is, how well the Six Sigma design principles and tools are practiced by the DFSS project teams. Intensity and constancy of purpose beyond the norm are required to constantly improve deployment. Rapid deployment of DFSS plus commitment, training, and practice characterize winning deploying companies.

In the predeployment phase, the deployment leadership should create a compelling business case for initiating, deploying, and sustaining DFSS as an effort. They need to raise general awareness about what DFSS is, why the company is pursuing it, what is expected of various people, and how it will benefit the company. Building the commitment and alignment among executives and deployment champions to aggressively support and drive deployment throughout the designated functions of the company is a continuous activity. Empowerment of leaders and DFSS operatives to effectively carry out their respective roles and responsibilities is key to success. A successful DFSS deployment requires the following prerequisites in addition to senior leadership commitment previously discussed

Establish Deployment Structure.[1] The first step taken by the senior deployment leader is to establish a deployment team to develop strategies and oversee deployment. With the help of the deployment team, the leader is responsible for designing, managing, and delivering the successful deployment of the initiative throughout the company, locally and globally. He or she needs to work with the human resources department to develop policy to ensure that the initiative becomes integrated into the culture, which may include integration with internal leadership development

[1]See Yang & El-Haik (2003) for product DFSS perspectives.

programs, career planning for black belts and deployment champions, a reward and recognition program, and progress reporting to the senior leadership team. In addition, the deployment leader needs to provide training, communication (as a single point of contact to the initiative), and infrastructure support to ensure consistent deployment.

The critical importance of the team overseeing the deployment to ensure smooth and efficient rollout cannot be overemphasized. This team puts the DFSS deployment effort in the path to success, whereby the proper individuals are positioned and support infrastructures are established. The deployment team is on the deployment forward edge, assuming the responsibility for implementation. In this role, team members perform a company assessment of deployment maturity, conduct a detailed gap analysis, create an operational vision, and develop a cross-functional Six Sigma deployment plan that spans human resources, information technology (IT), finance, and other key functions. Conviction about the initiative must be expressed at all times, even though in the early stages there is no physical proof of the company's specifics. They also accept and embody the following deployment aspects:

- Visibility of the top-down leadership commitment to the initiative (indicating a push system).
- Development and qualification of a measurement system with defined metrics to track the deployment progress. The objective here is to provide a tangible picture of deployment efforts. Later, a new set of metrics that target effectiveness and sustainment needs to be developed in maturity stages (end-of-deployment phase).
- Establish a goal-setting process in order to focus the culture on changing the process by which work gets done rather than adjusting current processes, leading to quantum rates of improvement.
- Strict adherence to the devised strategy and deployment plan.
- Clear communication of success stories that demonstrate how DFSS methods, technologies, and tools have been applied to achieve dramatic operational and financial improvements.
- Provide a system that will recognize and reward those who achieve success.

Deployment structure is not limited to the deployment team overseeing deployment both strategically and tactically, but also includes project champions, functional area deployment champions, owners of processes and designs for which solutions will be implemented, and master black belts (MBBs) who mentor and coach the black belts. All should have clearly defined roles and responsibilities with defined objectives. A premier deployment objective can be that the black belts are used as a task force to improve customer satisfaction, company image, and other strategic long-term objectives of the deploying company. To achieve such objectives, the deploying division should establish a deployment structure formed from the deployment directors, the centralized deployment team overseeing deployment, and master black belts (MBBs) with defined roles and responsibilities for long-term and short-term planning. The structure can take the form of a council meeting at definite, recurring times. We suggest using service DFSS to design the DFSS deployment process and strategy. The deployment team should:

- Develop a green belt structure of support to the black belts in every department.
- Cluster the green belts (GBs) as a network around the black belts for synergy and to increase the velocity of deployment.
- Ensure that the scopes of the projects are within control, that the project selection criteria are focused on the company's objectives, like quality, cost, customer satisfaction, delivery drivers, and so on.
- Hand off (match) the right scoped projects to black belts.
- Support projects with key up-front documentation like charters or contracts with financial analysis, highlighting savings and other benefits, efficiency improvements, customer impact, project rationale, and so on. Such documentation will be reviewed and agreed upon by primary stakeholders (deployment champions, design owners, black belts, and finance).
- Allocate the black belt resources optimally across many divisions of the company, targeting high-impact projects first, and create a long-term allocation mechanism to target a mix of DMAIC versus DFSS projects to be revisited periodically. In a healthy deployment, the number of DFSS projects should grow while the number of DMAIC projects should decrease over time. However, this growth in the number of DFSS projects should be managed. A growth model, such as an S-curve, can be modeled over time to depict this deployment performance. The initiating condition of how many and where DFSS projects will be targeted is a significant growth control factor. This is very critical aspect of deployment, in particular when the deploying company chooses not to separate the DMAIC and DFSS training tracks of the black belts and train the black belts in both methodologies.
- Available external resources will be used as leverage, when advantageous, to obtain and provide the required technical support.
- Promote and foster work synergy through the different departments involved in the DFSS projects.
- Maximize the utilization of the continually growing DFSS community by successfully closing most of the matured projects approaching the targeted completion dates.
- Keep leveraging significant projects that address company's objectives, in particular, the customer satisfaction targets.
- Maximize black belt certification turnover (set targets based on maturity).
- Achieve and maintain working relationships with all parties involved in DFSS projects, promoting an atmosphere of cooperation, trust, and confidence between them.

Other Deployment Operatives. A number of key people in the company are responsible for jump starting engagement in the company for successful deployment. The same people are also responsible for creating the momentum, establishing the culture, and driving DFSS through the company during the predeployment and deployment phases. This section describes who these people are in terms of their roles and responsibilities. The purpose is to establish clarity about what is expected of each deployment team member and to minimize the ambiguity that so often characterizes change initiatives that might otherwise be known as "the flavor-of-the month."

Deployment Champions. In the deployment structure, the deployment champion role is key. This position is usually given to an executive-ranked vice president assigned to various functions within the company (e.g., marketing, IT, communications, sales). Their tasks as part of the deployment team are to remove barriers within their functional area and to make things happen, review DFSS projects periodically to ensure that project champions are supporting their black belts' progress toward goals, assist with project selection, and serve as "change agents."

Deployment champions work full time at this assignment and should be at a level high enough to execute the top-down approach and the push system in the predeployment and deployment phases. They are key individuals with the managerial and technical knowledge required to create the focus and facilitate the leadership, implementation, and deployment of DFSS in designated areas of their respective organizations. In service DFSS deployment, they are tasked with recruiting, coaching, developing, and mentoring (but not training) black belts; identifying and prioritizing projects; leading product and process owners; removing barriers; providing the drumbeat for results; and expanding project benefits across boundaries via a mechanism of replication. Champions should develop a big-picture understanding of DFSS, deliverables, and tools to the appropriate level, and understand how DFSS fits within the service life cycle.

The deployment champion will lead his or her respective function's total quality efforts toward improving growth opportunities, quality of operations, and operating margins, among other functions, using service DFSS. This leader will have a blend of business acumen and management experience, and have a passion for process improvement. The deployment champions need to develop and grow a master black belt training program for the purpose of certifying and deploying homegrown future master back belts throughout deployment.

In summary, the deployment champion is responsible for broad-based deployment, establishing a common language, and effecting culture transformation by weaving Six Sigma into the company DNA, making it their own consistent, teachable point of view.

Project Champions. The project champions are accountable for the performance of black belts and the selection, scoping, and successful completion of black belt projects. The project champions remove roadblocks for black belts within their area of control and ensure timely completion of projects. The following considerations should be the focus of the deployment team project champions as they lay down their strategy:

- What does a DFSS champion need to know to be effective?
- How does one monitor impact and progress in projects?
- What are the expectations of senior leadership, the black belt population, and others?
- What are the expectations relative to the timeline for full adoption of DFSS into the development process?
- The playbook (reference) for the project champions.
- "Must have" versus "nice to have" tools. Learn DFSS project application.
- How does one use champions as "change agents"?
- Complete a FMEA exercise in their area, identifying deployment failure modes, ranking, and corrective actions. The FMEA will focus on potential failure modes in project execution.
- Plan for DFSS implementation: develop a timely deployment plan within their area of control, carry out project selection, select project resources and project pipeline.
- Develop guidelines, reference, and checklist (cheat sheet) for champions to help them understand (and force) compliance with service DFSS project deliverables.

The roles and responsibilities of a champion in project execution are a vital dimension of successful deployment that needs to be iterated in the deployment communication plan. Champions should develop their teachable point of view or resonant message.

A suggested deployment structure is presented in Figure 4.1.

Design/Process Owner. This operative is the owner of the process or design in which the DFSS project results and conclusions will be implemented. As owner of the design entity and resources, his or her buy-in is critical and he or she must be engaged early on. In predeployment, the design/process owners may be overwhelmed with the initiative and wondering why a black belt was assigned to fix their process. They need to be educated, consulted on project selection, and made responsible for the implementation of project findings. They should be tasked with the sustainment of project gains by tracking project success metrics after full implementation. Typically, they should serve as team members on the project, participate in reviews, and push the team to find permanent innovative solutions. In the deployment and postdeployment phases, process owners should be the first in line to staff their projects with "belts."

Master Black Belt (MBB). A master black belt should possess expert knowledge of the full Six Sigma tool kit and have proven experience with DFSS. As this is a full-

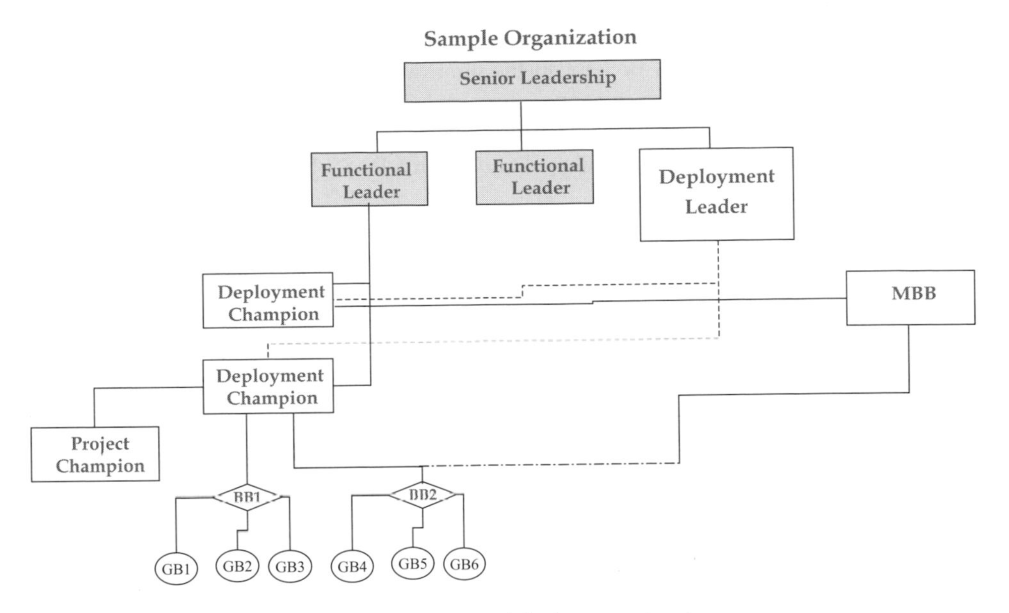

Figure 4.1 Suggested deployment structure.

time assignment, he or she should also have experience in training, mentoring, and coaching black belts, green belts, champions, and the leadership. Master black belts are ambassadors for the business and the DFSS initiative, people who will be able to go work in a variety of business environments with varying levels of Six Sigma penetration. A master black belt is a leader with a good command of statistics as well as the practical ability to apply Six Sigma in an optimal manner for the company. Knowledge of lean operation is also required to drive the initiative very fast. The MBB should be adaptable to the deployment phase requirements.

Some businesses trust them with managing the deployment and objectives of large projects. MBBs also need to get involved with project champions in project scoping, and coach the senior teams in each key function.

Black Belts.[2] Black belts are the critical resource of deployment as they initiate projects, apply service DFSS tools and principles, and close them with tremendous benefits. Selected for technical proficiency, interpersonal skills, and leadership ability, a black belt is an individual who solves difficult business issues for the last time. Typically, the black belts have a service life in the deployment phase of a couple of years. Nevertheless, their affect as disciples of service DFSS when they finish their service life (postdeployment for them) and move on as the next generation of leaders cannot be trivialized. It is recommended that a fixed popula-

[2]Although black belts may work in many areas, as seen in the previous section, we chose to separate them in one separate section due to their significant deployment role.

tion of black belts (usually computed as a percentage of affected functions where service DFSS is deployed) be kept in the pool during the designated service life. This population is not static, however. It is replenished every year by new blood. Repatriated black belts, in turn, replenish the disciple population, and the cycle continues until sustainment is achieved and service DFSS becomes the way of doing design business.

Black belts will learn and understand service methodologies and principles and find application opportunities within the project, cultivate a network of experts, train and assist others (e.g., green belts) in new strategies and tools, leverage surface business opportunities through partnerships, and drive concepts and methodology into the way of doing work.

The deployment of black belts is a subprocess within the deployment process itself with the following steps: (1) black belt identification, (2) black belt project scoping, (3) black belt training, (4) black belt deployment during the service life, and (5) black belt repatriation into the mainstream.

The deployment team prepares designated training waves or classes of service black belts to apply DFSS and associated technologies, methods, and tools on scoped projects. Black belts are developed by project execution, training in statistics and design principles with on-the-project application, and mentored reviews. Typically, with a targeted quick cycle time, a black belt should be able to close at least three projects a year. Our observations indicate that black belt productivity, on the average, increases after his/her training projects. Following their training focused, descoped project, black belt projects can get more complex and evolve into cross-function, supply chain, and customer projects.

The black belts are the leaders of the future. Their visibility should be apparent to the rest of the organization and they should be cherry-picked to join the service DFSS program with "leader of the future" statue. Armed with the right tools, processes, and DFSS principles, black belts are the change agents the deploying company should utilize to achieve its vision and mission statements. They need to be motivated and recognized for their good efforts while mentored at both the technical and leadership fronts by the master black belt and the project champions. Oral and written presentation skills are crucial for their success. To increase the effectiveness of the black belts, we suggest building a black belt collaboration mechanism for the purpose of maintaining structures and environments to foster individual and collective learning of initiative and DFSS knowledge, including initiative direction, vision, and prior history. In addition, the collaboration mechanism, whether virtual or physical, could serve as a focus for black belt activities to foster team building, growth, and inter- and intrafunction communication and collaboration. Another important reason for establishing such a mechanism is to ensure that the deployment team gets its information in an accurate and timely fashion to prevent and mitigate failure modes downstream of deployment and postdeployment phases. Historical knowledge might include lessons learned, best practices sharing, and deployment benchmarking data.

Table 4.1 summarizes the roles and responsibilities of the deployment operatives presented in this section. Figure 4.2 depicts the growth curve of the Six Sigma de-

Table 4.1 Deployment operative roles summary

Project champions	Manage projects across company Approve the resources Remove the barriers Create Vision
Master black belts	Review project status Teach tools and methodology Assist the champion Develop local deployment plans
Black belts	Train their teams Apply the methodology and lead projects Drive projects to completion
Green belts	Same as black belts (but done in conjunction with other full-time job responsibilities)
Project teams	Implement process improvements Gather data

ployment operatives. It is the responsibility of the deployment team to shape the duration and slopes of these growth curves subject to the deployment plan. The pool of black belts is replenished periodically. The 1% rule (one black belt per 100 employees) has been adopted by several successful deployments. The number of MBBs is a fixed percentage of the black belt population. Current practice ranges from 10 to 20 black belts per MBB.

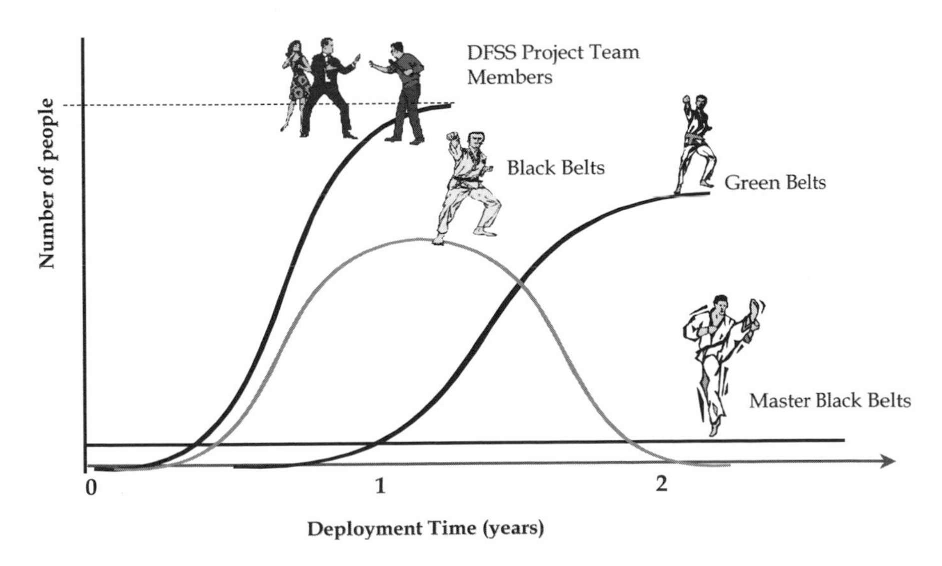

Figure 4.2 Deployment operatives growth curves.

Green Belt. A green belt is an employee of the deploying company who has been trained in Six Sigma and will participate on project teams as part of their full-time job. The green belt's knowledge of Six Sigma is less than a black belt's. The green belt's business knowledge of his or her company is a necessity to ensure success of the process improvement task. The green belt employee plays an important role in executing the Six Sigma process in day-to-day operations by completing smaller-scope projects. Black belts should be networked around green belts to support and coach them. Green belt training is not for awareness. The deployment plan should enforce certification while tracking their project status as control mechanisms for deployment. Green belts, like black belts, should be closing projects as well.

In summary, green belts are employees trained in Six Sigma methodologies who conduct or contribute to projects that require Six Sigma application. Following successful completion of training, green belts will be able to participate in larger projects being conducted by black belts, lead small projects, and apply Six Sigma tools and concepts to daily work.

Communication Plan. To ensure the success of service DFSS, the deployment team should develop a communication plan that highlights the key steps as service DFSS is being deployed. In doing so, they should target the audience that will receive necessary communication at various points in the deployment process with identifiable possible media of communication deemed most effective by the company. The deployment team should outline the overriding communication objectives at each major phase of service DFSS deployment and provide a high-level, recommended communication plan for each of the identified communicators during company DFSS initialization.

As service DFSS is deployed in a company, we recommend that various people communicate certain messages at certain times. For example, at the outset of deployment, the CEO should send a strong message to the executive population that the corporation is adopting service DFSS, why it is necessary, who will be leading the effort both at leadership and deployment team levels, why their commitment and involvement is absolutely required, and other important items. The CEO also sends, among other communiqués to other audiences, a message to the deployment champions, explaining why they have been chosen, what is expected of them, and how they are empowered to enact their respective roles and responsibilities.

Several key people will need to communicate key messages to key audiences as DFSS is initiated, deployed, and sustained; for example, the training and development leader, finance leader, HR leader, IT leader, project champions, deployment champions (functional leaders), mangers and supervisors, black belts, and green belts to name a few. Every leader involved in DFSS processes must be totally committed to the cause to avoid derailment of cultural evolution driven by DFSS. Every leader must seek out information from the deployment team to validate their conviction to the process.

To further assist in effective communications, the leader and others responsible for communicating DFSS deployment should delineate who delivers messages to whom during the predeployment phase. It is obvious that certain people have primary communication responsibility during the initial stages of Six Sigma deployment, including the CEO, Service DFSS deployment leader, and deployment champions. The company communications leader plays a role in supporting the CEO, deployment leader, and others leaders as they formulate and deliver their communiqués in support of predeployment. The communication plan should include the following minimum communiqués:

- Why the company is deploying DFSS, along with several key points about how Six Sigma supports and is integrated with company's vision, including other business initiatives.
- Set of financial targets, operational goals, and metrics that will be providing structure and guidance to the DFSS deployment effort, to be done with consideration of the targeted audience.
- A breakdown of where DFSS will be focused in the company; a roll-out sequence by function, geography, product or other scheme; a general time frame for how quickly and aggressively DFSS will be deployed.
- A firmly established and supported long-term commitment to the DFSS philosophy, methodology and anticipated results.
- Specific managerial guidelines to control the scope and depth of deployment for a corporation or function.
- Review and examine key performance metrics to ensure the progressive utilization and deployment of DFSS.
- Commitment of part-time and full-time deployment champions, project champion, and full-time black belt resources.

Service DFSS Project Sources. The successful deployment of the DFSS initiative within a company is tied to projects derived from the company's breakthrough objectives, multigeneration planning, growth and innovation strategy, and chronic, pressing redesign issues. Such service DFSS project sources can be categorized as retroactive or proactive sources. In either case, an active measurement system should be in place for both internal and external critical-to-satisfaction (CTS's) metrics, sometimes called the "big *Y*s." The measurement system should pass a Gauge R&R study in all big *Y* metrics.

How should big *Y*s be defined? This question underscores why we need to decide early who is the primary customer (internal and external) of our potential DFSS project. What is the big *Y* (CTS) in customer terms? It does us no good, for example, to develop a delivery system to shorten flow processes if the customer is mainly upset with quality and reliability. Likewise, it does us little good to develop a project to reduce tool breakage if the customer is actually upset with inventory cycle losses. It pays dividends to later project success to know the big *Y*. No big *Y* (CTS) simply means no project! Potential projects with hazy big *Y* definitions are setups for black belt failure. Again, it is unacceptable to not know the big *Y*s of top problems (retroactive project sources) or those of proactive project sources aligned with the annual objectives, growth and innovation strategy, benchmarking, multigeneration process and product planning, and technology road maps.

On the proactive side, black belts will be claiming projects from a multigenerational service/process plan or from the big *Y*s replenished, prioritized project pipeline. Green belts should be clustered around these key projects for the deploying function or business operations and tasked with assisting the black belts, as suggested by Figure 4.3.

We need some useful measure of big *Y*s, in variable terms,[3] to establish the transfer function, $Y = f(x)$. The transfer function is the means for determining customer satisfaction, or other big *Y*s, and can be identified by a combination of design mapping and design of experiment (if transfer functions are not available or cannot be derived). A transfer function is a mathematical relationship in the associated mapping, linking controllable and uncontrollable factors. Transfer functions can be derived, empirically obtained from a DOE, or regressed using historical data. In some cases, no closed mathematical formula can be obtained and the DFSS team can resort to modeling. In the DFSS project road map, there is a transfer function for every functional requirement, for every design parameter, for every process variable, and, ultimately, for every big *Y*.

Sometimes, we find that measurement of the big *Y* opens windows to the mind with insights powerful enough to solve the problem immediately. It is not rare to find customer complaints that are very subjective and unmeasured. The black belt needs to find the best measure available to his/her project big *Y* to help describe the variation faced and to support $Y = f(x)$ analysis. The black belt may have to develop a measuring system for the project to be true to the customer and big *Y* definition!

We need measurements of the big *Y* that we can trust. Studying problems with false measurements leads to frustration and defeat. With variable measurements, the issue is handled as a straightforward Gauge R&R question. With attribute or other subjective measures, it is an attribute Measurement system analysis (MSA) issue. It is tempting to ignore MSA of the big *Y*, but this is not a safe practice. More that 50% of the black belts we coached encounter MSA problems in their projects. This issue in the big *Y* measurement is probably worse because little thought is conventionally given to MSA at the customer level. The black belts should make every effort to assure themselves that their big *Y* measurement is error minimized. We need to be able to establish a distribution of *Y* from which to model or draw samples for $Y = f(x)$ study. The better the measurement of the big *Y*, the better the black belt can see the distribution contrasts needed to yield or confirm $Y = f(x)$.

What is the value to the customer? This should be a moot point if the project is a top issue. The value decisions have already been made. Value is a relative term with numerous meanings. It may be cost, appearance, status, but the currency of value must be decided on. In Six Sigma, it is a common practice to ask that each project generate average benefits greater than $250,000. This is seldom a problem in top projects that are aligned to business issues and opportunities.

[3]The transfer function will be weak and questionable without it.

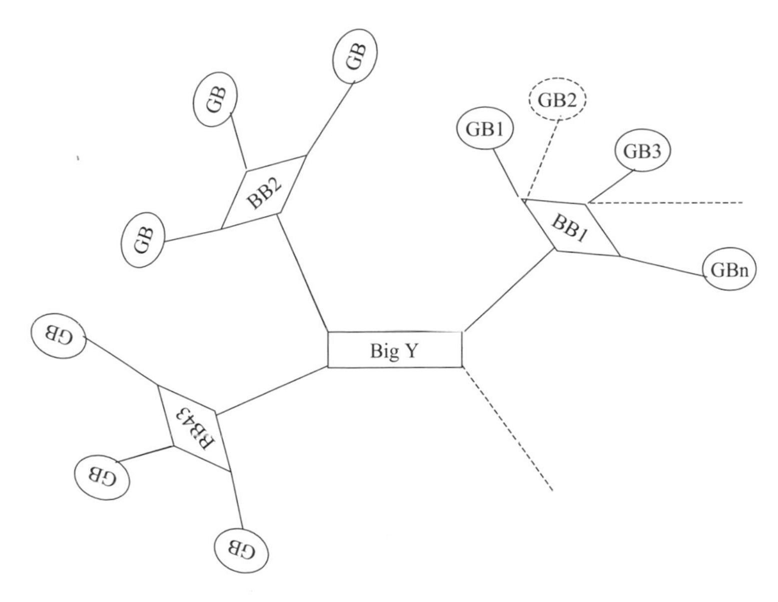

Figure 4.3 Green belt (GB) and black belt (BB) clustering scheme.

The black belt together with the finance individual assigned to the project should decide on a value standard and do a final check to ascertain that the potential project value is greater than the minimum. High-value projects are not necessarily harder than low-value projects. Projects usually hide their level of complexity until solved. Many low-value projects are just as difficult to complete as high-value projects, so the deployment champions should leverage their effort by value. Deployment management, including the local master black belt, has the lead in identifying redesign problems and opportunities as good potential projects. The task, however, of going from a potential to an assigned Six Sigma project belongs to the project champion. The deployment champion selects a project champion, who then carries out the next phases. The champion is responsible for the project scope, black belt assignment, ongoing project review, and, ultimately, the success of the project. This is an important and responsible position and must be taken very seriously. A suggested project initiation process is depicted in Figure 4.4.

It is a significant piece of work to develop a good project but black belts, particularly those already certified, have a unique perspective that can be of great assistance to the project champions. Green belts, as well, should be taught fundamental skills useful in developing a project scope. Black belt and green belt engagement is key to helping champions fill the project pipeline, investigate potential projects, prioritize them, and develop achievable project scopes. It is the observation of many skilled problem solvers that adequately defining the problem and setting up a solution strategy consumes the most time on the path to a successful project. The better we define and scope a project, the faster the deploying company and its customer base will benefit from the solution. That is the primary Six Sigma objective.

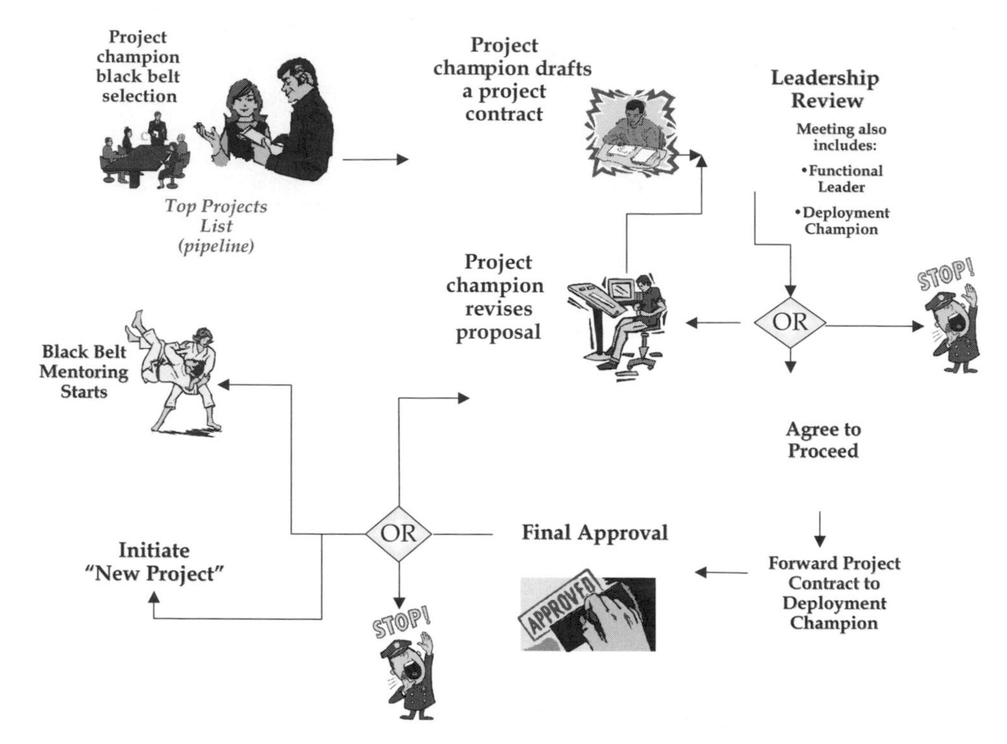

Figure 4.4 Service DFSS project initiation process.

It is the responsibility of management and the deployment and project champions, with the help of the design/process owner, to identify both retroactive and proactive sources of DFSS projects that are important enough to assign the company's limited, valuable resources to find a Six Sigma solution. Management is the caretaker of the business objectives and goals. They set policy, allocate funds and resources, and provide the personnel necessary to carry out the business of the company. Individual black belts may contribute to the building of a project pipeline but the overview is entirely management's responsibility.

It is expected that an actual list of projects will always exist and be replenished frequently as new information or policy directions emerge. Sources of information from which to populate the list include all retroactive sources, support systems such as warranty systems, internal manufacturing and production systems related to problematic metrics such as scrap and rejects, customer repairs/complaints database, and many others. In short, the information comes from the strategic vision and annual objectives, multigeneration service plans, customer surveys or other engagement methods, and the daily business of deployment champions, and it is their responsibility to approve what gets into the project pipeline and what does not. In general, service DFSS projects usually come from processes that have reached their ultimate capability (entitlement) and are still problematic, or those targeting new process designs.

In the case of retroactive sources, projects derive from problems that champions agree need solutions. Project levels can be reached by applying the "five why" technique (See Figure 4.5) to dig into root causes prior to the assignment of the black belt.

A scoped project will always provide the belt a good starting point and reduce the identify phase cycle time within the ICOV DFSS approach. They must prioritize because the process of going from a potential project to a properly scoped black belt project requires significant work and commitment. There is no business advantage in spending valuable time and resources on something with low priority. Usually, a typical company scorecard may include metrics relative to safety, quality, delivery, cost, and environment. We accept these as big sources (buckets), yet each category has a myriad of it's own problems and opportunities that can quickly drain resources if champions do not prioritize. Fortunately, the Pareto principle applies so we can find leverage in the significant few. It is important to assess each of the buckets according to the 80–20 principles of Pareto. In this way, the many are reduced to a significant few that still control over 80% of the problem in question. These need review and renewal by management routinely as the business year unfolds. The top project list emerges from this as a living document.

From the individual bucket Pareto lists, champions again must give us their business insight to plan an effective attack on the top issues. Given key business objectives, they must look across the several Pareto diagrams, using the 80–20 principle, and sift again until we have few top issues on the list with the biggest impact on the business. If the champions identify their biggest problem elements well, based on management business objectives and the Pareto principle, then how could any manager or supervisor in their right mind refuse to commit resources to achieve a solution? Solving any problems but these yields only marginal improvement.

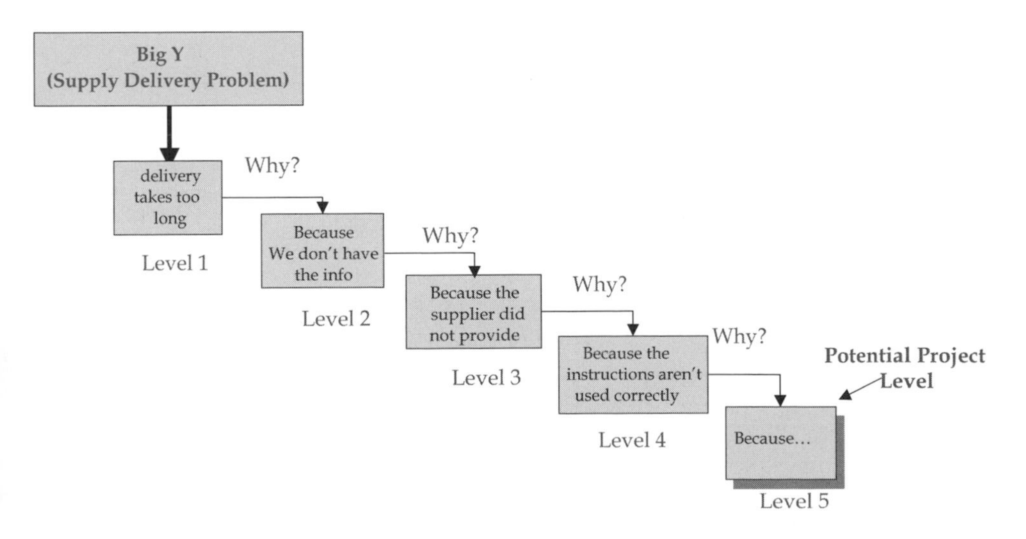

Figure 4.5 The "five why" scoping technique.

Resource planning for black belts, green belts, and other personnel is visible and simplified when they are assigned to top projects on the list. Opportunities to assign other personnel such as project team members are clear in this context. Your local deployment champion and/or master black belt need to manage the list. Always remember that a project focused on the top problems is worth a lot to the business. All possible efforts must be exerted to scope problems and opportunities into projects that black belts can drive to a Six Sigma solution. The process steps in Figure 4.6 help us turn a problem into a scoped project.

A critical step in the process is to define the customer. This is not a question that can be taken lightly! How can customers, either internal or external to the business, be satisfied if the black belt is not sure who they are? The black belt and his or her team must know customers to understand their needs and preferences. Never make guesses or assumptions about what your customers need—ask them. Several customer interaction methods will be referenced in the next chapters. For example, the customer of a service project on improving company image is the buyer of the service. On the other hand, if the potential project is to reduce tool breakage in a manufacturing process, then the buyer is too far removed to be the primary customer. Here the customer is more likely the process owner or other business unit manager. Certainly, if we reduce tool breakage, then we gain efficiency that may translate to cost or availability satisfaction, but this is of little help in planning a good project to reduce tool breakage.

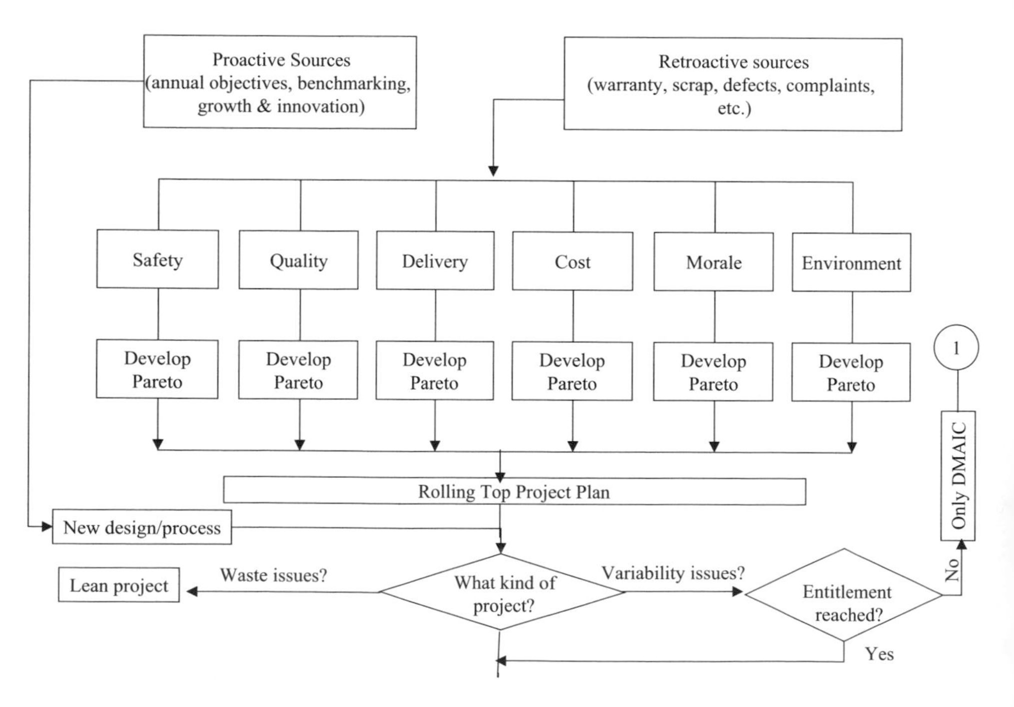

Figure 4.6 Six Sigma project identification and scoping process.

No customer, no project! Know your customer. It is especially unacceptable to not know your customer in the top project pipeline. These projects are too important to allow this kind of lapse.

Proactive DFSS Project Sources: Multigeneration Planning. A multigeneration plan is concerned with developing a timely design evolution of a service and for finding optimal resource allocation. An acceptable plan must be capable of dealing with uncertainty about future markets and availability of services when demanded by the customer. The incorporation of uncertainty into a resource-planning model of a service multigeneration plan is essential. For example, on the personal financial side, it was not all that long ago that a family was only three generations deep—grandparent, parent, and child. But as life expectancies increase, four generations are common and five generations are no longer unheard of. The financial impact of this demographic change has been dramatic. Instead of a family focused only in terms of its own finances, it may be necessary to deal with financial issues that cross generations. Whereas once people lived only a few years into retirement, now

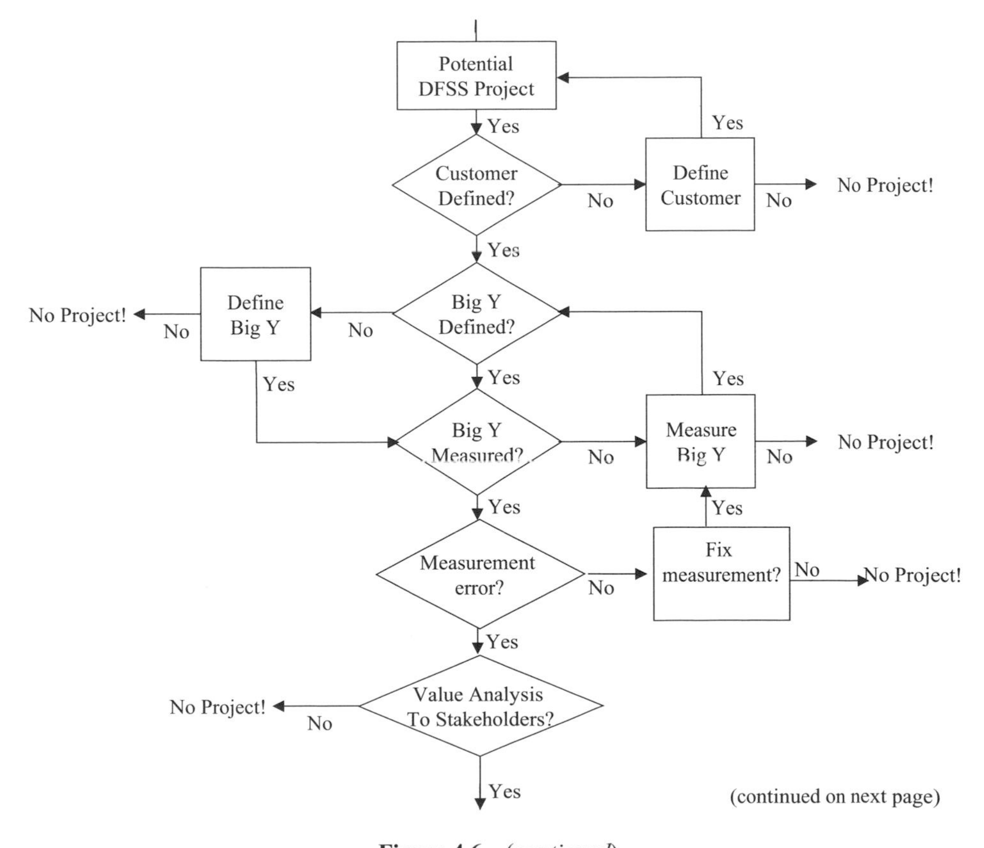

(continued on next page)

Figure 4.6 (*continued*)

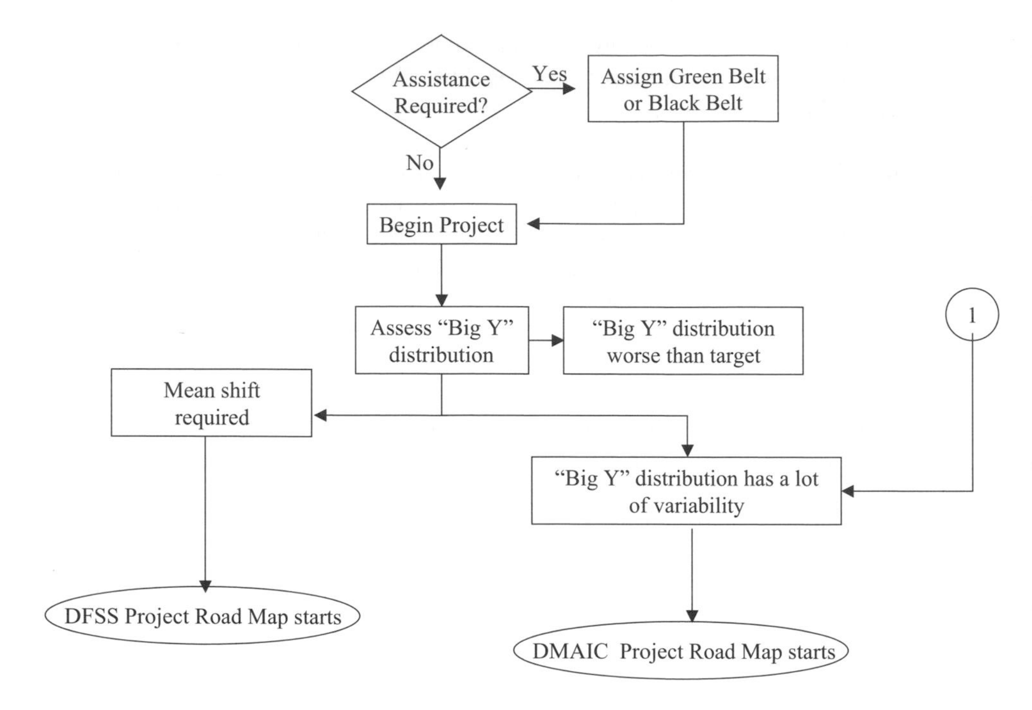

Figure 4.6 (*continued*)

they live 30 years or more. If the parents cannot take care of themselves, or they cannot afford to pay for high-cost, long-term care either at home or in a facility, their children may need to step forward. A host of financial issues are involved such as passing on the estate, business succession, college versus retirement, life insurance, and lending money. These are only a smattering of the many multigenerational financial issues that may arise.

Service design requires multigeneration planning that takes into consideration demand growth, the level of coordination in planning, and resource allocation among functions within a company. The plan should take into consideration uncertainties in demand, technology, and other factors by means of defining strategic design generations, which reflect gradual and realistic possible evolutions of the service of interest. The decision analysis framework needs to be incorporated in order to quantify and minimize risks for all design generations. Advantages associated with generational design in mitigating risks, financial support, economies of scale, and reductions of operating costs are key incentives for growth and innovation.

The main step is to produce generation plans for service design CTSs and functional requirements or other metrics with an assessment of uncertainties about achieving them. One key aspect for defining the generation is to split the plan into periods in which flexible generations can be defined. The beginning of generational periods may coincide with milestones or relevant events. For each period, a generational plan gives an assessment of how each generation should perform against an adopted set of metrics. For example, a company generational plan for its ERP system may be as depicted in Figure 4.7. This is a multigenerational plan that lays out the key metrics and the enabling technologies and processes by time horizon.

Training. To jump-start the deployment process, DFSS training is usually outsourced in the first year or two of deployment. The deployment team needs to devise a qualifying scheme for training vendors once their strategy is finalized and approved by the senior leadership of the company. Specific training session content for executive leadership, champions, and black belts should be planned with the selected vendor's heavy participation. This facilitates a coordinated effort, allowing better management of the training schedule and more prompt service. In this phase, simple guidelines for training deployment champions, project champions, and any other individuals whose scope of responsibility intersects with the training function needs to be discussed. Attendance is required all day of each day of training. In order to get the full benefit of the training course, each attendee needs to be present for all of the material that is presented. Each training course should be carefully developed and condensed into the shortest possible period by the vendor. Missing any part of a course will result in a diminished understanding of the covered topics and, as a result, may severely delay the progression of projects.

	Generation 0 "As Is"	Generation 1 120 days	Generation 2 6 - 12 months
VISION		Use DFSS to create standard process with scalable features that provide a framework to migrate to future states	Evolve process into SAP environment and drive 20% productivity improvement
METRICS			
Touch Time	Unknown	L – 40 hrs M – 20 hrs S – 10 hrs	Same
Cycle Time	Manual, 1 – 20 weeks	Manual, 3 - 10 days	Automated
Win Rate	Unknown	Measured	Automated
Accuracy	Unknown	Accuracy	Accuracy
Completeness	Unknown	Completeness	Completeness
Win Rate	Unknown	Win Rate	Win Rate
Compliance	Hope	Planned	Mistake proofed
Auditable/ Traceable	Hope	Planned	Mistake proofed
SCOPE		Service 1	Service 2

Figure 4.7 ERP bid and proposal process (service design multigeneration plan example).

Existence of a Service Program Development Management System. Our experience is that a project road map is required for successful DFSS deployment. The road map works as a compass, leading black belts to closure by laying out the full picture of the DFSS project. We would like to think of this road map as a recipe that can be further tailored to the customized application within the company's management system that spans the design life cycle (see Chapter 1). Usually, the service DFSS deployment team has two choices at this point:

1. Develop a new business process management system (BPMS; see Chapter 8) to include the proposed DFSS roadmap. Examples are advanced product quality planning (APQP) in the automotive industry, integrated product and process design (IPPD) in the aerospace industry, the Project Management Institute (PMI) process, and James Martin in software industry. It is the experience of the authors that many companies lack such universal discipline in a practical sense. This choice is suitable for companies practicing a variety of BPMS and hoping that alignment will evolve. The BPMS should span the service life cycle presented in Chapter 1
2. Integrate with the current BPMS by laying this roadmap over it and synchronizing when and where needed.

In either case, the DFSS project will be paced at the speed of the leading program from which the project was derived in the BPMS. Initially, highly leverage projects should target subsystems that the business and the customer are sensitive to. A sort of requirement flowdown or cascading method should be adopted to identify these subsystems. Later, when DFSS becomes the way of doing business, system-level DFSS deployment becomes the norm and the issue of synchronization with the BPMS will diminish eventually. Actually, the BPMS will be crafted to reflect the DFSS learning experience that the company gained over years of experience.

4.3.3 Deployment

This phase is the period of time in which champions are trained and select initial black belt projects, as well as when the initial wave of black belts are trained and projects are completed that yield significant soft and hard operational benefits. The training encompasses most of the deployment activities in this phase and is discussed in the following section. Additionally, this phase includes the following assignments of the deployment team:

- Reiterate to key personnel their responsibilities at critical points in the deployment process.
- Reinforce the commitment of project champions and black belts to aggressively execute selected improvement projects. Mobilize and empower both populations to effectively carry out their respective roles and responsibilities.

- Recognize exemplary performance in execution and in culture at the project champion and black belt levels.
- Inform the general employee population about the tenets of Six Sigma and the deployment process.
- Build information packets for project champions and black belts that contain administrative, logistical, and other information they need to execute their responsibilities at given points in time.
- Document and publicize successful projects and the positive consequences for the company and its employees.
- Document and distribute project savings data by business unit, product, or other appropriate areas of focus.
- Hold Six Sigma events or meetings with all employees at given locations where leadership is present and involved and where such topics are covered.

4.3.3.1 Training. The critical steps in DFSS training are (1) determining the content and outline, (2) developing the materials, and (3) deploying training classes. In doing so, the deployment team and their training vendor of choice should be very cautious about cultural aspects and weave the culture change into the soft side of the initiative training. Training is the significant mechanism within deployment that, in addition to equipping trainees with the right tools, concepts and methods, will expedite deployment and help shape a data-driven culture. This section will present a high-level perspective of the training recipients and what type of training they should receive. The training recipients are arranged below by level of complexity.

Senior Leadership. Training for senior leadership should include an overview of business and financial benefits of implementation, benchmarking of successful deployments, and specific training on tools to ensure successful implementation.

Deployment Champions. Training for deployment champions is more detailed than that provided to senior leadership. Topics would include the DFSS concept, methodology, and "must-have" tools and processes to ensure successful deployment within their function. A class focused on how to be an effective champion as well as their roles and responsibilities is often beneficial.

Master Black Belts. Initially, experienced master black belts are hired from the outside to jump-start the system. Additional homegrown MBBs may need to have additional training beyond their black belt training. Training for master black belts must be rigorous about the concept, methodology, and tools, as well as include detailed statistics training, computer analysis, and other tool application. Their training should include soft and hard skills to get them to a level of proficiency compatible with their roles. On the soft side, topics like strategy, deployment lessons learned, their roles and responsibilities, presentation and writing skills, leadership and resource management, and critical success factors benchmarking history and outside deployment are included. On the hard side, typical training may go into the theory of topics like DOE, ANOVA, axiomatic design, theory of inventive problem solving (TRIZ), hypothesis testing of discrete random variables, and lean tools.

Black Belts. As project leaders, the black belts will implement the DFSS methodology and tools within a function on projects aligned with the business objectives. They lead projects, institutionalize a timely project plan, determine appropriate tool use, perform analyses, and act as the central point of contact for their projects. Training for black belts includes detailed information about the concept, methodology and tools. Depending on the curriculum, the duration is usually between three to six weeks on a monthly schedule. Black belts will be assigned a training-focused, descoped project that has ample opportunities for tool application to foster learning while delivering deployment objectives. The weeks between the training sessions will be spent on gathering data, forming and training their teams, and applying concepts and tools where necessary. DFSS concepts and tools flavored by some soft skills are the core of the curriculum. Of course, DFSS training and deployment will be synchronized with the service development process already adopted by the deploying company. We provide in Chapter 5 of this book a suggested service DFSS project road map to serve as a design algorithm for the Six Sigma team. The road map will work as a compass, leading black belts to closure by laying out the full picture of a typical DFSS project.

Green Belts. The green belts may also take training courses developed specifically for black belts where there needs to be more focus. Short-circuiting theory and complex tools to meet the allocated short training time (usually less than 50% of black belt training period) may dilute many subjects. Green belts can resort to their black belt network for help on complex subjects and for coaching and mentoring.

4.3.3.2 Six Sigma Project Financial Aspects. In general, DFSS project financials can be categorized as *hard* or *soft* savings and are mutually calculated or assessed by the black belt and the assigned financial analyst to the project. The financial analyst assigned to a DFSS team should act as the lead in quantifying the financials related to the project "actions" at the initiation and closure phases, assist in identification of "hidden factory" savings, support the black belt on an ongoing basis, and if financial information is required from areas outside his/her area of expertise, he/she needs to direct the black belt to the appropriate contacts, follow up, and ensure that the black belt receives the appropriate data. The analyst, at project closure, should ensure that the appropriate stakeholders concur with the savings. This primarily affects processing costs, design expense, and nonrevenue items for rejects not directly led by black belts from those organizations. In essence, the analyst needs to provide more than an audit function.

"Hard savings" are defined as measurable savings associated with improvements in repairs, rework, absorptions, writeoffs, concessions, inspections, material costs, warranty, labor savings (will be achieved/collectable through work rebalances), revenue associated with reductions in customer dissatisfaction, cash flow savings (i.e., inventory), and other values of lost customer satisfaction. Hard savings are calculated against present operating levels, not against a budget or a plan. They represent the bottom-line savings that directly affect the company's income statement and cash flow and are the result of measurable product, service, and process improvements. The effect on company financial statements will be determined off-line by the appropriate company office. "Soft" savings are less direct in nature and include projects that open floor space (as a side benefit), may allow for the location of future operations, reduce vehicle weight, may enable other design actions to delete expensive light weight materials, and lead to cost avoidance. Cost avoidance is usually confused with cost savings. For example, employing an ATM instead a teller to reduce operating cost is avoidance rather than saving.

The financial analyst should work with the black belt to assess the projected annual financial savings based on the information available at that time (e.g., scope, expected outcome). This is not a detailed review, but a rough order-of-magnitude approval. These estimates are expected to be revised as the project progresses and more accurate data become available. The project should have the potential to achieve the annual target, usually $250,000. The analyst confirms the business rationale for the project where necessary.

Yang and El-Haik (2003) developed a scenario of black belt target cascading that can be customized to different applications. It is based on project cycle time, number of projects handled simultaneously by the black belt, and their importance to the organization.

4.3.4 Postdeployment Phase

This phase spans the period of time during which subsequent waves of black belts are trained, when the synergy and scale of Six Sigma builds to a critical mass, and when additional elements of DFSS deployment are implemented and integrated.

In what follows, we present some thoughts and observations that were gained through our deployment experience of Six Sigma and, in particular, DFSS. The purpose is to determine factors that lead to maintaining and expanding the momentum of DFSS deployment so that it becomes sustainable.

This book presents the service DFSS methodology that exhibits the merging of many tools at both the conceptual and analytical levels and penetrates dimensions like characterization, optimization, and validation by integrating tools, principles, and concepts. This vision of DFSS is a core competency in a company's overall technology strategy to accomplish its goals. An evolutionary strategy that moves the deployment of the DFSS method toward the ideal culture is discussed. In the strategy, we have identified the critical elements, needed decisions, and deployment concerns.

The literature suggests that more innovative methods fail immediately after initial deployment than at any other stage. Useful innovation attempts that are challenged by cultural change are not directly terminated, but are allowed to fade slowly and silently. A major reason for the failure of technically viable innovations is the inability of leadership to commit to an integrated, effective, cost-justified, and evolutionary program for sustainability that is consistent with company's mission. The DFSS deployment parallels in many aspects the technical innovation challenges from the cultural perspective. The DFSS initiatives are particularly vulnerable if they are too narrowly conceived, built upon only one major success mechanism, do not fit the larger organizational objectives. The tentative top-down deployment approach works where the top leadership support is the significant driver. However, this approach can be strengthened when built around superior mechanisms like DFSS as a design approach and the methodologies are attractive to designers who want to become more proficient in their jobs.

Although there is a need to customize the deployment strategy, it should not be rigid. The strategy should be flexible enough to meet expected improvements. The deployment strategy itself should be DFSS driven and responsive to anticipated changes. It should be insensitive to expected swings in the financial health of company and should be attuned to the company's objectives on a continuous basis.

The strategy should consistently build coherent linkages between DFSS and daily design business. For example, engineers and architects need to see how all of the principles and tools fit together, complement one another, and build toward a coherent whole process. DFSS needs to be seen, initially, as an important part, if not the central core, of an overall effort to increase technical flexibility.

4.3.4.1 DFSS Sustainability Factors. In our view, DFSS possesses many inherent sustaining characteristics that are not offered by other current design practices. Many deign methods, some called best practices, are effective if the design is at a low level and needs to satisfy minimum number of functional requirements, for example, a component or a process. As the number of the requirements increases (the design becomes more complex), the efficiency of these methods decreases. In addition, these methods are hinged on heuristics and developed algorithms (e.g., design for processability; see Chapter 10), limiting their application across the different development phases.

The process of design can be improved by constant deployment of DFSS, which begins from a different premise, namely, the principle of design. The design axioms and principles are central to the conception part of DFSS. As will be defined in Chapter 8, axioms are general principles or truths that cannot be derived, except when there are no counterexamples or exceptions. Axioms are fundamental to many engineering disciplines; examples are thermodynamics laws, Newton's laws, and the concepts of force and energy. Axiomatic design provides the principles to develop a good design systematically and can overcome the need for customized approaches.

In a sustainability strategy, the following attributes would be persistent and pervasive features:

- A deployment measurement system that tracks the critical-to-deployment requirements and failure modes, and implements corrective actions.
- Continued improvement in the effectiveness of DFSS deployment by benchmarking other successful deployment elsewhere.

- Enhanced control (over time) over the company's objectives via selected DFSS projects that lead to rapid improvement.
- Extended involvement of all levels and functions.
- Embedding DFSS into the everyday operations of the company.

The prospectus for sustaining success will improve if the strategy yields a consistent day-to-day emphasis of recognizing that DFSS represent a cultural change and a paradigm shift, and allows the necessary time for projects success. Several deployments found it very useful to extend their DFSS initiative to key suppliers and extending these beyond the component level to subsystem- and system-level projects. Some call these projects intraprojects when they span different areas, functions, and business domains. This ultimately will lead to integrating the DFSS philosophy as a superior design approach within the BPMS and align the issues of funding, timing, and reviews to the embedded philosophy. As a side bonus of the deployment, conformance to narrow design protocols will start fading away. In all cases, sustaining leadership and managerial commitment to adopting appropriate, consistent, relevant, and continuing reward and recognition mechanisms for black belts and green belts is critical to the overall sustainment of the initiative.

The vision is that DFSS as a consistent, complete, fully justified, and usable process should be expanded to other areas. The deployment team should keep an eye on the changes that are needed to accommodate altering a black belt's tasks from individualized projects to broader-scope, intrateam assignments. A prioritizing mechanism for future projects that targets the location, size, complexity, involvement of other units, type of knowledge to be gained, and potential for fit within the strategic plan should be developed.

Another sustaining factor lies in providing relevant on-time training and opportunities for competency enhancement of black belts and green belts. The capacity to continue learning and the alignment of rewards with competency and experience must be fostered. Instituting an accompanying accounting and financial evaluation that enlarges the scope of consideration of the impact of the project on both fronts—hard and soft savings—is a lesson learned. Finance and other resources should be moved upfront toward the beginning of the design cycle in order to accommodate the DFSS methodology.

If the DFSS approach is to become pervasive as a central culture underlying a development strategy, it must be linked to larger company objectives. In general, the DFSS methodology should be linked to:

1. The societal contribution of the company in terms of developing more reliable, efficient, and environmentally friendly products, processes, and services.
2. The goals of the company including profitability and sustainability in local and global markets.
3. The explicit goals of management embodied in company mission statements, including such characteristics as greater design effectiveness, efficiency, cycle time reduction, responsiveness to customers and the like.

4. A greater capacity for the deploying company to adjust and respond to customers and competitive conditions.
5. The satisfaction of mangers, supervisors, and designers.

A deployment strategy is needed to sustain the momentum achieved in the deployment phase. The strategy should show how DFSS allows black belts and their teams to respond to a wide variety of externally induced challenges and how complete deployment of DFSS will fundamentally increase the yield of company operations and its ability to provide a wide variety of design responses. DFSS deployment should be a core competence of a company. DFSS will enhance the variety and quality of design entities and design processes. These two themes should be continuously stressed in strategy presentations to more senior leadership. As deployment proceeds, the structures and processes used to support deployment will also need to evolve. Several factors need to be considered and built into the overall sustainability strategy. For example, the future strategy and plan for sustaining DFSS needs to incorporate more modern learning theory on the usefulness of the technique for green belts and other members at the time they need the information. Once DFSS deployment has been achieved, we suggest that the DFSS community (black belts, green belts, master black belts, champions, and deployment directors) commit to the following:

- Support their company image and mission as a highly motivated producer of world class, innovative products, processes, or services that lead in quality and technology and exceed customer expectations in satisfaction and value.
- Take pride in their work and in the contribution they make internally and externally.
- Constantly pursue "do it right the first time" as a means of reducing the cost to their customers and company.
- Strive to be recognized as a resource vital to both current and future development programs and management of operations.
- Establish and foster a partnership with subject-matter experts and the technical community in their company.
- Treat DFSS lessons learned as a corporate source of returns and savings by replicating solutions and processes in other relevant areas.
- Promote the use of DFSS principles, tools, and concepts where possible in both project and day-to-day operations and promote the data-driven decision culture, the crest of Six Sigma culture.

4.4 BLACK BELT AND DFSS TEAM: CULTURAL CHANGE

The first step in DFSS is to create an environment of teamwork. One thing the black belt will eventually learn is that team members have very different abilities, motivations, and personalities. For example, there will be some team members who are pioneers and others want to vanish. If black belts allow the latter behavior, those team members become dead weight and a source of frustration. The black belt must not let this happen. When team members vanish, it may not be entirely their fault. Take someone who is introverted and finds it stressful to talk in a group. They like to think things through before they start talking. They consider others' feelings and do seek a way to participate. It is the extroverts' responsibility to consciously include the introvert, to not talk over their heads or take the floor away from them. If the black belt wants the team to succeed, he or she has to accept that they must actively manage others. One of the first things the black belt should do as a team member is make sure every member knows every other member beyond name introduction. It is important to get an idea about what each person is good at, and what resources they can bring to the project.

One thing to realize is that when teams are new, each individual is wondering about their identity within the team. Identity is a combination of personality, competencies, behavior, and position in the organization chart. The black belt needs to push for another dimension of identity, that of belonging to the same team with the DFSS project as the task on hand. Vision is, of course, key. Besides the explicit DFSS project-phased activities, what are the real project goals? A useful team exercise, which is also a deliverable, is to create a project charter, with a vision statement, among themselves and with the project stakeholders. The charter is basically a contract that says what the team is about, what their objectives are, what they are ultimately trying to accomplish, where they will get resources, and what kinds of benefits will be gained as a return on their investment on closing the project. The best charters are usually those that gain from each member's input. A vision statement may also be useful. Each member should separately figure out what they think the team should accomplish, then get together see if there are any common elements out of which they can build a single, coherent vision statement that each person can commit to. The reason why it is helpful to use common elements of members' input is that this capitalizes on the common direction and motivates the team going forward.

It is a critical step in the DFSS project endeavor to establish and maintain a DFSS project team that has a shared vision. Teamwork fosters the Six Sigma transformation and instills the culture of execution and pride. It is difficult for teams to succeed without a leader, the black belt, who should be equipped with several leadership qualities acquired by experience and through training. It is a fact that there will be team functions that need to be performed, and he or she could do all of them, or split up the job among pioneer thinkers within the team. One key function is that of facilitator. The black belt will call meetings, keep members on track, and pay attention to team dynamics. As a facilitator, the black belt makes sure that the team focuses on the project, encourages the participation of all members, prevents personal attacks, suggests alternative procedures when the team is stalled, and summarizes and clarifies the team's decisions. In doing so, the black belt should stay neutral until the data starts speaking, and stop meetings from running too long, even if they are going well, or people will try to avoid coming next time. Another key function is that of liaison. The black belt will serve as liaison between the team and the project stakeholders for most of the work in progress. Finally, there is the project management function. As a manger of the DFSS project, the black belt organizes the project plan and sees to it that it is implemented. He or she needs to be able to take a whole project task and break it down into scoped and bounded activities with clearly defined deliverables to be handed out to team members as assignments. The black belt has to be able to budget time and resources and get members to execute their assignments at the right time.

Team meetings can be very useful if done right. One simple thing that helps a lot is having an updated agenda. Having a written agenda, the black belt will make it possible for the team to steer things back to the project activities and assignments. The written agenda serves as a compass.

There will be many situations in which the black belt needs to provide feedback to other team members. It is extremely important to avoid any negative comments that would seem to be about the member personally, rather than their work or their behavior. It is very important that teams assess their own performance from time to time. Most teams have good starts, and then drift away from their original goals and eventually collapse. This is much less likely to happen if from time to time the black belt asks everyone how they are feeling about the team, and takes the performance pulse of the team and compares it to the project charter. It is just as important that the black belt to *maintain* the team to continuously improve its performance. This function, therefore, is an ongoing effort throughout the project's full life cycle.

The DFSS teams emerge and grow through systematic efforts to foster continuous learning, shared direction, interrelationships, and a balance between intrinsic motivators (a desire that comes from within) and extrinsic motivators (a desire stimulated by external actions). Winning is usually contagious. Successful DFSS teams foster other teams. Growing synergy arises from ever-increasing numbers of motivated teams and accelerates improvement throughout the deploying company. The payback for the small, up-front investment in team performance can be enormous.

DFSS deployment will shake up many guarded and old paradigms. People's reaction to change varies from denial to pioneering, passing through many stages in between. In this area, the objective of the black belt is to develop alliances for his or her efforts as he or she progresses. Yang and El-Haik (2003) defined the different stages of change (Figure 4.8). The Six Sigma change stages are linked by what is called the "frustration curves." We suggest that the black belt draw such a curve periodically for each team member and use some or all of the strategies listed below to move his team members to the positive side, the "recommitting" phase. What about Six Sigma culture? What we are finding to be powerful in cultural transformation is the premise that the company results wanted equals the culture wanted. Leadership must first identify objectives that the company must achieve. These objectives must be carefully defined so that the other elements such as employees' beliefs, behaviors, and actions support them. A company has certain initiatives and actions that they must maintain in order to achieve the new results, but to achieve Six Sigma results, certain things must be stopped while others must be started (e.g., deployment). These changes will require a behavioral shift that people must make in order

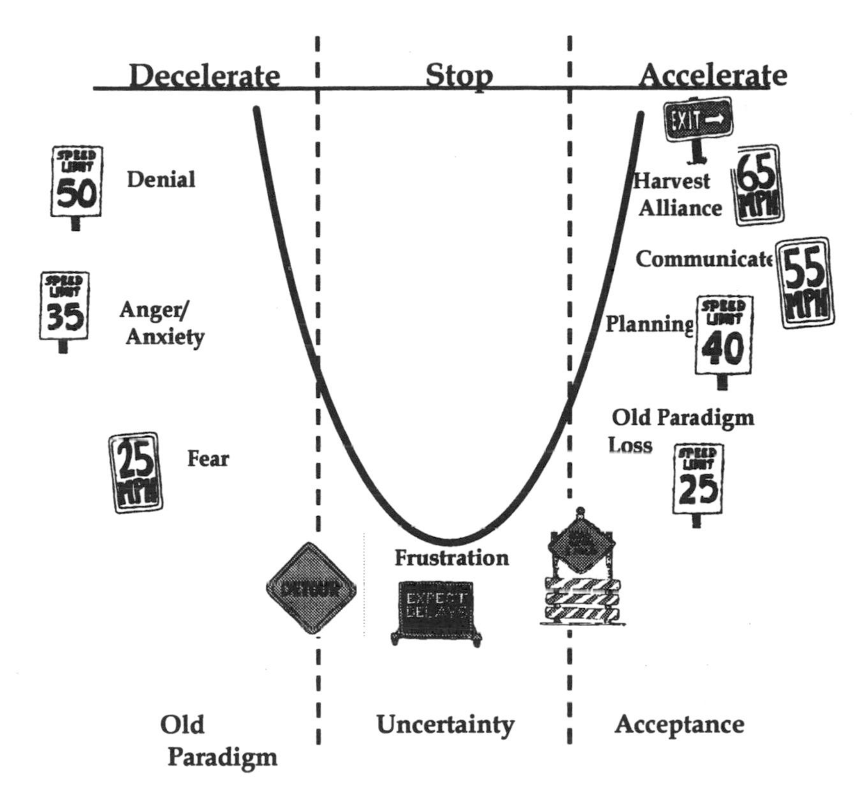

Figure 4.8 The "frustration curve."

for the Six Sigma cultural transition to evolve. True behavior change will not occur, let alone last, unless there is an accompanying change in the beliefs of leadership and the deployment team. Beliefs are powerful in that they dictate action plans that produce desired results. Successful deployment benchmarking (initially) and experiences (later) determine the beliefs, and beliefs motivate actions. So, ultimately, leaders must create experiences that foster beliefs in people. The bottom line is that for Six Sigma data-driven culture to be achieved, the company cannot operate with its old set of actions, beliefs, and experiences, otherwise the results it will get are the results it is currently having. Experiences, beliefs, and actions have to change.

The biggest impact on the culture of a company is the initiative of the founders themselves, starting from the top. The new culture is just maintained by the employees once the transition is complete. They keep it alive. Leadership sets up structures (the deployment team) and processes (the deployment plan) that consciously perpetuate the culture. A new culture means a new identity and new direction—doing things the Six Sigma way.

In implementing large-scale change through Six Sigma deployment, the effort enables the company to identify and understand the key characteristics of the current culture. Leadership together with the deployment team then develop the Six Sigma culture characteristics and the deployment plan of "how to get there." Companies with major internal conflicts or those faced with accelerated changes in business strategy are advised to move with more caution in their deployment.

Several topics that are vital to deployment success should be considered from cultural standpoint. These include:

- Elements of cultural change in the deployment plan
- Assessing resistance
- Ways to handle resistance to change relative to culture
- Types of leaders and leadership needed at different points in the deployment effort
- How to communicate effectively when very little is certain initially
- Change readiness and maturity measurement or assessment

A common agreement between the senior leadership and deployment team should be reached on major deployment priorities and timing relative to cultural transformation and those areas where further work is needed.

At the team level, there are several strategies a black belt could use to his or her advantage in order to deal with team change in the context of Figure 4.8. To help reconcile differences, the black belt needs to listen with empathy, acknowledge difficulties, and define what is out of scope and what is not. To help stop the old paradigm and reorient the team to the DFSS paradigm, the black belt should encourage redefinition, utilize management to provide structure and strength, rebuild a sense of identity, gain a sense of control and influence, and encourage opportunities for creativity. To help recommit the team to the new paradigm, he or she should reinforce the new beginning, provide clear purpose, develop a detailed plan, be consistent in the spirit of Six Sigma, and celebrate success.

5

SERVICE DFSS PROJECT ROAD MAP

5.1 INTRODUCTION

This chapter is written primarily to present the service DFSS project road map in order to support the service black belt and green belt and his or her team and the functional champion in the project execution mode of deployment. The design project is the core of the DFSS deployment and has to be executed consistently using a road map that lays out the DFSS principles, tools, and methods within an adopted gated design process (see Chapter 3, Section 3.5). From a high-level perspective, this road map provides the immediate details required for a smooth and successful DFSS deployment experience.

The chart presented in Figure 5.1 depicts the road map proposed. The road map objective is to develop Six Sigma service solutions entities with unprecedented levels of fulfillment of customer wants, needs, and delight over its total life cycle (see Chapter 1, Section 1.3). The service DFSS road map has four phases identify, charcterize, optimize, and validate, (ICOV) in seven developmental stages. Stages are separated by milestones called the *tollgates*. Coupled with design principles and tools, the objective of this chapter is to combine all of these factors into a comprehensive, implementable sequence in a manner that enables deployment companies to systematically achieve desired benefits by executing projects. In Figure 5.1, a *design stage* constitutes a collection of design activities and can be bounded by entrance and exit tollgates. A tollgate, denoted as TG, represents a milestone in the service design cycle and has some formal meaning defined by the company's own service development coupled with management recognition. The ICOV stages are averages of the authors' studies of several deployments. They need not be adopted blindly, but may be customized to reflect the deployment of interest. For example, in industry, service production cycles and volume are factors that can contribute to the shrinkage or elongation of some of the phases. Generally, the life cycle of a service or a process starts with some form of idea generation, whether in a freely invented format or in a more disciplined format such as multigeneration service planning and growth strategy.

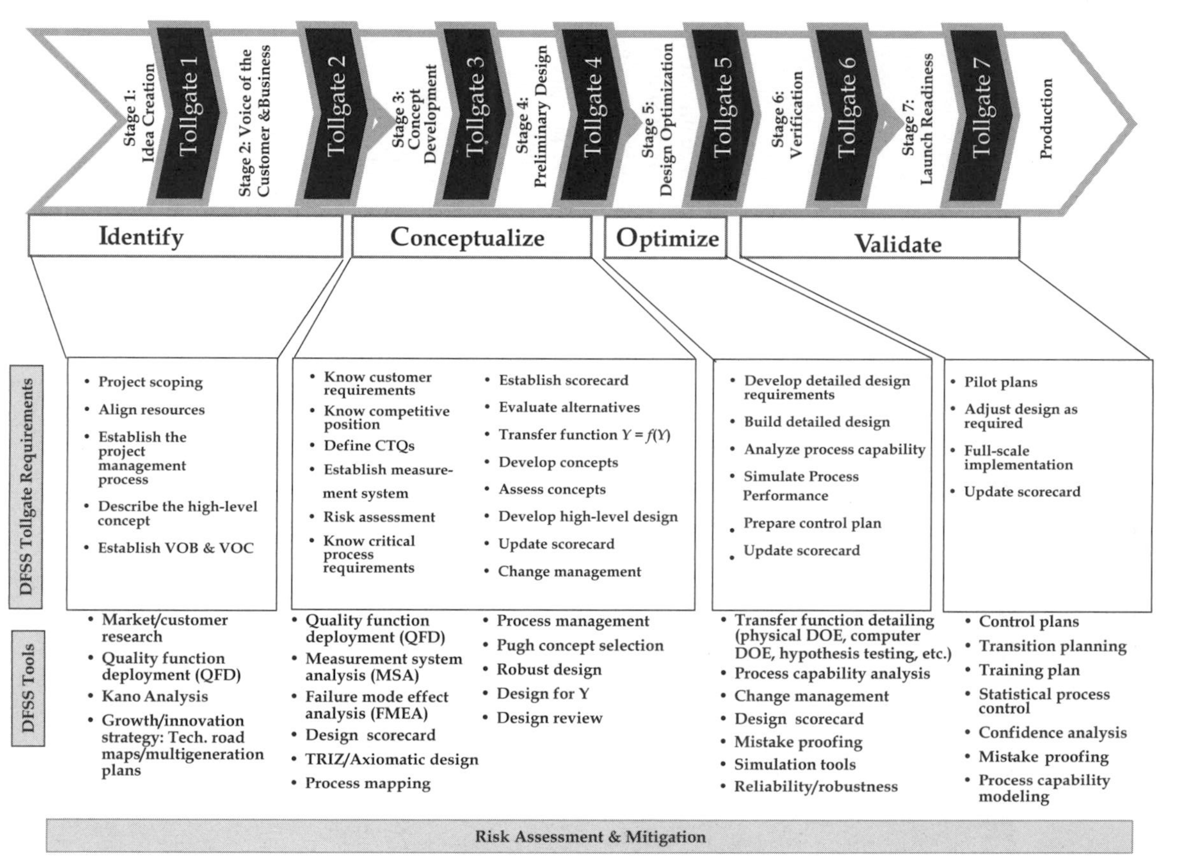

Stage 1: Idea Creation
Tollgate 1
Stage 2: Voice of the Customer &Business
Tollgate 2
Stage 3: Concept Development
Tollgate 3
Stage 4: Preliminary Design
Tollgate 4
Stage 5: Design Optimization
Tollgate 5
Stage 6: Verification
Tollgate 6
Stage 7: Launch Readiness
Tollgate 7
Production
Identify
Conceptualize
Optimize
Validate
DFSS Tollgate Requirements
• Project scoping
• Align resources
• Establish the project management process
• Describe the high-level concept
• Establish VOB & VOC
• Know customer requirements
• Know competitive position
• Define CTQs
• Establish measurement system
• Risk assessment
• Know critical process requirements
• Establish scorecard
• Evaluate alternatives
• Transfer function Y = f(Y)
• Develop concepts
• Assess concepts
• Develop high-level design
• Update scorecard
• Change management
• Develop detailed design requirements
• Build detailed design
• Analyze process capability
• Simulate Process Performance
• Prepare control plan
• Update scorecard
• Pilot plans
• Adjust design as required
• Full-scale implementation
• Update scorecard
DFSS Tools
• Market/customer research
• Quality function deployment (QFD)
• Kano Analysis
• Growth/innovation strategy: Tech. road maps/multigeneration plans
• Quality function deployment (QFD)
• Measurement system analysis (MSA)
• Failure mode effect analysis (FMEA)
• Design scorecard
• TRIZ/Axiomatic design
• Process mapping
• Process management
• Pugh concept selection
• Robust design
• Design for Y
• Design review
• Transfer function detailing (physical DOE, computer DOE, hypothesis testing, etc.)
• Process capability analysis
• Change management
• Design scorecard
• Mistake proofing
• Simulation tools
• Reliability/robustness
• Control plans
• Transition planning
• Training plan
• Statistical process control
• Confidence analysis
• Mistake proofing
• Process capability modeling
Risk Assessment & Mitigation

Figure 5.1 Service DFSS project road map.

Prior to starting on the DFSS road map, the black belt team needs to understand the rationale of the project. We advise they assure the feasibility of continuing the project by validating the project scope, the project charter, and the project resource plan (see Chapter 4, Section 4.3.2, Service DFSS Project Sources). A session with the champion should take place once the matching between the black belt and project charter is done. The objective is to make sure that everyone is aligned with the objectives and to discuss the next steps.

In service DFSS deployment, we will emphasize the synergistic service DFSS crossfunctional team. A well-developed team has the potential to design winning Six Sigma solutions. The growing synergy, which arises from ever-increasing numbers of successful teams, accelerates deployment throughout the company. The payback for up-front investments in team performance can be enormous. Continuous vigilance on the part of the black belt at improving and measuring team performance throughout the project life cycle will be rewarded with ever-increasing capability and commitment to deliver winning design solutions. Given time, there will be a transition from resistance to embracing the methodology, and the company culture will be transformed.

5.2 THE SERVICE DESIGN FOR SIX SIGMA TEAM[1]

It is well known that services intended to serve the same purpose in the same market may be designed and produced in radically different varieties. For example, compare your lodging experience at different hotel chains or your mortgage experience when shopping for a home. Why is it that two services function and feel so different? From the perspective of the design process, the obvious answer is that the design entity derives from a series of decisions, and that different decisions made at the tollgates in the process result in such differentiation. This is common sense; however, it has significant consequences. It suggests that a design can be understood, not only in terms of the design process adopted, but also in terms of the decision-making process used to arrive at it. Measures to address both sources of design variation need to be institutionalized. We believe that the adoption of the ICOV DFSS process presented in this chapter will address one issue—the consistency of development activities. For service design teams, this means that the company structures used to facilitate coordination during the project execution have an effect on the core of the design process. In addition to coordination, the primary intent of an organizational design structure is to control the decision-making process. It is logical to conclude, then, that we must consider the design implications of the types of organizational structures we deploy the ICOV process in to manage design practice. When flat organizational structures are adopted by design teams, members must negotiate design decisions among themselves, because a top-down approach to decision making may not be available.

[1]Adapted from David A. Owens, http://216.239.57.104/search?q=cache:mG3a-1wdz4sJ:mba.vanderbilt.edu/david.owens/Papers/2000%2520Owens%2520Status%2520in%2520Design%2520Teams%2520DMJ.pdf+motivating+design+teams&hl=en

Members of a service design team negotiating decisions with each other during design projects is obviously such a practice. A common assumption seems to be that these decision-making negotiations proceed in a reasonable manner, this being a basic premise of concurrent service design. Patterns and outcomes of decision making are best explained as a dynamic behavior of the teams. Even if two teams develop similar services using the same process, members of the otherwise comparable design teams may have varying levels of influence as decisions are made. The rank of different members of a design team can play a substantial role in team dynamics from the perspective of day-to-day decisions. It is the responsibility of the black belt to balance such dynamics in his or her team. As team leaders, black belts and master black belts need to understand that design teams must make decisions and that, invariably, some set of values must drive those decisions.

Decision making and team structure in companies that use hierarchical structures follow known patterns. Although day-to-day decision making is subject to team dynamics, the milestone decisions are not. In the latter, decisions are made based on formal rank. That is, decisions made by higher-ranking individuals override those made by lower-ranking individuals. Such authoritative decision-making patterns make sense as long as the rank of the decision maker is matched with expertise and appreciation of company goals. This pattern will also assure that those higher in rank can coordinate and align the actions of others with the goals of the company. We adopted this model for DFSS deployment in Chapter 4. Despite these clear benefits, a number of factors make this traditional form of hierarchical structure less attractive, particularly in the context of the design team. For example, risk caused by increased technological complexity of the services being designed, market volatility, and other factors make it difficult to create a decision-making structure for day-to-day design activities. To address this problem, we suggest a flatter, looser structure that empowers team members, black belts, and master black belts to use their own expertise when needed in day-to-day activities. In our view, an ideal design team should consist of team members who represent every phase of a service life cycle. This concurrent structure combined with the road map will assure a company-consistent, minimal design process variation, and successful DFSS deployment. This approach allows information to flow freely across the bounds of time and distance, in particular for geographically wide-spread companies. It also ensures that representatives of later stages of the life cycle have the same influence in making design decisions as do those representatives of earlier stages (e.g., customer service, maintenance, vendors, aftermarket). While benefits such as these can result from a flattened structure, it need not be taken to the extreme. It is apparent that having no structure means the absence of a sound decision-making process. Current practice indicates that a design project is far from a rational process of simply identifying day-to-day activities and then assigning the expertise required to handle them. Rather, the truly important design decisions are more likely to be subjective decisions made based on judgments, incomplete information, or personally biased values, even though we strive to minimize these gaps in VOC and technology roadmapping. In milestones, the final say over decisions in a flat design team remains with the champions or tollgate (TG) approvers. It must not happen at random but, rather, in a mapped fashion.

Our recommendation is twofold. First, a deployment company should adopt a common design process that is customized with their design needs and has the flexibility to adapt the DFSS process to obtain design consistency and to assure success. Second, choose flatter, looser design team structures that empower team members to use their own expertise when needed. This practice works best in companies servicing advanced development work in high-technology domains.

A cross-functional synergistic design team is one of the ultimate objectives of any deployment effort. The black belt needs to be aware of the fact that full participation in design is not guaranteed simply because members are assigned into a team. The structural barriers and interests of others on the team are likely to be far too formidable as the team travels down the ICOV DFSS process.

The success of service development activities depends on the performance of the team that is fully integrated, that is, one having members internal and external (suppliers and customers) groups. Special efforts may be necessary to create a multifunctional DFSS team that collaborates to achieve a shared project vision. Roles, responsibilities, membership, and resources are best defined up front, collaboratively, by the teams. Once the team is established, however, it is just as important to maintain the team to continuously improve its performance. This first step, therefore, is an ongoing effort throughout the service DFSS ICOV cycle of planning, formulation, and production.

The primary challenge for a design organization is to learn and improve faster than their competitors. Lagging competitors must go faster to catch up. Leading competitors must go faster to stay in front. A service DFSS team should learn rapidly, not only about what needs to be done, but about how to do it—how to pervasively implement DFSS process.

Learning without application is really just gathering information, not learning. No company becomes premier by simply knowing what is required, but rather by practicing, by training day in and day out, and by using the best contemporary DFSS methods. The team needs to monitor competitive performance using benchmarking of services and processes to help guide directions of change and employ lessons learned to help identify areas for improvement. In addition, they will benefit from deploying program and risk management best practices throughout the project life cycle (Figure 5.1). This activity is key to achieving a winning rate of improvement by avoiding or eliminating risks. The team is advised to continuously practice design principles and systems thinking, that is, thinking in terms profound knowledge of the total service.

5.3 SERVICE DESIGN FOR SIX SIGMA ROAD MAP

In Chapter 3, we learned about the ICOV process and the seven developmental stages spaced by bounding tollgates indicating a formal transition between entrance and exit. As depicted in Figure 5.2, tollgates or design milestone events include reviews to assess what has been accomplished in the current developmental stage and preparation for the next stage. The service design stakeholders, including the pro-

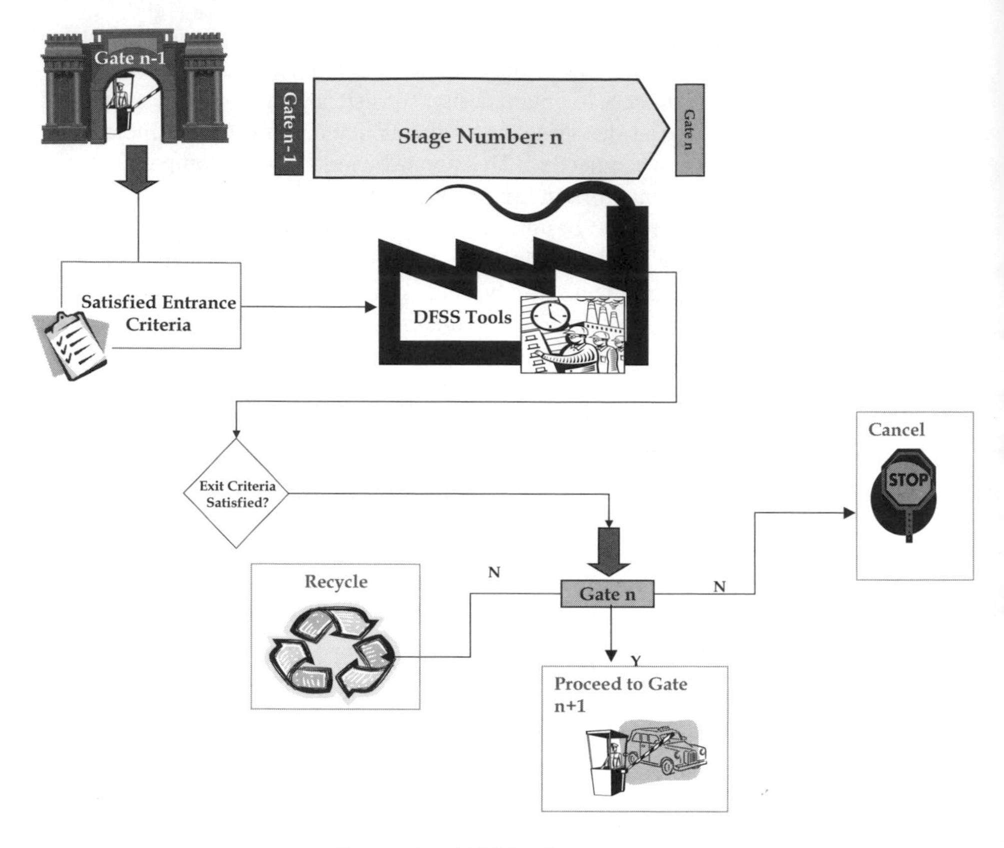

Figure 5.2 DFSS tollgate process.

ject champion, process owner, and deployment champion, conduct tollgate reviews. In a tollgate review, three options are available to the champion or his/her designated tollgate approver:

- Proceed to next stage of development
- Go back for further clarification on certain decisions
- Cancel the project

In tollgate (TG) reviews, work proceeds when the exit criteria (required decisions) are met. Consistent exit criteria from each tollgate blend both service DFSS deliverables due to the application of the approach itself and the business unit or function-specific deliverables as needed.

In this section, we will first expand on the ICOV DFSS process activities by stage and comment on the applicable key DFSS tools and methods that were baselined in Chapter 3. Subsections covering each phase are presented below.

5.3.1 Service DFSS Phase I: Identify Requirements

This phase includes two stages: idea creation (Stage 1) and voice of the customer and business (Stage 2).

Stage 1: Idea Creation

Stage 1 Entrance Criteria. Entrance criteria may be tailored by the deploying function for the particular program/project, provided that the modified entrance criteria, in the opinion of the function, are adequate to the support the exit criteria for this stage. They may include:

- A target customer or market
- A market vision, with an assessment of marketplace advantages
- An estimate of development cost
- Risk assessment

Tollgate "1"—Stage 1 Exit Criteria

- Decision to collect the voice of the customer to define customer needs, wants, and satisfaction
- Verification that adequate funding is available to define customer needs
- Identification of the tollgate keepers'[2] leader and the appropriate staff

Stage 2: Customer and Business Requirements Study

Stage 2 Entrance Criteria

- Closure of Tollgate 1: Approval of the gatekeeper is obtained
- A service DFSS project charter that includes project objectives, service design statement, big *Y* and other business levers, metrics, resources, and team members

These are almost the same criteria required for DMAIC projects. However, project duration is usually longer and initial cost is probably higher. The DFSS team, relative to DMAIC, typically experiences longer project cycle times. The goal here is either designing or redesigning a different entity, not just patching up the holes in an existing one. Higher initial cost is due to the fact that the value chain is being energized by process/service development and not by manufacturing or production arenas. There may be new customer requirements to be satisfied adding more cost to the developmental effort. For DMAIC projects, we may only

[2]A tollgate keeper is an individual or group who will assess the quality of work done by the design team and initiate a decision to approve, reject, cancel, or recycle the project to an earlier gate. Usually, the project champion(s) is tasked with this mission.

work on improving a very limited subset of the critical-to-satisfaction characteristics, the CTSs or big *Y*s.

- Completion of a market survey to determine customer needs critical to satisfaction (CTS) or the voice of the customer (VOC)

In this step, customers are fully identified and their needs collected and analyzed with the help of quality function deployment (QFD) and Kano analysis (see Chapter 7). Then, the most appropriate set of CTSs or big *Y* metrics are determined in order to measure and evaluate the design. Again, with the help of QFD and Kano analysis, the numerical limits and targets for each CTS are established. The list of tasks in this step follows. A detailed explanation is provided in later chapters.

- Determine methods of obtaining customer needs and wants
- Obtain customer needs and wants and transform them into a list of the voice of customer (VOC)
- Finalize requirements
- Establish minimum requirement definitions
- Identify and fill gaps in customer-provided requirements
- Validate application and usage environments
- Translate the VOC to CTSs as critical-to-quality, critical-to-delivery, critical-to-cost, and so on
- Quantify CTSs or big *Y*s
- Establish metrics for CTSs
- Establish acceptable performance levels and operating windows
- Start flow-down of CTSs
- Make an assessment of required technologies
- Make a project development plan (through TG2)
- Conduct risk assessment
- Align with business objectives—voice of the business (VOB) relative to growth and innovation strategy

Tollgate "2"—Stage 2 Exit Criteria

- Assess of market opportunity
- Command a reasonable price or be affordable
- Commit to development of the conceptual designs
- Verify that adequate funding is available to develop the conceptual design
- Identify the gatekeepers' leader (gate approver) and the appropriate staff
- Continue flow-down of CTSs to functional requirements

5.3.1.1 Identify Phase Road Map. DFSS tools used in this phase include (see Figure 5.1):

- Market/customer research
- Quality function deployment (QFD): Phase I
- Kano analysis
- Growth/innovation strategy

5.3.1.2 Service Company Growth and Innovation Strategy: Multigeneration Planning.[3] Even within best-in-class companies, there is the need and opportunity to strengthen and accelerate progress. The first step is to establish a set of clear and unambiguous guiding growth principles as a means to characterize company position and focus. For example, growth in emerging markets might be the focus abroad, whereas effectiveness and efficiency of resource usage within the context of enterprise productivity and sustainability may be the local position. Growth principles and vision at a high level are adequate to find agreement, focus debate within the zone of interest, and exclude or diminish nonrealistic targets. The second key step is to assess the current know-how and solutions of the services portfolio in the context of these growth principles. An inventory is made of what the senior leadership team knows they have, and how they integrate that into the set of guiding growth principles. The third step is to establish a vision of the ultimate state for the company. The final step is to develop a *multigeneration plan* to focus the research, product development, and integration efforts in planned steps to move toward that vision. The multigeneration plan is key because it helps the deploying company stage progress in realistic developmental stages, one DFSS project at a time, but always with an eye on the ultimate vision.

In today's business climate, successful companies must be efficient and market-sensitive to beat their competitors. By focusing on new services, companies can create custom solutions to meet customer needs, enabling customers to keep in step with new service trends and changes that affect them. As the design team engages the customers (through surveys, interviews, focus groups, etc.) and starts the QFD, they gather competitive intelligence. This information helps increase design team awareness of competing services or how they stack up competitively with a particular key customer. By doing this homework, the team identifies potential gaps in their development maturity. Several in-house tools to manage the life cycle of each service product from the cradle to the grave need to be developed, including the multigeneration plan and a customized version of the ICOV DFSS process, if required. The multigeneration plan evaluates the market size and trends, service position, competition, and technology requirements. This tool provides a means to easily identify any gaps in the portfolio while directing the DFSS project road map. The multigeneration plan needs to be supplemented with a decision-analysis tool used to determine the financial and strategic value of potential new applications over a medium time horizon. If the project passes this decision-making step, it can be lined up with others in the Six Sigma project portfolio for start of schedule.

[3]Scott H. Hutchins, http://216.239.57.104/search?q=cache:WTPP0iD4WTAJ:cipm.ncsu.edu/symposium/docs/Hutchins_text.doc+product+multi-generation+plan&hl=en

5.3.1.3 ***Research Customer Activities.*** This is usually done by the service planning departments (service and process) or market research experts, who should be on the DFSS team. The black belt and his team start by brainstorming all possible customer groups of the product. Using the affinity diagram method to group the brainstormed potential customer groups, categories of markets, user types, or service and process applications types will emerge. From these categories, the DFSS team should work toward a list of clearly defined customer groups from which individuals can be selected.

External customers might be drawn from customer centers, independent sales organizations, regulatory agencies, societies, and special interest groups. Merchants and, most importantly, the end user, should be included. The selection of external customers should include both existing and loyal customers, recently lost customers, and newly acquired customers within the market segments. Internal customers might be drawn from production, functional groups, facilities, finance, employee relations, design groups, and distribution organizations. Internal research might assist in selecting internal customer groups that would be most instrumental in identifying wants and needs in operations and service operations.

The ideal service definition, in the eye of the customer, may be extracted from customer engagement activities. This will help turn the knowledge gained from continuous monitoring of consumer trends, competitive benchmarking, and customer likes and dislikes into a preliminary definition of an ideal service. In addition, it will help identify areas for further research and dedicated efforts. The design should be described from a customer's viewpoint (external and internal) and should provide the first insight into what a good service should look like. Concept models and design studies using TRIZ (see Chapter 9) and Axiomatic Design[4] (see Chapter 8) are good sources for evaluating consumer appeal and areas of likes or dislikes.

The array of customer attributes should include all customer and regulatory requirements and social and environmental expectations. It is necessary to understand requirement and prioritization similarities and differences in order to understand what can be standardized and what needs to be tailored.

5.3.2 Service DFSS Phase 2: Characterize Design

This phase spans the following two stages: concept development (Stage 3) and preliminary design (Stage 4).

Stage 3: Concept Development

Stage 3 Entrance Criteria

- Closure of Tollgate 2: Approval of the gatekeeper is obtained.
- Define system technical and operational requirements. Translate customer requirements (CTSs or big *Y*s) to service/process functional requirements. Cus-

[4]Download the Axiomatic Design Software "Acclaro DFSS Light®" from ftp://ftp.wiley.com/public/sci_tech_med/six_sigma. See Section 8.5. Acclaro software is protected under copyright and patents pending. Acclaro is a registered trademark of Axiomatic Design Solutions, Inc.

tomer requirements (CTSs) give us ideas about what will satisfy the customer but they usually cannot be used directly as the requirements for product or process design. We need to translate customer requirements to service and process functional requirements. Another phase of QFD can be used to develop this transformation. Axiomatic design principles will be also very helpful for this step.

- Select a process or service conceptual design.
- Trade off alternate conceptual designs with the following steps. (1) Generate design alternatives. After the determination of the functional requirements for the new design entity (service or process), we need to characterize (develop) design entities that are able to deliver those functional requirements. In general, there are two possibilities. The first is that if the existing technology or known design concept is able to deliver all the requirements satisfactorily, then this step becomes almost a trivial exercise. The second possibility is that if the existing technology or known design is not able to deliver all requirements satisfactorily, then a new design concept has to be developed. This new design should be "creative" or "incremental," reflecting the degree of deviation from the baseline design, if any. The TRIZ method (see Chapter 9) and axiomatic design (see Chapter 8) will be helpful in generating many innovative design concepts in this step. (2) Evaluate design alternatives. Several design alternatives might be generated in the last step. We need to evaluate them and make a final determination as to which concept will be used. Many methods can be used in design evaluation, including the Pugh concept selection technique, design reviews, and FMEA. After design evaluation, a winning concept will be selected. During the evaluation, many weaknesses of the initial set of design concepts will be exposed and the concepts will be revised and improved. If we are designing a process, process management techniques will also be used as an evaluation tool.
- Functional, performance, and operating requirements are allocated to service design components (subprocesses).
- Do development cost estimate (tollgates 2 through 5).
- Do target product/service unit production cost assessment.
- With regard to the market,

  Determine profitability and growth rate
  Do supply chain assessment
  Do time-to-market assessment
  Do share assessment

- Do overall risk assessment.
- Create a project management plan (tollgates 2 through 5) with schedule and test plan.
- Create a team member staffing plan.

Tollgate "3"—Stage 3 Exit Criteria

- Assess that the conceptual development plan and cost will satisfy the customer base.
- Decide whether the service design represents an economic opportunity (if appropriate).
- Verify that adequate funding will be available to perform preliminary design.
- Identify the tollgate keeper and the appropriate staff.
- Write an action plan to continue flow-down of the design functional requirements.

Stage 4: Preliminary Design

Stage 4 Entrance Criteria

- Closure of tollgate 3: approval of the gatekeeper is obtained.
- Flow-down of system functional, performance, and operating requirements to subprocesses and steps (components).
- Document design data package with configuration management[5] at the lowest level of control.
- Development-to-production operations transition plan published and in effect.
- Subprocesses (steps) functionality, performance, and operating requirements verified.
- Development testing objectives complete under nominal operating conditions.
- Test with design parametric variations under critical operating conditions. Tests might not utilize the intended operational production processes.
- Perform design, performance, and operating transfer functions.
- Do reports documenting the design analyses, as appropriate.
- Create a procurement strategy (if applicable).
- Decide whether to make or buy.
- Arrange for sourcing (if applicable).
- Perform risk assessment.

Tollgate "4"—Stage 4 Exit Criteria:

- Acceptance of the selected service solution/design.
- Agreement that the design is likely to satisfy all design requirements.
- Agreement to proceed with the next stage of the selected service solution/design.

[5]A systematic approach to defining design configurations and managing the change process.

- An action plan to finish the flow-down of the design functional requirements to design parameters and process variables.

DFSS tools used in this phase:

- QFD
- TRIZ/axiomatic design
- Measurement system analysis (MSA)
- Failure mode effect analysis (FMEA)
- Design scorecard
- Process mapping
- Process management
- Pugh concept selection
- Robust design
- Design for X
- Design reviews

5.5.3 Service DFSS Phase 3: Optimize the Design

This phase spans the Stage 5 only, the "Design Optimization" stage.

Stage 5: Design Optimization

Stage 5 Entrance Criteria

- Closure of tollgate 4: approval of the gatekeeper is obtained.
- Design documentation defined: the design is complete and includes the information specific to the operations processes (in the opinion of the operating functions).
- Design documents are under the highest level of control.
- Formal change configuration is in effect.
- Operations are validated by the operating function to preliminary documentation.
- Demonstration test plan put together that must demonstrate functionality and performance under operational environments
- Risk assessment performed.

Tollgate "5"—Stage 5 Exit Criteria

- Reach agreement that functionality and performance meet the customers' and business requirements under the intended operating conditions.
- Decide to proceed with verification test of a pilot built to preliminary operational process documentation.

- Perform analyses to document that the design optimization meets or exceeds functional, performance, and operating requirements.
- Optimized transfer functions. DOE (design of experiments) is the backbone of process design and redesign improvement. It represents the most common approach to quantify the transfer functions between the set of CTSs and/or requirements and the set of critical factors, the *X*s, at different levels of design hierarchy. DOE can be conducted by hardware or software (e.g., simulation). From the subset of vital few *X*s, experiments are designed to actively manipulate the inputs to determine their effect on the outputs (big *Y*s or small *y*s). This phase is characterized by a sequence of experiments, each based on the results to the previous study. "Critical" variables are identified during this process. Usually, a small number of *X*s account for most of the variation in the outputs.

The result of this phase is an optimized service entity with all functional requirements released at Six Sigma performance level. As the concept design is finalized, there are still a lot of design parameters that can be adjusted and changed. With the help of computer simulation and/or hardware testing, DOE modeling, Taguchi's robust design methods, and response surface methodology, the optimal parameter settings will be determined. Usually, this parameter optimization phase will be followed by a tolerance-optimization step. The objective is to provide a logical and objective basis for setting requirements and process tolerances. If the design parameters are not controllable, we may need to repeat Stages 1–3 of service DFSS.

The DFSS tools used in this phase are:

- Transfer function detailing (physical DOE, computer DOE, hypothesis testing, etc.)
- Process capability analysis
- Design scorecard
- Simulation tools
- Mistake-proofing plan
- Robustness assessment (Taguchi methods: parameter and tolerance design)

5.3.4 Service DFSS Phase 4: Validate the Design

This phase spans the following two stages: verification (Stage 6) and launch readiness (Stage 7).

Stage 6: Verification

Stage 6 Entrance Criteria

- Closure of tollgate 5: Approval of the gatekeeper is obtained
- Risk assessment

Tollgate (TG) "6"—Stage 6 Exit Criteria. After the parameter and tolerance design is finished, we will move to the final verification and validation activities, including testing. The key actions are:

- The pilot tests are audited for conformance with design and operational process documentation.
- Pilot test and refining: No service should go directly to market without first being subjected to piloting and refining. Here we can use design failure mode effect analysis (DFMEA, see Chapter 11) as well as pilot and small-scale implementations to test and evaluate real-life performance.
- Validation and process control: In this step we will validate the new entity make sure that the service, as designed, meets the requirements, and establish process controls in operations to ensure that critical characteristics are always produced to the specifications of the Optimize phase.

Stage 7: launch Readiness

Stage 7 Entrance Criteria

- Closure of tollgate 6: approval of the gatekeeper is obtained
- The operational processes have been demonstrated
- Risk assessment has been performed
- All control plans are in place.
- Final design and operational process documentation has been published
- The process is achieving or exceeding all operating metrics
- Operations have demonstrated continuous operation without the support of the design development personnel
- Planned sustaining development personnel are transferred to operations:

  Optimize, eliminate, automate, and/or control the vital few inputs defined in the previous phase

  Document and implement the control plan

  Sustain the gains identified

  Reestablish and monitor long-term delivered capability

- A transition plan is in place for the design development personnel
- Risk assessment is performed

Tollgate (TG) "7" Exit Criteria

- The decision is made to reassign the DFSS black belt
- Full commercial rollout and handover to new process owner: As the design entity is validated and process control is established, we will launch full-scale commercial rollout and the new designed service together with the supporting

operations processes can be handed over to design and process owners, complete with requirements settings and control and monitoring systems.

- Closure of tollgate 7: Approval of the gatekeeper is obtained

DFSS tools used in this phase are:

- Process control plan
- Control plans
- Transition planning
- Training plan
- Statistical process control
- Confidence analysis
- Mistake-proofing
- Process capability modeling

5.4 SUMMARY

In this chapter, we presented the service design for Six Sigma road map. The road map is depicted in Figure 5.1, which highlights at a high level the identify, characterize, optimize, and validate phases, and the seven service development stages (idea creation, voice of the customer and business, concept development, preliminary design, design optimization, verification, and launch readiness). The road map also recognizes the tollgates, design milestones at which DFSS teams update the stakeholders on development status and ask for decisions to be made whether to approve going to the next stage, going back to an earlier stage or cancelling the project altogether.

The road map also highlights the most appropriate DFSS tools by ICOV phase. It indicates where it is most appropriate to start tool usage. These tools are presented in Chapters 6 and 7.

6

SERVICE DFSS TRANSFER FUNCTION AND SCORECARDS

6.1 INTRODUCTION

In its simplest form, a *transfer function* is a mathematical representation of the relationship between the input and output of a system or a process. For example, putting a speaker in a room will make the speaker sound different than it does in a nonenclosed space. This is caused by reflection, absorption, and resonance of the room. This change is called the transfer function.

Speed of answering in a call center is dependent on call volume, number of service representatives, and time spent on each call. The speed of answering is a function of these three factors and their relationship is called a transfer function.

All systems have a transfer function or a set of transfer functions. Typically, the most noticeable change in artificial (man-made) systems is caused by excitation by an energy source. An output response or a design functional requirement is inherent because the system is like a box, an enclosed entity that promotes reaction to energy excitement. But the transfer function is more than just a change in a system's or process's status; it also can cause major changes in the system's physical embodiment. As designers, DFSS teams usually determine the transfer function(s) of their designed service entity through their design activities. They can map the activities and optimize the output functional requirements, and then build the system or process and take advantage of their knowledge of the transfer function. So what is a transfer function?

A transfer function describes the relationship between design parameters or process variables ($x_1, x_2, \ldots, x_n$) and a functional requirement (y), denoted by the relationship $y = f(x_1, x_2, \ldots, x_n)$. The variable y is the dependent output variable of a process or a system of interest to the customer. In service design, the transfer function is used to design and monitor a process to detect if it is out of control, or if symptoms are developing within a process. Once derived or quantified through design of experiment (DOE, see Chapters 12 and 13), a transfer function $y = f(x_1, x_2, \ldots, x_n)$ can be developed to define the relationship of process variables that can lead to a plan for the control a process. The response y is an output measure or a functional or design requirement of a process, such as productivity or customer satisfaction. The transfer function explains the transformation of the inputs into the output with x's being process inputs or process steps that are involved in producing the service. For example, in a service call center, customer satisfaction (y) is a function (f) of the service time (x_1), waiting time (x_2), phone system (x_3), the associate's knowledge (x_4), and so on. All of these x's can be defined, measured and used for optimizing (y).

Optimization in the service DFSS context means shifting the mean of y (μ_y) to the target (T_y) and minimizing its variance (σ_y^2). Design transfer functions are tied to design mappings and design hierarchy. These are explained below.

6.2 DESIGN MAPPINGS

The DFSS project road map recognizes three types of mapping, as depicted in Figure 6.1:

1. Customer mapping (from customer attributes to functional requirements)
2. Functional mapping (from functional requirements to design parameters)
3. Process mapping (from design parameters to process variables)

The service DFSS road map of Chapter 5 is focused on providing a solution framework for design processes in order to produce healthy conceptual entities with Six Sigma quality potential. The mapping from the customer attributes domain to the functional requirements domain (ys) is conducted using quality function deploy-

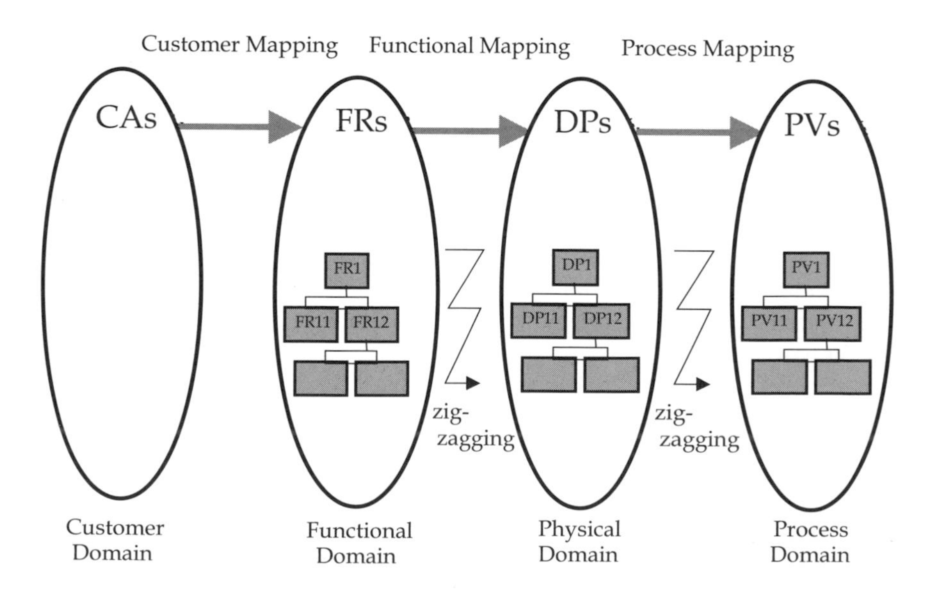

Figure 6.1 Design mappings.

ment (QFD) as described in Chapter 7. QFD represents a mapping from the raw customer attributes to the functional requirements.

Functional mapping is done in the following phase, where the Six Sigma conceptual potential of the functional requirements, the *y*s, are set. This mapping phase presents a systematic approach for establishing the capability at the conceptual level in the service entity by cascading requirements into a hierarchy while deploying design best practices such as design principles. The application of design principles, in particular those promoted to axioms, is the team's insurance against repeating already known vulnerabilities. The axiomatic design[1] method and FAST techniques (see Chapter 8) are two prime tools for functional mapping and analysis. A call center design-mapping example is depicted in Figure 6.2.

Process mapping is a mapping from the design parameters domain to the process variables domain. Process variables are parameters that can be dialed (adjusted or changed) to deliver design parameters that in turn satisfy a functional requirement. Graphical process mapping, IDEF[2] family, and value stream mapping (Chapter 8) are techniques used to map a design process.

In addition to design mapping, Figure 6.1 also conveys the concept of design hierarchy such as process, subprocess, step, and so on. At a given level of design hierarchy, there exists a set of requirements defined as the minimum set of needed requirements at that level. Defining acceptable functional requirements may involve several iterations when a limited number of logical questions in the mapping process are employed.

6.2.1 Functional Mapping

Functional mapping is design mapping between functional requirements and design parameters. Functional mapping can be represented graphically or mathematically, depicting the input–output or cause-and-effect relationships of functional elements. In its graphical form, block diagrams are used to capture the mapping and are composed of nodes connected by arrows depicting such relationships. A block diagram should capture all design elements within the DFSS project scope and ensure correct flow-down to critical parameters. Functional mapping represented mathematically using mapping matrices; matrices belonging to the same hierarchical level are clustered together. Design hierarchy is built by decomposing a design into a number of simpler design matrices that collectively meet the high-level functional requirements identified in the first mapping. The collection of design matrices or blocks forms the conceptual design blueprint and provides a means to track the chain of effects for design changes as they propagate across the design. The decomposition starts by the identification of a minimum set of functional requirements that deliver the design

[1]Download the Axiomatic Design Software "Acclaro DFSS Light®" from ftp://ftp.wiley.com/public/sci_tech_med/six_sigma. See Section 8.5. Acclaro software is protected under copyright and patents pending. Acclaro is a registered trademark of Axiomatic Design Solutions, Inc.

[2]IDEF is a family of mapping techniques that were developed by the U.S. Air Force. They are discussed in Chapter 8. IDEF stands for Integrated computer aided manufacturing (ICAM) DEFinition (ICAM definition).

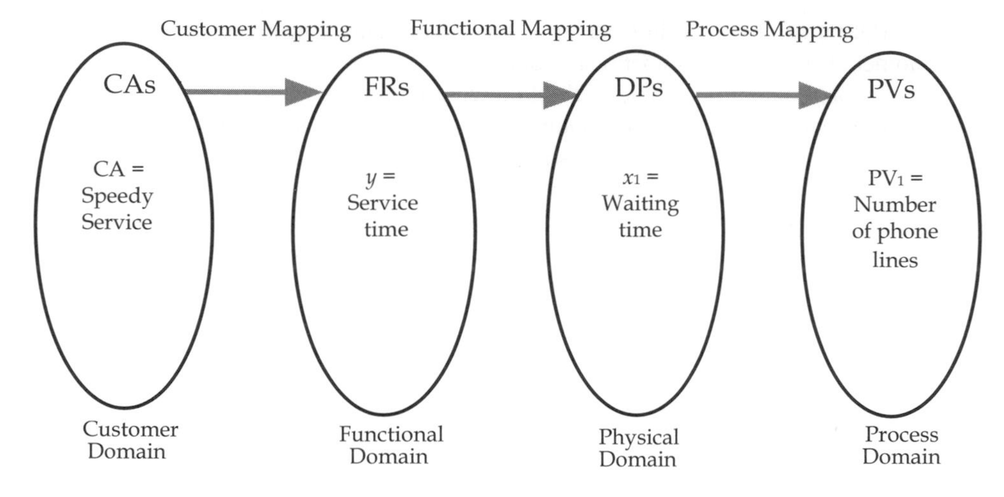

Figure 6.2 Call center design mapping.

tasks as defined by the customer. Both the design principles and the team's creativity guide the decomposition heuristic process of functional definition through logical questions (see Chapter 8). The efficient use of design principles can gear the synthesis and analysis design activities to produce vulnerability-free solutions.

6.2.2 Process Mapping

Functional mappings are conceptual representations at different hierarchal levels. These matrices are not stand-alone, and a complete solution entity for the design project needs to be delivered in some physical form. We call this embodiment. The ultimate goal of design mapping is to facilitate design detailing when the relationships are identified in the form of transfer functions or other models. Design mapping, a design analysis step, should be conducted *first,* prior to design synthesis activities.

A detailed transfer function is not only useful for design optimization but also for further design analysis and synthesis activities. For example, the functional requirement of a given subprocess can be the input signal of another subprocess delivering another requirement, and so on. These relationships create the design hierarchy.

Similar to functional mapping, process mappings can also be depicted graphically or mathematically. Refer to Chapter 8 for further material. Next, we will discuss the steps necessary to conduct a design mapping.

6.2.3 Design Mapping Steps

The following synthesis steps can be used in both functional and process mapping. We will use the functional mapping technique for illustration.

1. Obtain the high-level functional requirements using quality function deployment (QFD, see Chapter 7).

2. Define boundaries from the project scope.
3. Conduct the mapping, using the techniques from Chapter 8, to the lowest possible level and identify the transfer functions at every level. The lowest level represents the standard design parameters or process variables. For example, in service DFSS projects, forms, standards, and procedures are at the lowest level.
4. Define the respective hierarchical levels of design mapping.
5. Within a level, for each mapping, and for each requirement (y), determine the potential mapped-to-design parameters and process variables.
6. Select and use a consistent mapping technique (see Chapter 8) for both functional and process mappings for compatible analysis.
7. Aggregate the chains of mappings in every hierarchical level into an overall mapping. You will produce something that resembles Figure 6.1 for your project.

6.3 DESIGN SCORECARDS AND TRANSFER FUNCTION[3]

In a world in which data is plentiful and data storage is relatively inexpensive, it is a painful irony that most design companies are unable to use that data to make better decisions. They are rich in data but poor in knowledge. Hidden within those megabytes of information is useful knowledge about customers, services, and processes that, if uncovered, could result in significant cost savings or revenue enhancement. Transfer function models uncover hidden relationships and improve design decision making. DFSS teams can use transfer function models effectively to cleanse, augment, and enhance historical knowledge.

A transfer function is a means of optimization and design detailing and is usually documented in a scorecard. A transfer function should be treated as a living entity within the DFSS road map that passes through its own life cycle stages. A transfer function is first identified using functional mapping and then detailed by derivation, modeling, or experimentation. Design transfer functions belonging to the same hierarchical level in the design should be recorded in the scorecard of that hierarchical level. A scorecard is used to record and optimize the transfer functions. The transfer functions are integrated with other DFSS concepts and tools like process mapping, DOE, and DFMEA. Functional mapping and process mapping are the premier activities performed in the characterize phase of the DFSS road map (see Chapter 5). Both types of design activities as well their corresponding techniques are covered in Chapter 8.

Through several transfer function detailing options such as DOE, transfer functions can be approximated by an algebraic equation of a polynomial additive model

[3]Usually, transfer function models refer to a statistical technique used in forecasting a variable over time. The technique is similar to regression analysis and is often referred to as dynamic regression analysis. Transfer function models are most often used as a relatively simple and effective time-series forecasting technique and is useful in forecasting demand and capacity in logistic problems.

and augmented with some modeling and experimental error. This approximation is valid in any local area or volume of the design space from which it was obtained. In other words, we can infer within the tested space but not predict outside the limits of testing. An error term is usually present to represent the difference between the actual transfer function and the predicted one.

Transfer function additivity is extremely desirable in all design mappings. As the magnitude of the error term reduces, the transfer function additivity increases, as it implies less interaction.[4] In additive transfer functions, the significance of a design parameter is relatively independent from the effect of other parameters. Service solution entities that are designed following axiomatic design methods (see Chapter 8) will have additive transfer functions that can be optimized easily, thus reducing the DFSS project cycle time. From an analysis standpoint, this additivity is needed in order to employ statistical analysis techniques like robust design, DOE, and regression.

The transfer functions in the functional and process mappings are usually captured in design scorecards. Design scorecards document and assess quantitatively the DFSS project progress, store the results of the learning process, and show all critical elements of a design and their performance. Their benefits include documenting transfer functions and design optimization, predicting final results, and enabling communication among all stakeholders in the project while evaluating how well the design is supported by production processes.

The set of transfer functions of a given design are the means for optimizing customer satisfaction. Transfer functions can be mathematically derived, empirically obtained from a DOE, or regressed using historical data. In several cases, no closed mathematical formula can be obtained and the DFSS team can resort to mathematical and/or simulation modeling (see Chapter 14). In DFSS, there should be a transfer function for every functional requirement (y) depicting the functional mapping and for every design parameter depicting the process mapping. The transfer functions at a given hierarchy level are recorded in one scorecard.

6.3.1 DFSS Scorecard Development

The scorecard identifies which design parameters contribute most to the variation and mean in the response of the transfer function and the optimized design point. Tightening tolerances may be appropriate for those parameters that affect the output most significantly. The team has the freedom to customize the scorecard. During scorecard development, the team should remember that a scorecard has a hierarchical structure that parallels the concerned mapping. The number of scorecards will equal the number of hierarchal levels in the concerned mapping.

The extension of the DFSS methodology from a product to a service environment is facilitated by the design scorecard and life cycle elements mapping. Think about a simple database used in a transactional process. The life cycle is build, load, use, report, and update (see Figure 6.3). The common requirements of a typical de-

[4]Interaction is the cross product term of a transfer function. In a designed experiment, it occurs when the difference of a functional requirement between levels of one parameter or variable is not the same at all levels of the other factors. See Chapter 12 for more details.

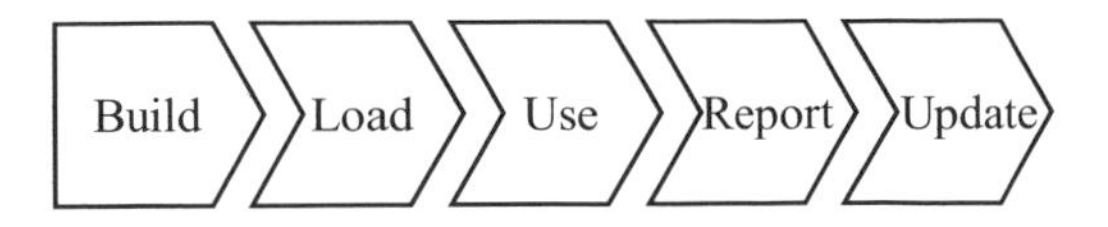

Figure 6.3 Scorecard database life cycle stages.

sign scorecard would be accuracy, completeness, and time requirements for inputting, reporting, and updating steps, as shown in Figure 6.4.

6.3.2 Transfer Function Life Cycle[5]

Transfer functions are living entities in the DFSS roadmap. The life cycle of a transfer function in the DFSS project passes through the following sequential stages:

1. Identification: Obtained by conducting the mapping between the design domains.
2. Conceptual treatment: Fix, add, replace, or delete some of the independent variables in the codomain to satisfy design principles and axioms (see Chapter 8).
3. Detailing: Achieved by finding the cause-and-effect (preferably mathematical) relationship between all variables in the concerned mapping domains. Detailing involves validating both the assumed relationship and the sensitivities of all independent parameters and variables, that is, $y = f(x_1, x_2, \ldots, x_n)$. Transfer functions can be categorized as (1) empirical as obtained from testing, (2) mathematical in the form of equations, inequalities, and other mathematical relations among design parameters and/or process variables that the describe behavior of design entity. For some cases, it is possible to analytically have equations in closed form. However, in many situations, closed form solutions are impossible and one has to resort to other methods such as simulation.
4. Optimization: After detailing, the dependent variables in the transfer functions are optimized by shifting their mean, reducing their variation, or both in the optimize phase of the DFSS road map. This can be achieved by adjusting $(x_1, x_2, \ldots, x_n)$ parameter means and variances. This optimization propagates to the customer domain via the established transfer function in the design mapping, resulting in increased satisfaction.
5. Validation: The transfer function is validated in both mappings.
6. Maintenance, by controlling all significant independent variables, after the optimization, in-house or outside.

Stages 1–6 are experienced in a DFSS project.

7. Disposal. The design transfer functions are disposed or reach entitlement in delivering high-level functional requirements when either new customer at-

[5]See Yang and El-Haik (2003).

	Quantity	Defects	DPU	RTY	Zlt	Zst
Top Level						
Input	100	6	0.06	0.941765	1.569761	3.069761
Report	2	3	1.5	0.22313	-0.76166	0.738336
Maintain	5	6	1.2	0.67032	0.440798	1.940798
TOTAL				0.140858	-1.07647	0.423529

	Quantity	Defects	DPU	RTY	Zlt	Zst
Input Level						
Accuracy	100	3	0.03	0.970446	1.887383	3.387383
Completeness	100	1	0.01	0.99005	2.328215	3.828215
Time to input	100	2	0.02	0.980199	2.057868	3.557868
	100	6		0.941765	1.569761	3.069761

	Quantity	Defects	DPU	RTY	Zlt	Zst
Report Level						
Accuracy	2	1	0.5	0.606531	0.270288	1.770288
Completeness	2	1	0.5	0.606531	0.270288	1.770288
Process Time	2	1	0.5	0.606531	0.270288	1.770288
	2	3		0.22313	-0.76166	0.738336

	Quantity	Defects	DPU	RTY	Zlt	Zst
Maintain Level						
Completeness	5	0	0	1	6	7.5
Accuracy	5	0	0	1	6	7.5
Time to Perform	5	2	0.4	0.67032	0.440798	1.940798
	5	6		0.67032	0.440798	1.940798

Figure 6.4 Design scorecard for database.

tributes that cannot be satisfied with the current design emerge or when the mean or controlled variance of the requirements are no longer acceptable by the customer. This stage is usually followed by the evolution of new transfer functions to satisfy the needs that have emerged.

8. Evolution of a new transfer function. Per TRIZ (see Chapter 9), an evolution usually follows certain basic patterns of development. Evolutionary trends of the functional performance of a certain design can be plotted over time and have been found to evolve in a manner that resembles an S curve (Figure 6.5).

The following are some possible sources of "detailed" transfer functions:

1. Direct documented knowledge like equations derived from laws of business (e.g., profit = price – cost, interest formulas) Transfer function variability optimization and statistical inference can be done using simulation, usually via the Monte Carlo method or Taylor series expansions. The transfer functions obtained this way are very much dependent on the team's understanding of and competency with their design and the discipline of knowledge it represents.

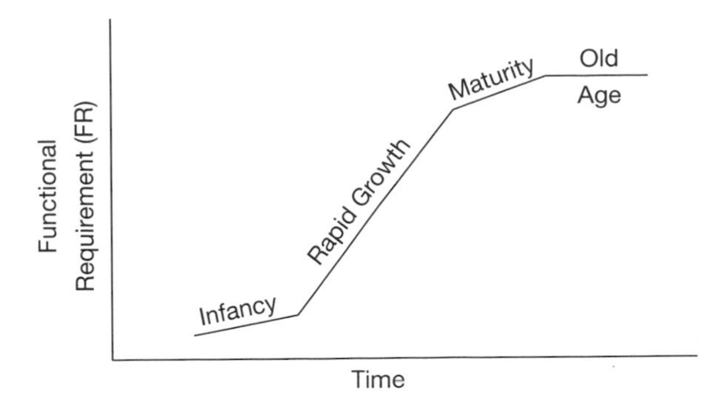

Figure 6.5 The S curve of FR evolution.

2. Transfer functions can be obtained by sensitivity analysis to estimate the derivatives, either with the prototype process, the baseline design, or a credible mathematical model. The sensitivity analysis involves changing one parameter or variable, requantifying the transfer function (y) to obtain the result, and then comparing the change of both sides to each other. The process is repeated for all independent parameters or variables ($x_1, x_2, \ldots, x_n$).
3. Design of Experiment (DOE) is another source of transfer function detailing. In many ways, a DOE is another form of sensitivity analysis. DOE analysis runs the inputs through their possible experimental ranges, not through incremental area or volume.

   The resultant predictive equation from a DOE analysis using statistical packages such as Minitab™, for example, is in essence a transfer function. The black belt may use the derivative of the predictive equation to estimate sensitivities. Special DOE techniques are more appropriate due to the richness of information they yield. These include the response surface method (Yang & El-Haik, 2003), factorial experimentation, and robust design (see Chapters 12 and 13).

   A DOE can be run using physical entities or mathematically. Additionally, simulation models can be obtained using techniques such as the Monte Carlo method (see Section 6.8) or products such as Simul8®, SigmaFlow®, Witness®, and ProModel®. Mathematical models are descriptive; they provide a representation that is a good and realistic substitute for the real system while retaining characteristics of importance. We simulate behavior of real systems by conducting virtual experiments on analytical models in lieu of experiments on real systems. Simulation saves cost and effort and provides a replica of reality for further investigation, including if–then scenarios.

   In simulation, the team needs to define design parameters and/or process variables. The simulation model then samples a number of runs from these distributions and forms an output distribution. The statistical parameters, such as the mean and variance, of this distribution are then estimated. Afterward, statistical inference analysis can be used.

4. Regression transfer equations are obtained when the functional requirements are regressed over all input variables/parameters of interest. Multiple regressions coupled with multivariate analysis of variance (MANOVA) and covariance (MANCOVA) are typically used to estimate parameter (variable) effects. These subjects are beyond the scope of this book.

6.4 TRANSFER FUNCTION

In this section we provide a view of the role mathematics plays in describing how design elements change in a transfer function. We graph the kind of functions that arise in service design, with major concentration on polynomials up to quadratic terms. Polynomial transfer functions are continuous and differentiable functions, and they represent the most common components of many useful models. A non-polynomial function can be approximated by polynomial terms using Taylor series expansion.

In graphing polynomials and other functions, we begin with the assignment of numerical coordinates to points in a plane. These coordinate are called Cartesian coordinates, in honor of Descartes, a French mathematician. They make it possible to graph algebraic equations in two variables (or parameters), lines and curves. Cartesian coordinates also allow the calculation of angles and distances and the writing of coordinate equations to describe the paths along which objects move and change. Most of us are familiar with the notions of x-axis, y-axis, and the point where the coordinates intersect, the origin. Motion from left to right along the x-axis is called motion in the positive x-direction. Along the y-axis, the positive direction is up, and the negative direction is down. When plotting data in the Cartesian coordinate plane whose parameters have different units of measures, the units shown on the two coordinate axes may have different interpretations. For example, if we plot the size of national debt at different months, the x-axis will show time in months and the y-axis will show billions of dollars.

If we track the prices of food, steel, or computers, we can watch their progress on graph paper by plotting points and fitting them to a curve. We extend the curve day by day as new prices appear. To what uses can we then put such a curve? We can see what the price was on any given date. We can see by the slope of the curve, the rate at which the prices are rising or falling. If we plot other data on the same sheet of paper, we can perhaps see what relation they have to the rise and fall of prices. The curve may also reveal patterns that can be help us forecast or affect the future with more accuracy than could someone who has not graphed the data. In Chapter 12, we introduce plotting one factor effect at a time (commonly known as mean effect plots) and the two-factor effects (commonly known as interaction effect plots), and so on. In addition, graphing helps make a connection between rates of change and slopes of smooth curves. One can imagine the slope as the grade[6] of a roadbed. Consider Equation (6.1):

[6]Civil engineers calculate the slope of a roadbed by calculating the ratio of the distance it rises or falls to the distance it runs horizontally, calling it the *grade* of the roadbed.

$$y = f(x_1) = a_0 + a_1x_1 \tag{6.1}$$

Where a_1x_1 is a linear term. This is an equation of a liner transfer function in which a_0 and a_1 are the y-intercept and slope, respectively. It is a polynomial of the first degree or a linear polynomial. In this case, the output y is called the dependent variable, whereas x is called the independent variable. This relationship can be exemplified by the inventory cost of an item that is replenished at a constant rate a_1, with an initial stocking cost of a_0. In this case, the y output is the cost of total units in the inventory since acquisitions, and x_1 is the stocked units (quantity) variable. The terms y, a_0, and a_1x_1 are in units of cost. Therefore, the unit of a_1, the slope, is the unit of inventory cost per unit of the item. Another example is revenue (y) as a function of sales. This function is graphed in Figure 6.6.

Suppose there is another item denoted by x_2 that is also stocked in the same inventory space, and we are interested in the overall inventory cost of both items. In this case, the cost can be given as

$$y = f(x_1, x_2) = a_0 + a_1x_1 + a_2x_2 \tag{6.2}$$

where $a_1x_1 + a_2x_2$ are linear terms. The plot of such a transfer function is given in Figure 6.7. As you may conclude, it is a plane. Note that this transfer function may be enhanced by cross product terms reflecting the interaction between parameters or factors x_1 and x_2. In Chapter 12, we will call this cross product term *interaction*. This may occur when the there is a limited inventory space and competing item stocks. Equation (6.2) then becomes

$$y = f(x_1, x_2) = a_0 + a_1x_1 + a_2x_2 + a_{12}x_1x_2 \tag{6.3}$$

where $a_1x_1 + a_2x_2$ are linear terms and $a_{12}x_1x_2$ is an interaction term.

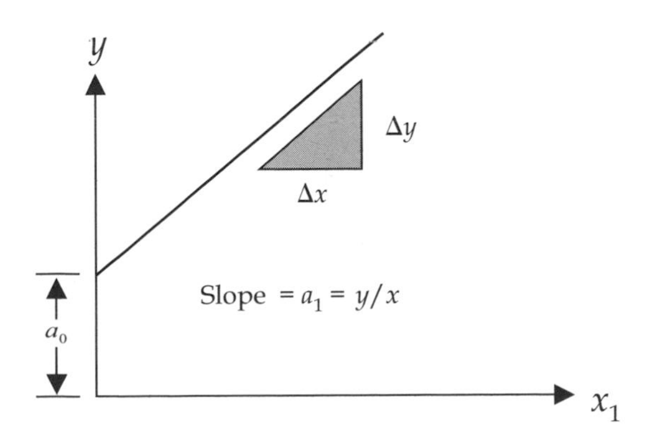

Figure 6.6 Linear transfer function.

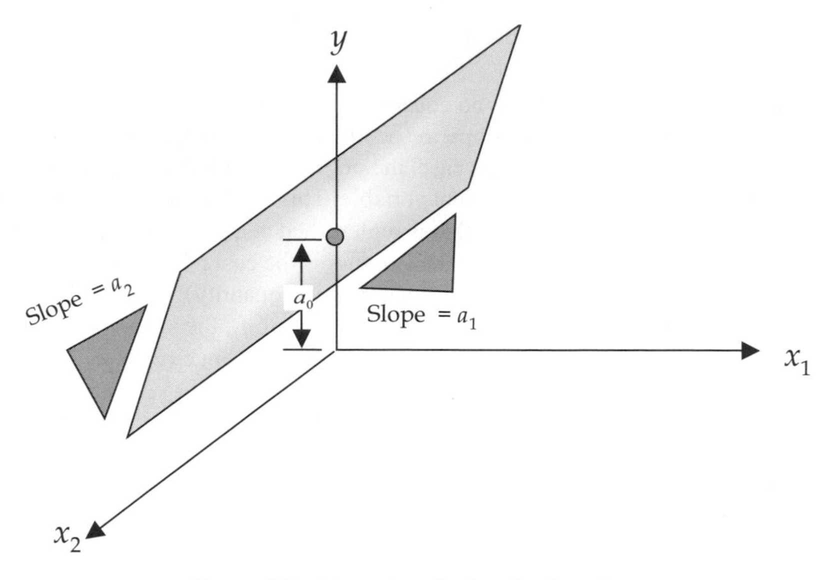

Figure 6.7 Linear transfer function in a plane.

We often encounter transfer functions that changes at a linear rate rather than a constant rate. That is, the instantaneous rate of change in the output y relative to a factor x_1 is linear, similar to Equation (6.1), which means the y is a quadratic function in the independent variables. In the inventory example, assume that there is only one item, say x_1, that is replenished linearly. The cost in this case is quadratic, as given in Equation (6.4):

$$y = f(x_1) = a_0 + a_1 x_1 + a_{11} x_1^2 \tag{6.4}$$

where $a_1 x_1$ is a linear term and $a_{11} x_1^2$ is a quadratic term.

Quadratic transfer functions are often encountered in customer-satisfaction target setting in the context of design mapping (Figure 6.1). In this case the customer satisfaction (y) is highest when an influential independent factor (x) assumes a desired target value. For example, in a restaurant that collects customer feedback, in particular for lunch, the highest satisfaction is often encountered when the overall experience took on average one hour. Satisfaction is measured on a six-point scale from poor (1) to excellent (6). This will allow restaurant management to target employee responsiveness to customers, including friendly table visits and prompt delivery. More or less time results in decay in customer satisfaction (y), as depicted in Figure 6.8. By mathematical treatment, the customer satisfaction is given by Equation (6.5):

$$y = f(x) = \frac{|y_o - y_T|}{(-x_o^2 + 2x_o x_T - x_T^2)}(x - x_T)^2 + y_T \tag{6.5}$$

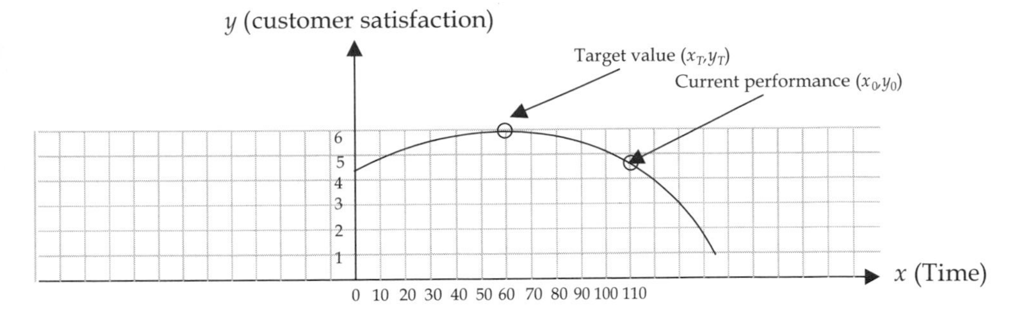

Figure 6.8 Quadratic transfer function example.

This high-level introduction of linear and quadratic polynomial transfer functions will allow the reader to grasp the forthcoming concepts in the book. Both represent building blocks of most encountered service transfer functions. Other forms are listed below. The reader can extend such forms to three (by introducing x_3) and four (by introducing x_4) independent factors (variables or parameters), and so on.

$$y = f(x_1, x_2) = a_0 + \underbrace{a_1 x_1 + a_2 x_2}_{\text{Linear terms}} + \underbrace{a_{12} x_1 x_2}_{\text{Interaction terms}} + \underbrace{a_{11} x_1^2 + a_{22} x_2^2}_{\text{Quadratic terms}} \tag{6.6}$$

$$y = f(x_1, x_2, x_3) = a_0 + \underbrace{a_1 x_1 + a_2 x_2 + a_3 x_3}_{\text{Linear terms}} + \underbrace{a_{12} x_1 x_2 + a_{13} x_1 x_3 + a_{23} x_2 x_3}_{\text{Interaction terms}} + \underbrace{a_{11} x_1^2 + a_{22} x_2^2 + a_{33} x_3^2}_{\text{Quadratic terms}} \tag{6.7}$$

When empirical methods are used to obtain a transfer function, the result is an approximation and not exact because of the inclusion of experimental error as well as the effect of unaccounted for design parameters, process variables, and noise factors. An error term will represent the difference between the actual transfer function and the empirical one. The notation $\hat{y}$ is used to reference an empirically obtained transfer function.

6.5 TRANSFER FUNCTIONS AND OPTIMIZATION

A transfer function that includes random parameters or variables, which have probabilistic uncertainty, must incorporate the parameters' or variables' uncertainty in order to be valid. A transfer function is like design DNA. The evaluation of transfer functions that contain random parameters or (independent) variables by point estimates alone does not include the uncertainty and, therefore, does not provide complete representation.

Without clear justification for preferring one value in the whole uncertainty domain, point estimates are not credible. Ignoring variability (uncertainty) may result in misleading conclusions. Therefore, transfer function detailing should take into consideration all possible sources of uncertainty, including uncontrollable factors (noise factors; refer to Chapter 13) to gain DFSS optimization credibility.

Optimization in the context of service DFSS of a requirement, say y, is carried out in two steps: first, minimizing its variance (σ_y^2), and second, shifting the mean (μ_y) to the target (T_y). This two-step optimization implies design robustness, and when coupled with Six Sigma targets means a Six Sigma design. In conducting optimization, design transfer functions are the workforce. Some of the transfer functions are readily available from existing knowledge. Others will require perhaps some intellectual or monetary capital to obtain them.

A key philosophy of DFSS project mapping is that during the optimization phase, *inexpensive parameters* can be identified and studied, and can be combined in a way that will result in performance that is insensitive to uncontrolled yet adversely influential factors (noise factors). The team's task is to determine the combined best settings (parameter targets) for each of the design parameters, which have been judged by the design team to have the potential to most improve the process or service. By varying the parameter target levels in the transfer function (design point), a region of nonlinearity can be identified. This area of nonlinearity is the most optimized setting for the parameter under study. Consider two settings or means of a design parameter (x), setting 1 (x^*) and setting 2 (x^{**}), having the same variance and probability density function (statistical distribution) as depicted in Figure 6.9. Consider, also, the given curve of a hypothetical transfer function,

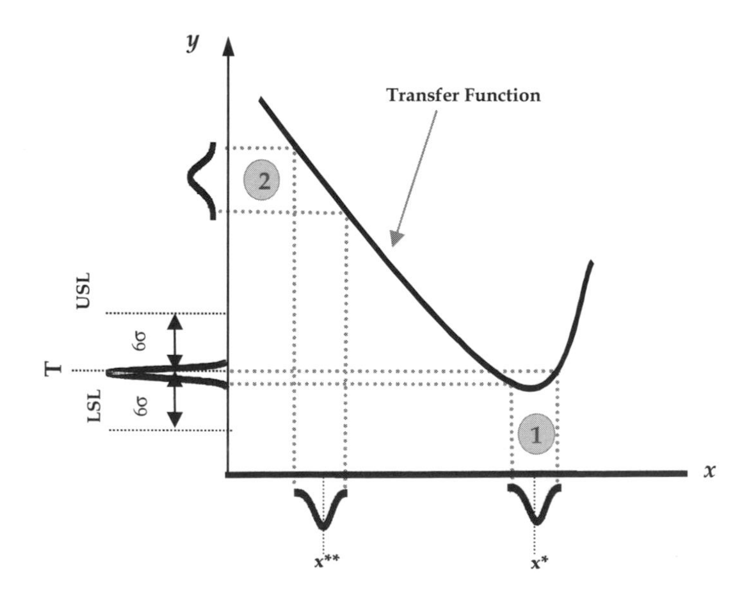

Figure 6.9 Transfer function and optimization.

which is in this case a nonlinear function in the design parameter, x. It is obvious that setting 1 produces less variation in the functional requirement (y) than setting 2 by capitalizing on nonlinearity.[7] Setting 1 (x^*) will also produce a lower quality loss, a concept that is entertained in Chapter 13. In other words, the design produced by setting 1 (x^*) is more robust (optimized) than that produced by setting 2. Setting 1 (x^*) robustness is evident in the amount of transferred variation through the transfer function to the y response in Figure 6.9.[8] When the distance between the specification limits is six times the standard deviation ($6\sigma_y$), a Six Sigma level optimized y is achieved. When all design functional requirements are released at this level, a Six Sigma design is obtained.

The black belt and the rest of the DFSS team need to detail the transfer functions in the context of the DFSS algorithm if they want to optimize, validate, and predict the exact performance of their project in the intended environment. However, the team should be cautious about the predictability of some of the transfer functions due to the stochastic effects of noise factors in the intended environment, which are particularly hard to predict. The team should explore all knowledge to obtain the transfer functions desired. For example, stack-up based on tolerances may contain descriptions of functionality that are based on a lumped-mass model and geometry. Transfer functions that are obtained by means other than direct documented knowledge are approximate and, for the purposes of the DFSS roadmap, are valid until proven wrong or disposed due to the evolution of other adopted concepts.

6.6 MONTE CARLO SIMULATION

Monte Carlo simulation is named for Monte Carlo, Monaco, where the primary attraction is gambling. Gambling implies the use of chance and probability, similar to the simulation mechanism of employing randomness. Games of chance, such as roulette wheels, dice, and slot machines, exhibit random behavior. When we use the word simulation, we refer to any analytical method meant to replicate a real-life process, especially when other transfer functions are too mathematically complex or too difficult to reproduce. Without the aid of simulation, a developed model will generally reveal a single outcome, the most likely or average scenario. Analysis uses simulation to automatically analyze the effect of varying inputs on outputs of the modeled process. One type of spreadsheet simulation is Monte Carlo simulation, which randomly generates values for variables over and over to simulate a process.

The random behavior in gambling games is similar to how Monte Carlo simulation selects variable values at random to simulate a model. When a die is rolled, you know that a 1, 2, 3, 4, 5, or 6 will come up, but you don't know which will come up for any particular trial. It's the same with the process variables that have a known

[7]In addition to nonlinearity, leveraging interactions between the noise factors and the design parameters is another popular empirical parameter design approach.

[8]Also the flatter quadratic quality loss function in Figure 13.3.

range of values but an uncertain value for any particular time or event (e.g., staffing needs, stock prices, inventory, phone calls per minute).

For each process variable (one that has a range of possible values), the DFSS team defines the possible values with a probability distribution. The type of distribution you select is based on the conditions surrounding that variable. Distribution types include normal, uniform, exponential, and triangular. To add this sort of function to an Excel spreadsheet, you would need to know the equation that represents this distribution. In commercially available software packages such as Crystal Ball®, these equations are automatically calculated for you. They can even fit a distribution to any historical data available.

A simulation calculates multiple scenarios of a model by repeatedly sampling values from the probability distributions for the process variables and use those values for the advancement of transactions in the process. Simulations can consist of hundreds or even thousands of trials in just a few seconds. During a single Monto Carlo trial, a simulation model randomly selects a value from the defined possibilities (the range and shape of the distribution) for each process variable and then recalculates the spreadsheet.

Monte Carlo simulation is a popular simulation method to evaluate the uncertainty of a functional requirement (y). It involves some sampling from the independent parameters or variables, and probability distributions to provide their values. A random-number generator is used to obtain the parameter value based on the probability density function. When all independent parameters or variables are quantified, the functional requirement is quantified to produce the numerical result. This process is repeated for a desired number of iterations. A histogram of the results is then built from the answer and can be used for statistical inference. The black belt needs to:

- Determine the applicable distribution of y.
- Estimate the parameters of that distribution.
- Use the distribution for optimization and inference. For example, assuming normality y, we can use the following upper z value, $z_y = (USL - \mu_y)/\sigma_y$, to calculate the defects per million.

The drawbacks of this method are:

- Simulation time may be an issue, in particular for some complex mathematical forms of transfer functions and large numbers of iterations.
- Identifying the proper probability distribution functions for the independent parameters and variables may be difficult due to lack of data or not understanding the underlying process.
- Randomness of the random-number generator may be an issue.

Tables of ready-to-use generated random numbers are available, especially in Monte Carlo simulations, where random sampling is used to estimate certain requirements. Also, most simulation software packages have the capability of automatic random number generation. The detail of sampling from probability distributions is not within the scope of this chapter.

Monte Carlo simulation models are time-independent (static) models that deal with a process of fixed state. In such spreadsheet-like models, certain variable values are changed by random generation and a certain measure or more are evaluated per such changes without considering the timing and the dynamics of such changes, a prime disadvantage that is handled by discrete event simulation models (see Chapter 14). In discrete event simulation, the time dimension is live.

6.7 SUMMARY

Design can be defined by a series of mappings between four domains. The DFSS project road map recognizes three types of mapping:

1. Customer mapping (from customer attributes to functional requirements)
2. Functional mapping (from functional requirements to design parameters)
3. Process mapping (from design parameters to process variables)

Transfer functions are mathematical relationships relating a design response, usually a functional requirement, with design parameters and/or process variables. Transfer functions are DFSS optimization vehicles. They facilitate the DFSS optimization of process outputs related to the service entity of interest by defining the true relationship between input variables and the output. Optimization in this context means minimizing the requirement variability and shifting its mean to some desired target value desired by the customer. Design scorecards are used to document the transfer function as well as the optimization calculations.

7

QUALITY FUNCTION DEPLOYMENT (QFD)

7.1 INTRODUCTION

In this chapter, we will cover the history of quality function deployment (QFD), describe the methodology of applying the QFD within the DFSS project road map (see Chapter 5), and apply the QFD to our supply chain example.

QFD is a planning tool that allows the flow-down of high-level customer needs and wants through to design parameters and then to process variables critical to fulfilling the high-level needs. By following the QFD methodology, relationships are explored between quality characteristics expressed by customers and substitute quality requirements expressed in engineering terms (Cohen 1988, 1995). In the context of DFSS, we call these requirements "*critical-to*" characteristics. These critical-to characteristics can be expanded along the dimensions of speed (*critical to delivery,* CTD), quality (CTQ), cost (CTC), as well as the other dimensions introduced in Figure 1.1. In the QFD methodology, customers define their wants and needs using their own expressions, which rarely carry any actionable technical terminology. The voice of the customer can be refined into a list of needs and wants that can be used as input to a relationship matrix called QFD's house of quality (HOQ) using the affinity diagram.

Knowledge of the customer's needs and wants is paramount in designing effective products and services by innovative and rapid means. Utilizing the QFD methodology allows the developer to attain the shortest development cycle while ensuring the fulfillment of the customer's needs and wants.

Figure 7.1 shows that teams that use QFD place more emphasis on responding to problems early in the design cycle. Intuitively, it incurs more effort, time, resources, and energy to implement a design change at production launch than at the concept phase because more resources are required to resolve problems than to prevent their occurrence in the first place.

With QFD, quality is defined by the customer. Customers want products and ser-

Service Design for Six Sigma. By Basem El-Haik and David M. Roy

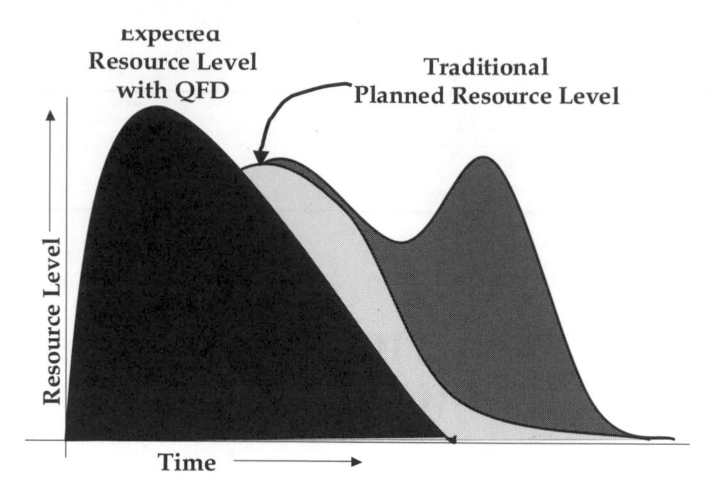

Figure 7.1 The time-phased effort for DFSS versus traditional design.

vices that throughout their lifetimes meet their needs and expectations at a value that exceeds cost.

QFD methodology links these needs through design and into process control. The QFD's ability to link and prioritize at the same time guides the design team in focussing its energy and resources with amazing focus.

In this chapter, we will provide the detailed methodology to create the four QFD "houses" and evaluate them for completeness and goodness, introduce the Kano model for the voice of the customer, and relate QFD with the DFSS road map introduced in Chapter 5.

7.2 HISTORY OF QFD

As we mentioned in Chapter 1, QFD was developed in Japan by Yoji Akao and Shigeru Mizuno in 1966 but not westernized until the 1980s. Their purpose was to develop a quality assurance method that would design customer satisfaction into a product before it was manufactured. For six years, the methodology was developed from the initial concept of Kiyotaka Oshiumi of Bridgestone Tire Corporation. Following the first publication of an article on quality deployment by Dr. Yoji Akao in 1972 (Akao, 1972) the pivotal development work was conducted at Kobe Shipyards for Mitsubishi Heavy Industry. The stringent government regulations for military vessels coupled with the large capital outlay forced the management at the shipyard to seek a method of ensuring upstream quality that would cascade down throughout all activities. The team developed a matrix that related all the government regulations, critical design requirements, and customer requirements to technical characteristics of how the company would achieve them. Within the matrix, the team depicted the importance of each requirement, which allowed for prioritization.

Following the successful deployment within the shipyard, Japanese automotive companies adopted the methodology to resolve the problem of rust in cars. Next, it was applied to car features and the rest, as we say, is history. In 1978, the detailed methodology was published in Japanese (Mizuno & Akao, 1978) and translated into English in 1994 (Mizuno & Akao, 1994).

7.3 QFD OVERVIEW

The benefits of utilizing the QFD methodology are mainly ensuring that high level customer needs are met, that the development cycle is efficient in terms of time and effort, and that the control of specific process variables is linked to customer wants and needs for continuing satisfaction.

In order to complete a QFD, three key conditions are required to ensure success. Condition 1 is that a multidisciplinary DFSS team is required to provide a broad perspective. Condition 2 is that more time is expended up front in the collecting and processing of the customer needs and expectations. Condition 3 is that the functional requirements defined in HOQ 2 (see below) be solution free.

This theory sounds logical and achievable; however, there are three realities that must be overcome in order to achieve success. Reality 1 is that the interdisciplinary DFSS team will not work together well in the beginning. Reality 2 is that there may be a prevalent culture of heroic problem solving in lieu of drab problem prevention. People get visibly rewarded and recognized for fire fighting but get no recognition for the problem prevention that drives a culture focused on correction rather than prevention. The final reality is that the team members and even customers will jump right to the solutions instead of following the details of the methodology and waiting until design requirements are specified.

7.4 QFD METHODOLOGY

Quality function deployment is accomplished by multidisciplinary DFSS teams using a series of matrixes, called "houses," to deploy critical customer needs throughout the phases of the design development. The QFD methodology is deployed through a four-phase sequence shown in Figure 7.2. The four phases are:

- Phase 1—critical-to-satisfaction planning (House 1)
- Phase 2—functional requirements planning (House 2)
- Phase 3—design parameters planning (House 3)
- Phase 4—process variable planning (House 4)

Each of these phases will be covered in detail within this chapter.

It is interesting to note that the QFD is linked to the voice of the customer (VOC) tools at the front end and then to design scorecards and customer satisfaction mea-

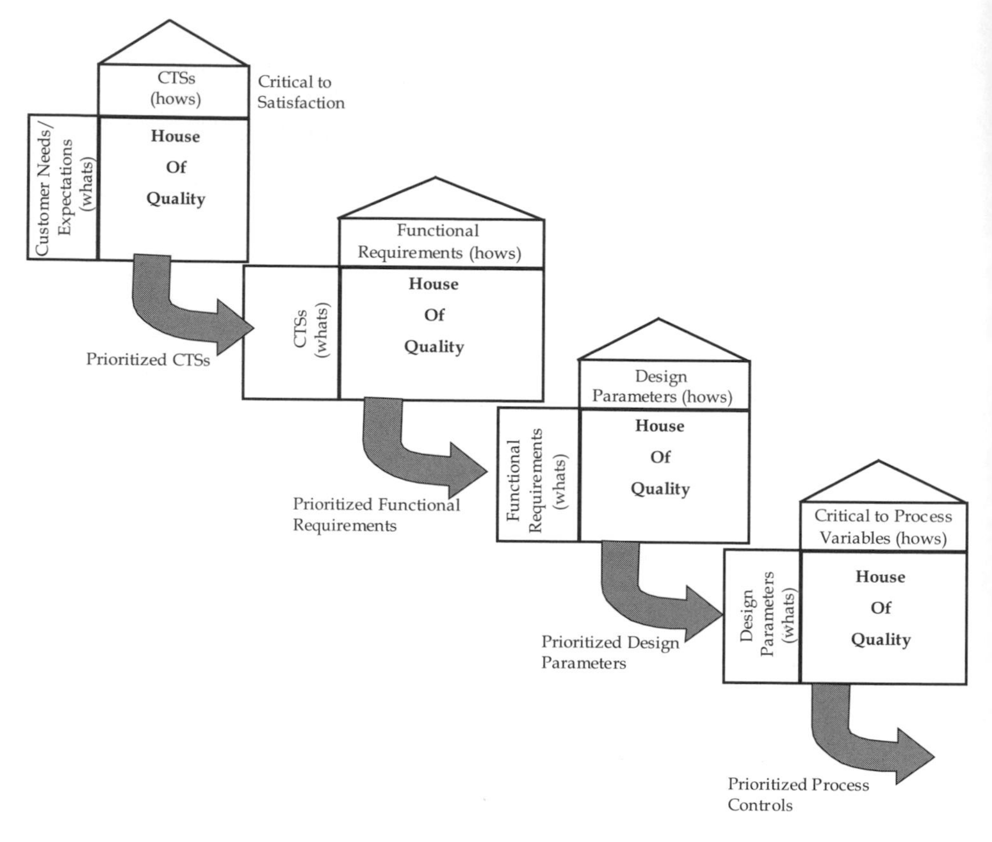

Figure 7.2 The four phases of QFD.

sures throughout the design effort. These linkages, along with adequate analysis, provide the feed-forward (requirements flow-down) and feed-backward (capability flow-up) signals that allow for the synthesis of design concepts (Suh, 1990).

Each of these four phases deploys the house of quality (HOQ), with the only content variation occurring in Room #1 and Room #3. Figure 7.3 depicts the generic HOQ. Going room by room we see that the input is into Room #1 in which we answer the question "What?". These "Whats" are either the results of VOC synthesis for HOQ 1 or a rotation of the Hows from Room #3 into the following HOQs. These Whats are rated in terms of their overall importance and placed in the Importance column.

Next, we move to Room #2 and compare our performance and the competitions' performance against the "whats" in the eyes of the customer. This is usually a subjective measure and is generally scaled from 1 to 5. A different symbol is assigned to the different providers so that a graphical representation is depicted in Room #2. Next, we must populate Room #3 with the "hows." For each "what" in Room #1 we ask "How can we fulfill this?" We also indicate which direction the improvement

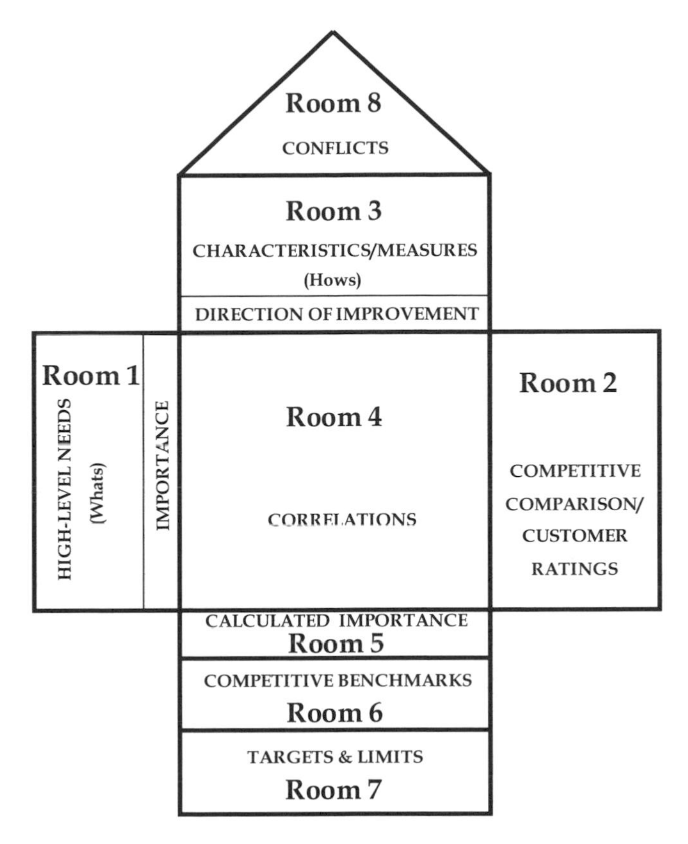

Figure 7.3 House of quality.

must be in to satisfy the "what:" maximize, minimize, or target. This classification is in alignment with robustness methodology (see Chapter 13) and indicates an optimization direction. In HOQ1, these become "How does the customer measure the 'what'?" In HOQ1 we call these critical-to-satisfaction (CTS) measures. In HOQ2, the "hows" are measurable and solution-free functions that are required to fulfill the "whats" of the CTSs. In HOQ3, the "hows" become design parameters (DPs), and in HOQ4 the "hows" become process variables (PVs). A word of caution: teams involved in designing new services or processes often times jump right to specific solutions in HOQ1. It is a challenge to stay solution free until HOQ3. There are some rare circumstances in which the VOC is a specific function that flows straight through each house unchanged.

Within Room #4, we assign the weight of the relationship between each "what" and each "how" using a 9 for strong, 3 for moderate, and 1 for weak. In the actual HOQ, these weightings will be depicted with graphical symbols, the most common being a solid circle for strong, an open circle for moderate, and a triangle for weak (Figure 7.4).

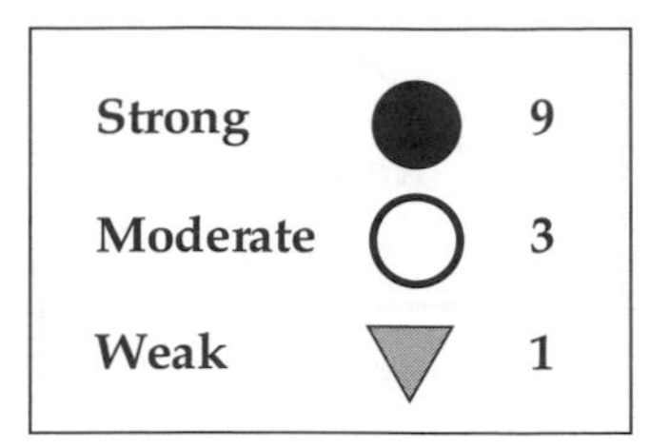

Figure 7.4 Rating values for affinities.

Once the relationship assignment is completed, by evaluating the relationship of every "what" to every "how," the calculated importance can be derived by multiplying the weight of the relationship and the importance of the "what" and summing for each "how." This is the number in Room #5. For each of the "hows" we can also derive quantifiable benchmark measures of the competition and our own performance; in the eyes of industry experts, this is what goes in Room #6. In Room #7, we can state the targets and limits of each of the "hows." Finally, in Room #8, often called the roof, we asses the interrelationships of the "hows." If we were to maximize one of the "hows," what happens to the others? If it were to improve, we would classify it as a synergy, whereas if it were to move away from the direction of improvement, it would be classified as a compromise. Wherever a relationship does not exist, it is just left blank. For example, if we wanted to improve customer intimacy by reducing the number of call handlers, then the wait time might degrade. This is clearly a compromise. Although it would be ideal to have correlation and regression values for these relationships, often they are just based on common sense or business laws.

This completes each of the eight rooms in the HOQ. The next step is to sort based on the importance in Room #1 and Room #5, and then evaluate the HOQ for completeness and balance.

7.5 HOQ EVALUATION

Completing the HOQ is the first important step; however, the design team should take the time to review their efforts for quality and checks and balances as well as design resource priorities. The following diagnostics can be utilized in the sorted HOQ:

1. Is there a diagonal pattern of strong correlations in Room #4? This will indicate good alignment of the "hows" (Room #3) with the "whats" (Room #1).
2. Do all "hows" in Room #3 have at least one correlation with the "whats" in Room #1?
3. Are there empty or weak rows in Room #4? This indicates unaddressed "whats" and could be a major issue. In HOQ1, this would indicate unaddressed customer wants or needs.

4. Are there empty columns in Room #4? This indicates redundant "hows" which don't fulfill any "whats."
5. Evaluate the highest score in Room #2. What should our design target be?
6. Evaluate the customer rankings in Room #2 versus the technical benchmarks in Room #6. If Room #2 values are lower than Room #6 values, the design team may need to work on changing the customer's perception. It could also mean that the correlation between the wants/needs and CTS is not correct.
7. Review Room #8 trade-offs for conflicting correlations. For strong conflicts/synergies, changes to one characteristic (Room #3) could effect other characteristics.

7.6 HOQ 1—"THE CUSTOMER'S HOUSE"

Quality function deployment begins with the voice of the customer (VOC), the first step required for HOQ 1. VOC can be collected by many methods and from many sources. Some common methods are historical research methods, focus groups, interviews, councils, field trials, surveys, and observations. Sources range from passive historical records of complaints, testimonials, warranty records, customer records, and call centers, to active customers, lost customers, and target customers. Stick with the language of the customer and think about how they speak when angered or satisfied; this is generally their natural language. These voices need to be prioritized and synthesized into rank order of importance. The two most common methods are the

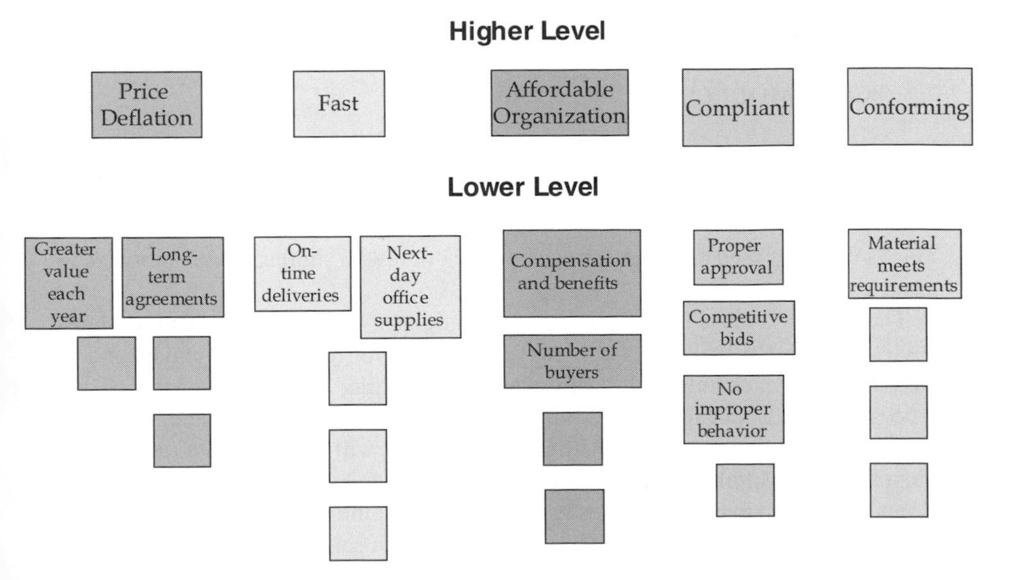

Figure 7.5 Affinity diagram for a supply chain.

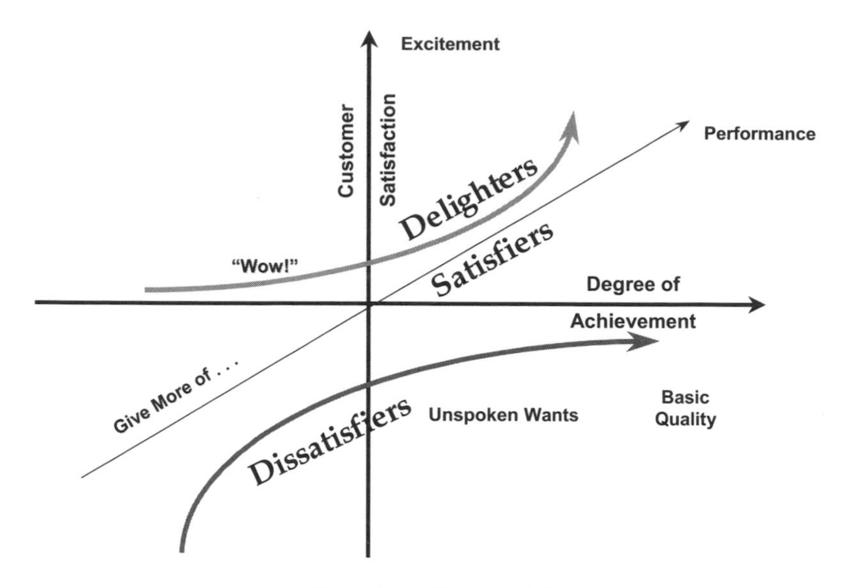

Figure 7.6 Kano model.

affinity diagram (see Figure 7.5) and Kano analysis. We will cover the Kano model (see Figure 7.6) before taking the prioritized CTSs into Room #1 of HOQ 2.

When collecting the VOC, make sure that it isn't the VOE (voice of the engineer or voice of the employee) or VOB (voice of the boss). Although the QFD is a robust methodology, if you start with a poor foundation, it will deteriorate throughout the process.

7.7 KANO MODEL

In the context of DFSS, customer attributes are potential benefits that the customer could receive from the design and are characterized by qualitative and quantitative data. Each attribute is ranked according to its relative importance to the customer. This ranking is based on the customer's satisfaction with similar design entities featuring that attribute.

The understanding of customer expectations (wants, needs), and delights ("wow" factors) by the design team is a prerequisite for further development and is, therefore, the most important action prior to starting the other functional and process mappings (Chapter 8). The fulfillment of these expectations and the provision of differentiating delighters (unspoken wants) will lead to satisfaction. This satisfaction will ultimately determine what products and services/processes the customer is going to endorse and consume or buy. In doing so, the design team needs to identify constraints that limit the delivery of such satisfaction. Constraints present opportunities to exceed expectations and create delighters. The identification of customer expectations is a vital step for the development of Six Sigma products and services the customer will buy in preference to those of the competitors. Noriaki Kano, a Japanese consultant, has developed a model relating design characteristics to customer satisfaction (Cohen, 1995). This model (see Figure 7.6) divides characteristics into categories—dissatifiers, satisfiers, and delighters—each of which affects customers differently.

"Dissatisfiers" are also known as basic, "must-be," or expected attributes, and can be defined as a characteristic that a customer takes for granted and that causes dissatisfaction when it is missing. "Satisfiers" are performance, one-dimensional, or straight-line characteristics, and are defined as something the customer wants and expects; the more, the better. "Delighters" are features that exceed competitive offerings in creating unexpected, pleasant surprises. Not all customer satisfaction attributes are equal from an importance standpoint. Some are more important to customers than others in subtly different ways. For example, dissatisfiers may not matter when they are met, but subtract from overall design satisfaction when they are not delivered.

When customers interact with the design team, delighters often surface that have not been independently conceived. Another source of delighters may emerge from team creativity, as some features have the unintended result of becoming delighters in the eyes of customers. Any design feature that fills a latent or hidden need is a delighter and, with time, becomes a want. A good example of this is the remote controls first introduced in televisions. Early on, these were differentiating delighters; today, they are common features in televisions, radios, and even automobile ignitions and door locks. Today, if you received a television without a remote control it would be a dissatisfier. Delighters can be sought in areas of weakness, competitor benchmarking, technical, social, and strategic innovation. Social aspects are becoming more important as educated buyers want to preserve the environment and human rights.

The design team should conduct a customer evaluation study. This is hard to do in creative design situations. Customer evaluation is conducted to assess how well the current or proposed design delivers on the needs and desires of the end user. The most frequently used method for this evaluation is to ask the customer (e.g., via a focus group or a survey) how well the design project is meeting each customer's expectations. In order to leap ahead of the competition, the team must also understand the evaluation and performance of their toughest competition. In the HOQ 1, the team has the opportunity to grasp and compare, side by side, how well the current, proposed, or competitive design solutions are delivering on customer needs.

The objective of the HOQ 1 Room 2 evaluation is to broaden the team's strategic choices for setting targets for the customer performance goals. For example, armed with meaningful customer desires, the team could aim their efforts at either the strengths or weaknesses of best-in-class competitors, if any. In another choice, the team might explore other innovative avenues to gain competitive advantages.

The list of customer wants and needs should include all from the customer as well as regulatory requirements and social and environmental expectations. It is necessary to understand requirements and prioritization similarities and differences in order to understand what can be standardized and what needs to be tailored.

Customer wants and needs, in HOQ1, social, environmental, and other company wants, can be refined to a matrix format for each identified market segment. The "customer importance rating" in Room 1 is the main driver for assigning priorities from both the customer and the corporate perspectives, as obtained through direct or indirect engagement with the customer.

The traditional method for conducting the Kano model is to ask functional and dysfunctional questions around known wants/needs or CTSs. Functional questions take the form of "How do you feel if the CTS is present in the product/service?" Dysfunctional questions take the form of "How do you feel if the CTS is not present in the product/service?"

Collection of this information is the first step, and then detailed analysis is required, which is beyond the scope of this book. For a good reference on processing the voice of the customer, see Burchill and Brodie (1997).

In addition to the traditional method it is possible to use qualitative assessment to put things into a Kano analysis plot and then have it validated by the end user. Figure 7.7 shows an example of this qualitative method.

In the Kano analysis plot, the y axis consists of the Kano model dimensions of must be, one-dimensional, and delighters. The top item, indefferent, is where the customer chooses opposite items in the functional and dysfunctional questions. The x axis is based on the importance of the CTSs to the customer. This type of plot can

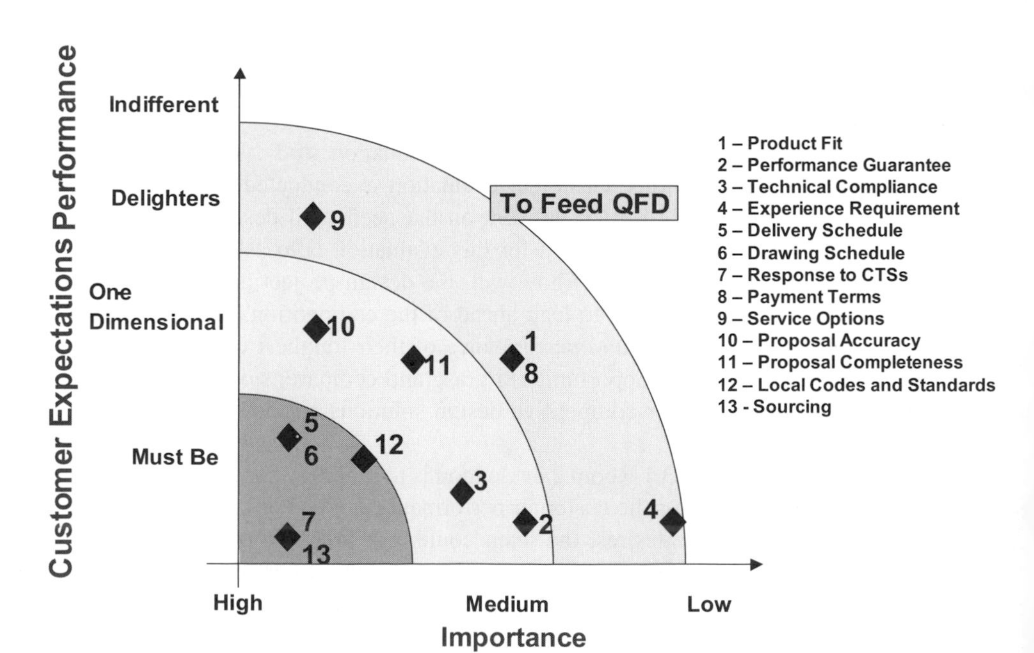

Figure 7.7 Qualitative Kano analysis plot for a sales inquiry process.

be completed from the Kano model or can be arranged qualitatively by the design team, but it must be validated by the customer or we will fall into the trap of voice of the engineer again.

7.8 QFD HOQ 2—"TRANSLATION HOUSE"

The critical-to-satisfaction (CTS) list is the set of metrics derived by the design team from the customer to answer the customer attributes list. The CTS list rotates into HOQ 2 Room #1 in this QFD phase. The objective is to determine a set of functional requirements (FRs) with which the critical-to-satisfaction requirements can be materialized. The answering activity translates customer expectations into requirements such as waiting time and number of mouse clicks for on-line purchasing service. For each CTS, there should be one or more FRs that describe a means of attaining customer satisfaction.

At this stage, only overall CTSs that can be measured and controlled need to be used. We will call these CTSs "Technical CTSs." As explained in Section 7.1, CTSs are traditionally known as substitute quality characteristics. Relationships between technical CTSs and FRs are often used to prioritize CTSs filling the relationship matrix of HOQ 2 rooms. For each CTS, the design team has to assign a value that reflects the extent to which the defined FRs contribute to meeting it. This value, along with the calculated importance index of the CTS, establishes the contribution of the FRs to the overall satisfaction and can be used for prioritization.

The analysis of the relationships of FRs and CTSs allows a comparison to other indirect information, which needs to be understood before prioritization can be finalized. The new information from Room 2 in the QFD HOQ needs to be contrasted with the available design information (if any) to ensure that reasons for modification are understood.

The purpose of the QFD HOQ 2 activity is to define the design functions in terms of customer expectations, benchmark projections, institutional knowledge, and interface management with other systems, and to translate this information into service technical functional requirement targets and specifications. This will facilitate the design mappings (see Chapter 8). Since the FRs are solution free, their targets and the specifications for them are flowed down from the CTs. For example, a CTS is for "speed of order" and the measure is hours to process, and we want order processing to occur within four hours. The functional requirements for this CTS, the "hows," could include *process design,* in which the *number of process steps* and the *speed of each step* would be the flow-down requirements to achieve "speed of order." Obviously, the greater the number of process steps, the shorter each step will need to be. Since at this stage we do not know what the process will be and how many steps will be required, we can allocate the sum of all process steps multiplied by their process time, not to exceed 4 hours. A major reason for customer dissatisfaction is that the service design specifications do not adequately link to customer use of the product or service. Often, the specification is written after the design is completed. It may also be a copy of outdated specifications. This reality may be attributed to the current planned design practices that do not allocate activities and resources in areas of importance to customers, and waste resources by spending too much time in activities that provide marginal value, a gap that is nicely filled by the QFD activities. The targets and tolerance setting activity in QFD Phase 2 should also be stressed.

7.9 QFD HOQ 3—"DESIGN HOUSE"

The functional requirements (FRs) are the list of solution-free requirements derived by the design team to answer the CTS array. The FRs list is rotated into HOQ 3 Room #1 in this QFD phase. The objective is to determine a set of design parameters which will fulfill the FRs. Again, the FRs are the "whats," and we decompose this into the "hows." This is the phase that most design teams want to jump right into so, hopefully, they will have completed the prior phases of HOQ 1 and HOQ 2 before arriving here. The design requirements must be tangible solutions

7.10 QFD HOQ 4—"PROCESS HOUSE"

The design parameters (DPs) are a list of tangible functions derived by the design team to answer the FRs array. The DPs list is rotated into HOQ 4 Room #1 in this QFD phase. The objective is to determine a set of process variables that, when controlled, ensure the DRs. Again, the DRs are the "whats," and we decompose this into the "hows."

7.11 QFD EXAMPLE

There was a business that purchased $3 billion of direct materials (materials that are sold to customers) and $113 million of indirect materials and services (materials and services that are consumed internally in order to support the customer-facing processes) and yet had no formal supply chain organization. The procurement function existed in each functional group, so Operations purchased what they needed as did Human Resources, Legal, and Information Systems. This distributed function resulted in a lack of power in dealing with suppliers, mostly supplier-favorable contracts, and some incestuous relationships (lavish trips and gifts given to the decision makers).

The corporate parent required that all procurement be consolidated into a formal supply chain organization with sound legal contracts, high integrity, and year-over-year deflation on the total spend. The design team was a small set of certified black belts and a master black belt who were being repatriated into a clean-sheet supply chain organization.

7.11.1 HOQ 1

The DFSS team will identify the customers and establish their wants, needs, delights, and usage profiles. In order to complete the HOQ, the team also needs to understand the competing performance and the environmental aspects the design will operate within. The design team realized that the actual customer base would include the corporate parent looking for results, a local management team that wanted no increased organizational cost, and the operational people who needed materials and services. Through interviews laced with a dose of common sense, the following customer wants and needs were derived:

- Price deflation
- Conforming materials
- Fast process
- Affordable process
- Compliant process
- Ease of use

The next step was to determine how these would be measured by the customer. The critical-to-satisfaction measures were determined to be:

- Percentage year-over-year price deflation
- Percentage defective deliveries
- Percentage late deliveries
- Cost per transaction
- Cost of organization/Cost of orders
- Speed of order (hours)
- Number of compliance violations
- Number of process steps

Next, the team used the QFD application, Decision Capture,[1] to assign the direction of improvement and importance of the wants/needs and the relationship between these and the CTSs. Figure 7.8 shows the results. We can see that price deflation is moderately related to percentage of defective deliveries and number of compliance violations, weakly related to percentage of late deliveries, and strongly related to year-over-year price deflation. We see that price deflation is extremely important and the direction of improvement is to maximize it.

In Figure 7.8 we can see that the importance of the CTSs has been calculated. The method of calculation is as follows for the top sorted CTS:

[1]Decision capture is a software application from International TechneGroup Inc. (www.decisioncapture.com).

Year-over-year price deflation = 9 (strong) × 5 (extremely important)
+ 1 (weak) × 5 (extremely important) + 3 (moderate) × 4 (very important)
+ 1 (weak) × 3 (somewhat important) = 65

The other columns are similarly calculated.

The completed HOQ1 is in Figure 7.9 and shows the competitive ratings of the wants/needs (Room 2) and the technical ratings of the CTSs (Room 6). For this pur-

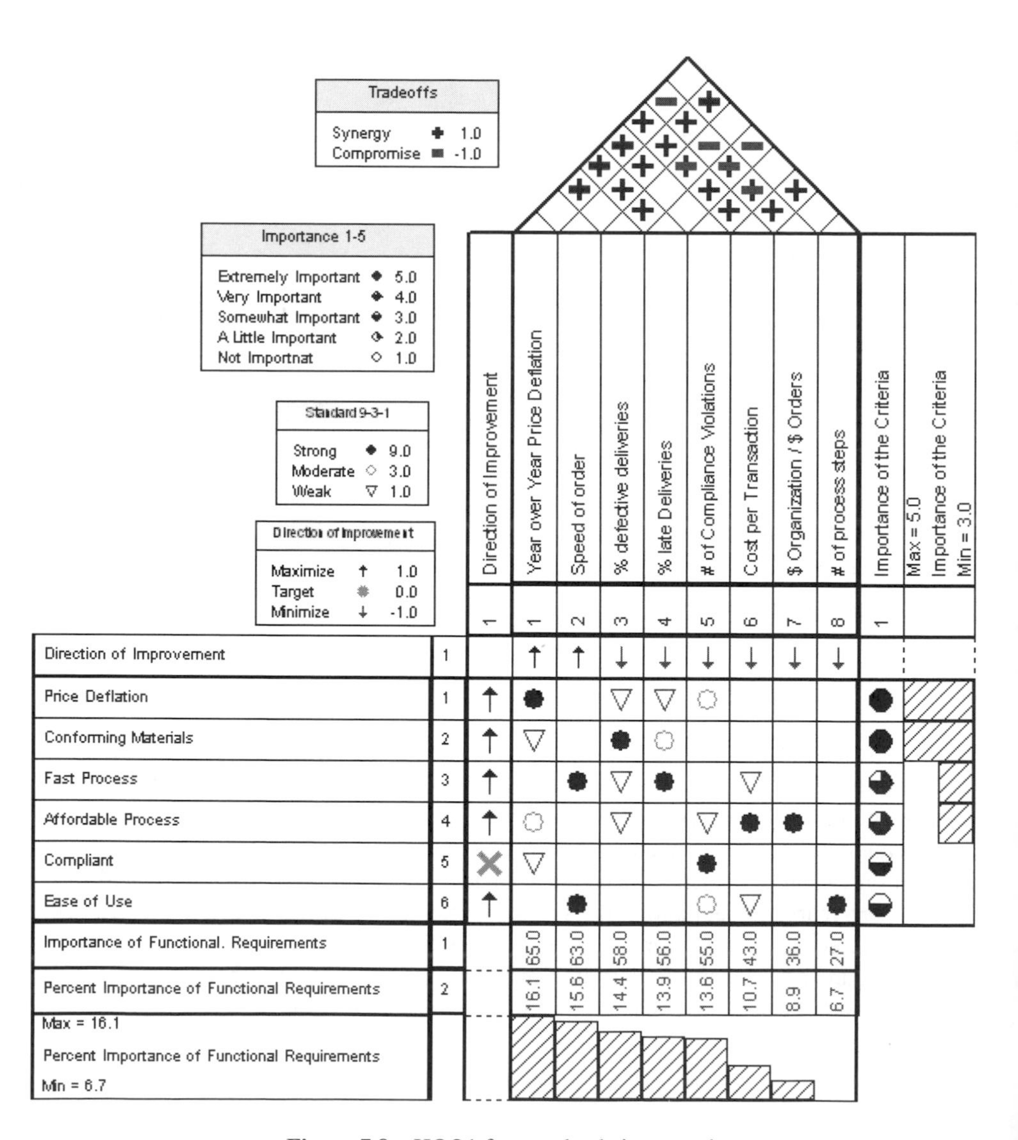

Figure 7.8 HOQ1 for supply chain example.

pose, the team used Competitor 1 as the incumbent process and used a sister business with a fully deployed supply chain organization as Competitor 2. The synergies and compromises have also been determined.

The next step would be to assess the HOQ for any "whats" with only weak or no relationships. In this example, no such case exists. Now we look for "hows" that are blank or weak and also see that there are none of these.

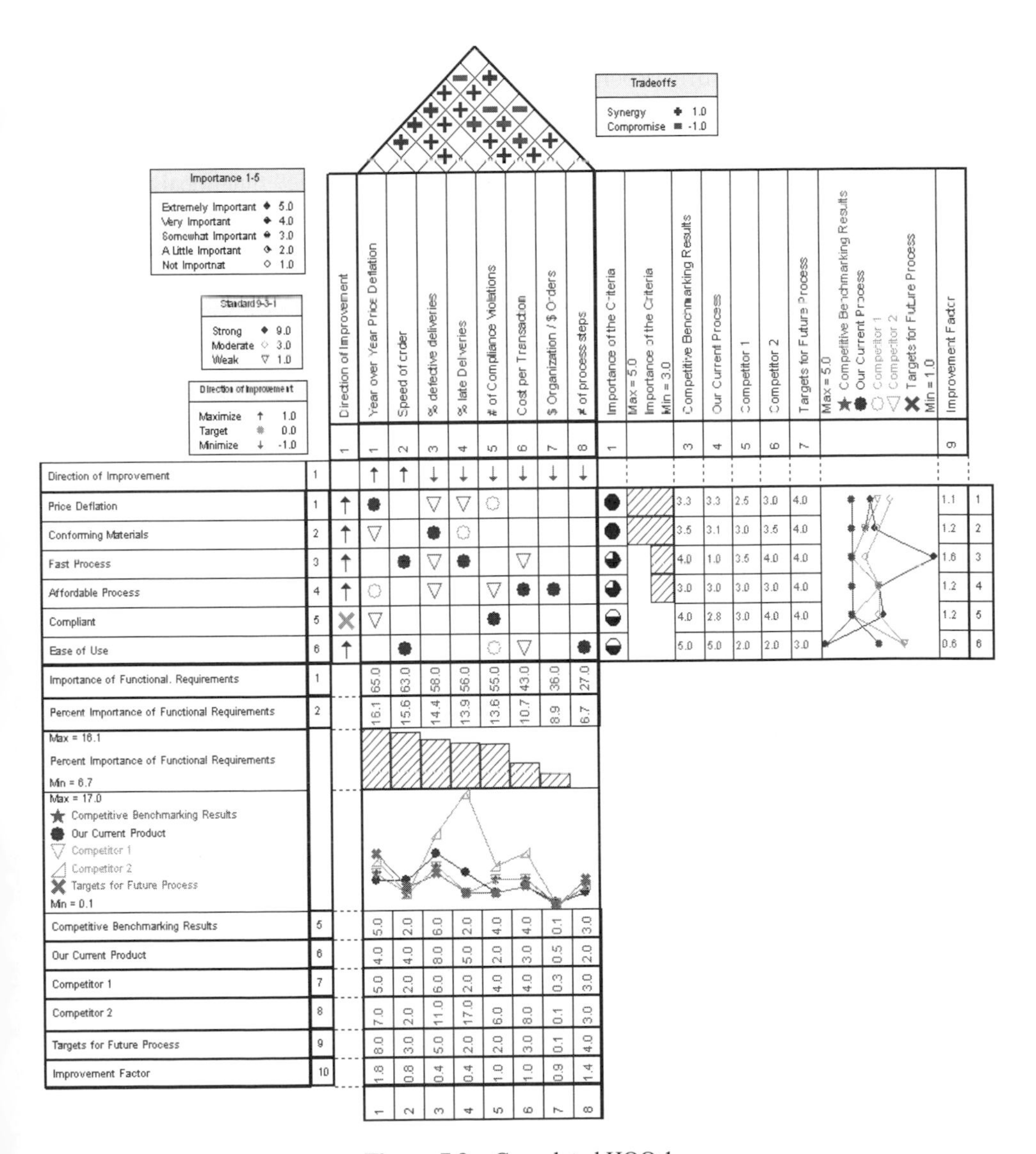

Figure 7.9 Completed HOQ 1.

7.11.2 HOQ 2

Now that the team feels confident about HOQ 1, they move on to HOQ 2. We will follow the seven-step process:

1. Define functional requirements (FRs) (Room 3)
2. Define direction of improvement (Room 3)
3. Define relationship between CTSs and FRs (Room 4)
4. Calculate the importance value (Room 5)
5. Sort importance values in descending order
6. Define the trade-offs (Room 8)
7. Evaluate the matrix for strengths and weaknesses

The first step is to decompose the CTSs into "solution-free" functional requirements. The first CTS is "percentage year-over-year price deflation." This is the "what," and we ask, "How can we do this?" First we need measures of deflation. Notice that this does not specify any specific means of measurement, and surely we need a measure if we are to convince the stakeholders of what we want to accomplish. In addition to this measure, we need methods for achieving deflation and resources to achieve deflation. So the first CTS can be fulfilled by:

- Measures of deflation
- Cost-reduction methods
- Resources to achieve deflation

Similarly, the other CTSs are decomposed into their functional requirements:

- Percentage of defective deliveries:
  - Measures of delivery quality
  - Delivered quality improvement methods
  - Resources to achieve delivery quality
- Percentage of late deliveries:
  - Measures of on-time delivery
  - Methods for improving on-time delivery
  - Resources to achieve on-time delivery
- Number of compliance violations:
  - Measures of compliance
  - Compliance training
  - Staffing methods
  - Compliance audits
- Cost per transaction:
  - Organizational cost
  - Systems process cost

- Purchased material and services cost divided by organizational cost:
  - Organizational design
  - Amount of spending controlled
  - Supplier support
- Speed of order:
  - Approval speed
  - Process speed
  - System availability
- Number of process steps:
  - Process design
  - Approval levels

We place these in the matrix (Room 3), define the direction of improvement, and then define the relationship between each CTS and FR (Room 4). We calculate the matrix and sort on the prioritized FRs, and then define the trade-offs on the roof (Room 8). Last, we evaluate the HOQ 2 for completeness and logic. Figure 7.10 shows the completed HOQ 2.

7.11.3 HOQ 3

Now that the team feels confident about HOQ 2, they move on to HOQ 3. We will again follow the seven-step process:

1. Define design parameters (DPs) (Room 3)
2. Define direction of improvement (Room 3)
3. Define relationship between FRs and DPs (Room 4)
4. Calculate the importance value (Room 5)
5. Sort importance values in descending order
6. Define the trade-offs (Room 8)
7. Evaluate the matrix for strengths and weaknesses

The first step is to decompose the FRs into the corresponding design parameters. The first FR is "resources," which are required for supplier development in delivery, quality and productivity. This is the "what," and we ask, "How can we do this?" First, we need to know the type of resources. In addition to this we need to know the quantity of resources. So the first FR can be fulfilled by:

- Type of resources
- Number of resources

Similarly, the other FRs are decomposed into their design parameters:

- Organizational design
  - Type of resources

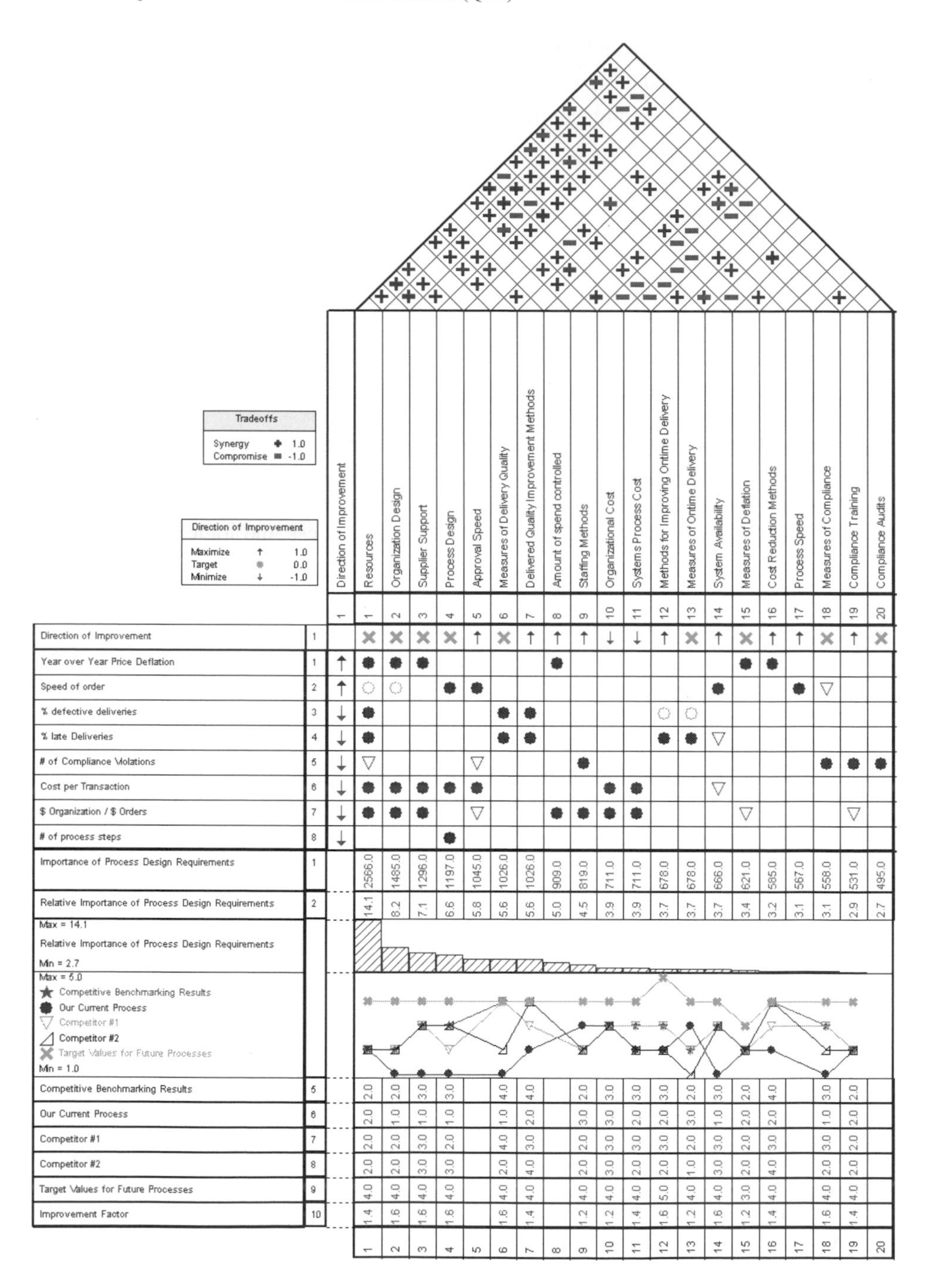

Figure 7.10 HOQ 2 example.

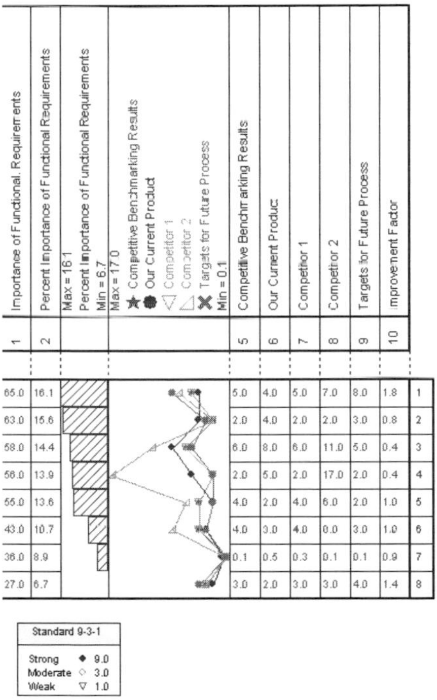

#		1	2	3	4	5	6	7	8
1	Importance of Functional. Requirements	65.0	63.0	58.0	56.0	55.0	43.0	36.0	27.0
2	Percent Importance of Functional Requirements	16.1	15.6	14.4	13.9	13.6	10.7	8.9	6.7
	Max = 16.1 Percent Importance of Functional Requirements Min = 6.7								
	Max = 17.0 Competitive Benchmarking Results Our Current Product Competitor 1 Competitor 2 Targets for Future Process Min = 0.1								
5	Competitive Benchmarking Results	5.0	2.0	6.0	2.0	4.0	4.0	0.1	3.0
6	Our Current Product	4.0	4.0	8.0	5.0	2.0	3.0	0.5	2.0
7	Competitor 1	5.0	2.0	6.0	2.0	4.0	4.0	0.3	3.0
8	Competitor 2	7.0	2.0	11.0	17.0	6.0	0.0	0.1	3.0
9	Targets for Future Process	8.0	3.0	5.0	2.0	2.0	3.0	0.1	4.0
10	Improvement Factor	1.8	0.8	0.4	0.4	1.0	1.0	0.9	1.4

Standard 9-3-1	
Strong	◆ 9.0
Moderate	◇ 3.0
Weak	▽ 1.0

Figure 7.10 (*continued*)

Quantity of resources
Location of resources

- Supplier support
  Type of support
  Frequency of support
  Magnitude of support
- Process design
  Intuitive
  Workflow
- Approval speed
  Approval levels
  Speed of approval levels
- Measures of delivery quality
  By supplier
  By commodity
  By part
  By buyer
- Methods for improving delivery quality
  Lean methods
  Variation reduction methods
  Contract delivery quality terms
- Amount of spend controlled
  Consolidated purchase amount
- Staffing methods
  Background checks
  Screening methods
- Organizational cost
  Compensation and benefits
  Travel and living
- Systems process cost
  Depreciation cost
  Telecommunication cost
  Mail cost
- Methods for improving on-time delivery
  Lean methods
  Variation reduction methods
  Contract on-time delivery terms
- Measures of on-time delivery
  By supplier

  - By commodity
  - By part
  - By buyer
- System availability
  - Uptime
  - Bandwidth
- Measures of deflation
  - By supplier
  - By commodity
  - By part
  - By buyer
- Cost reduction methods
  - Leveraged spend
  - Supplier deflation performance
  - Lean methods
  - Variation reduction methods
  - Contract cost-reduction terms
- Process speed
  - Number of process steps
  - Speed of each process step
  - Resource quantity
  - Resource skill
- Measures of compliance
  - By compliance category
  - By buyer
  - By process
- Compliance training
  - Target audience
  - Content
  - Frequency
- Compliance audits
  - Auditors
  - Audit method

We place these in the matrix (Room 3), define the direction of improvement, and then define the relationship between each FR and DP (Room 4). We calculate the matrix (Room 5), sort on the prioritized DPs, and then define the trade-offs on the roof (Room 8). Figure 7.11 shows the completed HOQ 3. Last, we evaluate the HOQ 3 for completeness and logic.

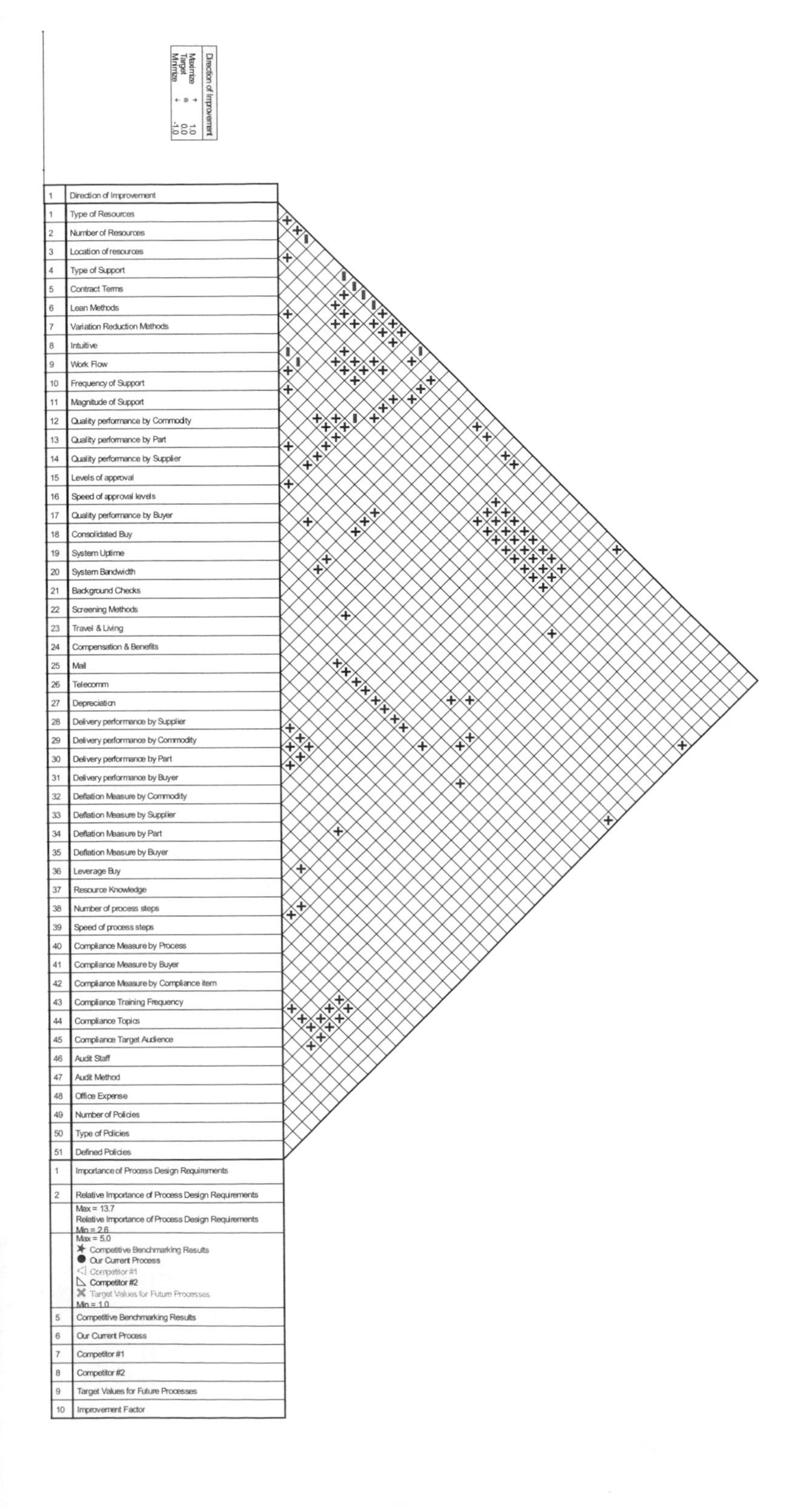

Direction of Improvement
Maximize 1.0
Target 0.0
Minimize -1.0
1 Direction of Improvement
1 Type of Resources
2 Number of Resources
3 Location of resources
4 Type of Support
5 Contract Terms
6 Lean Methods
7 Variation Reduction Methods
8 Intuitive
9 Work Flow
10 Frequency of Support
11 Magnitude of Support
12 Quality performance by Commodity
13 Quality performance by Part
14 Quality performance by Supplier
15 Levels of approval
16 Speed of approval levels
17 Quality performance by Buyer
18 Consolidated Buy
19 System Uptime
20 System Bandwidth
21 Background Checks
22 Screening Methods
23 Travel & Living
24 Compensation & Benefits
25 Mail
26 Telecomm
27 Depreciation
28 Delivery performance by Supplier
29 Delivery performance by Commodity
30 Delivery performance by Part
31 Delivery performance by Buyer
32 Deflation Measure by Commodity
33 Deflation Measure by Supplier
34 Deflation Measure by Part
35 Deflation Measure by Buyer
36 Leverage Buy
37 Resource Knowledge
38 Number of process steps
39 Speed of process steps
40 Compliance Measure by Process
41 Compliance Measure by Buyer
42 Compliance Measure by Compliance item
43 Compliance Training Frequency
44 Compliance Topics
45 Compliance Target Audience
46 Audit Staff
47 Audit Method
48 Office Expense
49 Number of Policies
50 Type of Policies
51 Defined Policies
1 Importance of Process Design Requirements
2 Relative Importance of Process Design Requirements
Max = 13.7
Relative Importance of Process Design Requirements
Min = 2.6
Max = 5.0
Competitive Benchmarking Results
Our Current Process
Competitor #1
Competitor #2
Target Values for Future Processes
Min = 1.0
5 Competitive Benchmarking Results
6 Our Current Process
7 Competitor #1
8 Competitor #2
9 Target Values for Future Processes
10 Improvement Factor

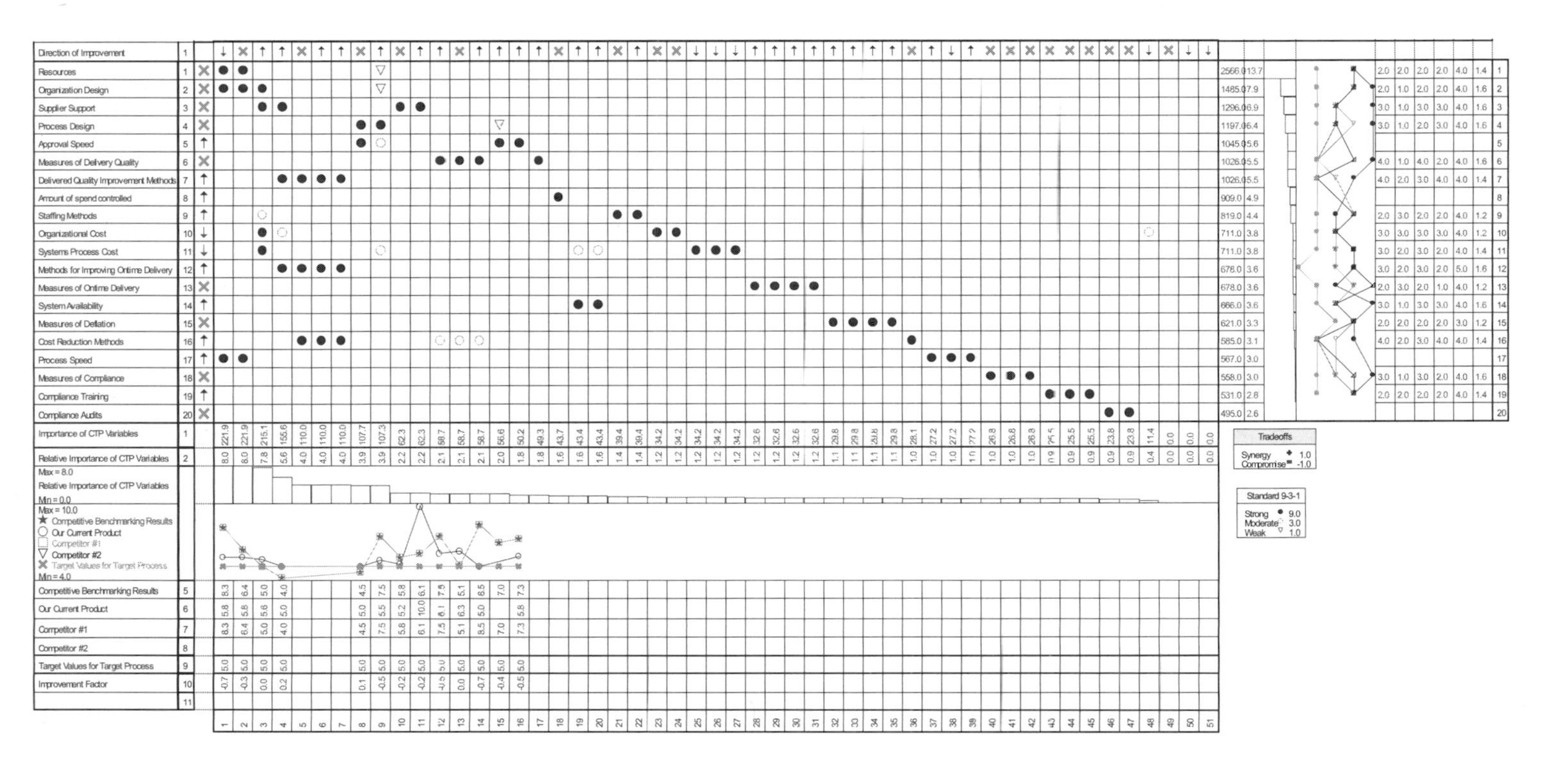

Figure 7.11 HOQ 3 example.

7.11.4 HOQ 4

Now that the team feels confident about HOQ 3, they move on to HOQ 4. We will follow the seven-step process one last time:

1. Define process variables (PVs) (Room 3)
2. Define direction of improvement (Room 3)
3. Define relationship between DPs and PVs (Room 4)
4. Calculate the importance value (Room 5)
5. Sort importance values in descending order
6. Define the trade-offs (Room 8)
7. Evaluate the matrix for strengths and weaknesses

The first step is to decompose the DPs into the corresponding process variables. The first DP is "type of resources." This is the "what," and we ask, "How can we do this?" First, we need commodity leaders. In addition to this we need buyers and clerical support. All of these positions need to be scaled by opportunity, so the process variables that we need in order to make these choices are dynamic and vary by activity and resource type. The first DP can be fulfilled by:

- Capacity model
- Volume per person

Similarly, the other DPs are decomposed into their process variables:

- Number of resources
  - Volume per person
  - Capacity model
- Location of resources
  - Travel cost per person
- Type of support
  - Problem resolution support
  - Productivity support
- Contract terms
  - Delivery performance terms
  - Quality performance terms
  - Productivity performance terms
- Lean methods
  - Lean material content
  - Number of persons trained in lean methods
  - Number of lean projects
- Variation-reduction methods

  Variation-reduction materials content
  Number of persons trained in variation-reduction methods
  Number of variation-reduction projects
- Intuitive
  Templates
  Self-help
- Workflow
  Percentage of workflow
- Frequency of support
  Date of activity
  Count of activity
- Magnitude of support
  Hours per supplier per year
- Quality performance by commodity
  Commodity class
  Quality measure
- Quality performance by part
  Part ID
  Quality measure
- Quality performance by supplier
  Supplier ID
  Quality measure
- Levels of approval
  Approval policies
- Speed of approval levels
  Time to approve by level
- Quality performance by buyer
  Buyer ID
  Quality measure
- Consolidated purchasing
  Supplier ID
  Annual spend by supplier
  Dollar amount of spend actively managed
  Number of suppliers
- System uptime
  Minutes available per schedule
- System bandwidth
  Transaction per minute
- Background checks

  Backgrounds performed
  Backgrounds with defects
- Screening methods
  Interview questions
- Travel and living expense
  Buyer ID
  Supplier ID
  Commodity class
  Trip approval
  Reason for travel
  Expense per month
- Compensation and benefits
  Buyer ID
  Expense per month
- Mail expense
  Cost per month
- Telecommunication expense
  Expense per month
  Number of telephones
- Depreciation expense
  Expense per month
- Delivery performance by supplier
  Supplier ID
  Delivery performance
- Delivery performance by commodity
  Commodity class
  Delivery performance
- Delivery performance by part
  Part ID
  Delivery performance
- Delivery performance by buyer
  Buyer ID
  Delivery performance
- Deflation measure by commodity
  Commodity class
  Deflation measure
- Deflation measure by supplier
  Supplier ID
  Deflation measure

- Deflation measure by part
  Part ID
  Deflation measure
- Deflation measure by buyer
  Buyer ID
  Deflation measure
- Leveraged purchases
  Annual spend by supplier
  Volume by part
- Resource knowledge
  Date of training
  Hours of training
  Type of training
  Defects per buyer
- Number of process steps
  Count of steps
- Speed of process steps
  Time to process per step
- Compliance measure by process
  Process step classification
  Count of compliance defects
- Compliance measure by buyer
  Buyer ID
  Count of compliance defects
- Compliance measure by compliance item
  Compliance category
  Count of compliance defects
- Compliance training frequency
  Compliance requirements by position
- Compliance items
  Compliance category
  Compliance requirements by position
- Compliance target audience
  Buyer ID
  Compliance requirements by position
- Audit staff
  Quantity
  Source
- Audit method

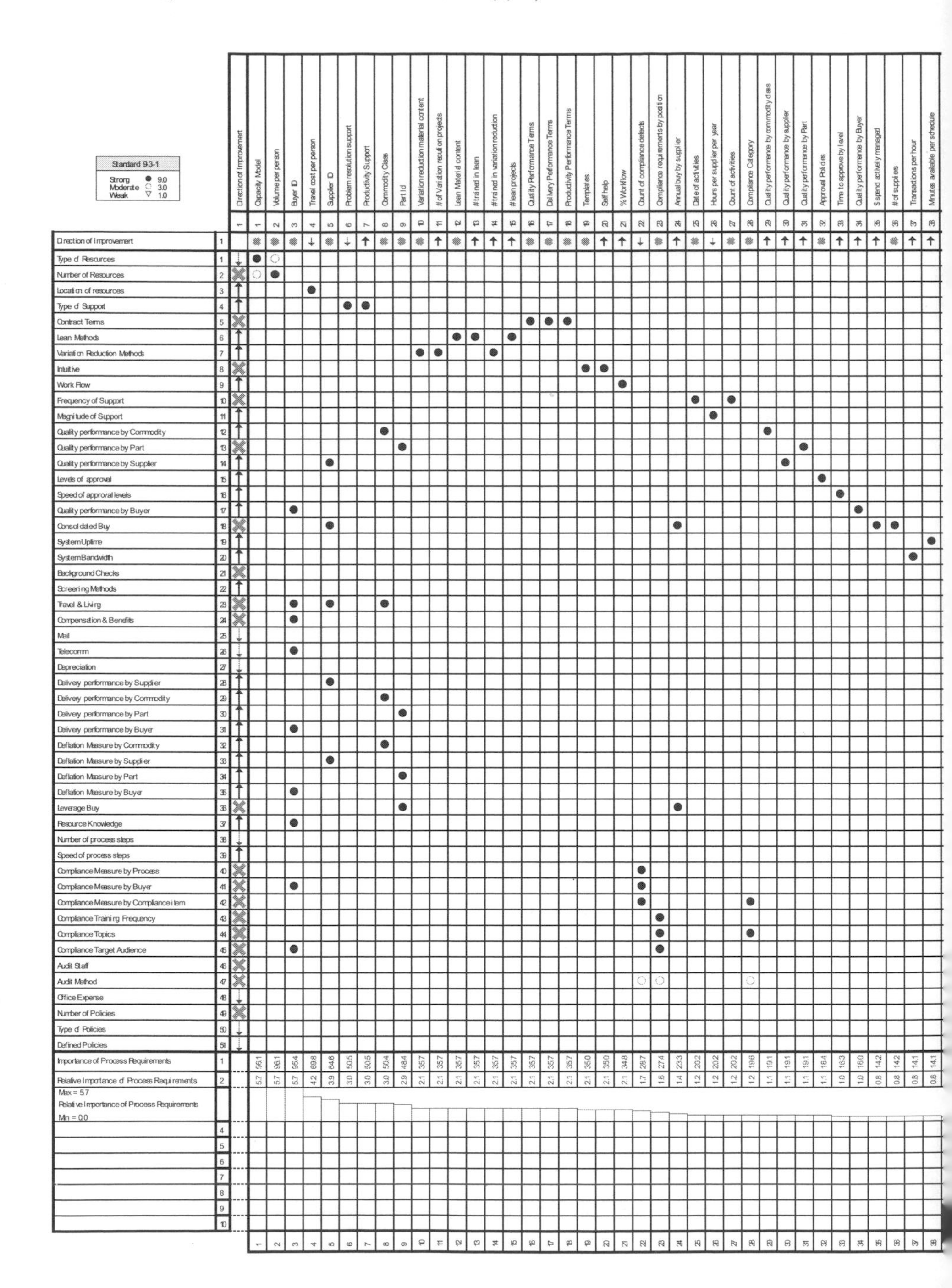

Figure 7.12 HOQ 4 example.

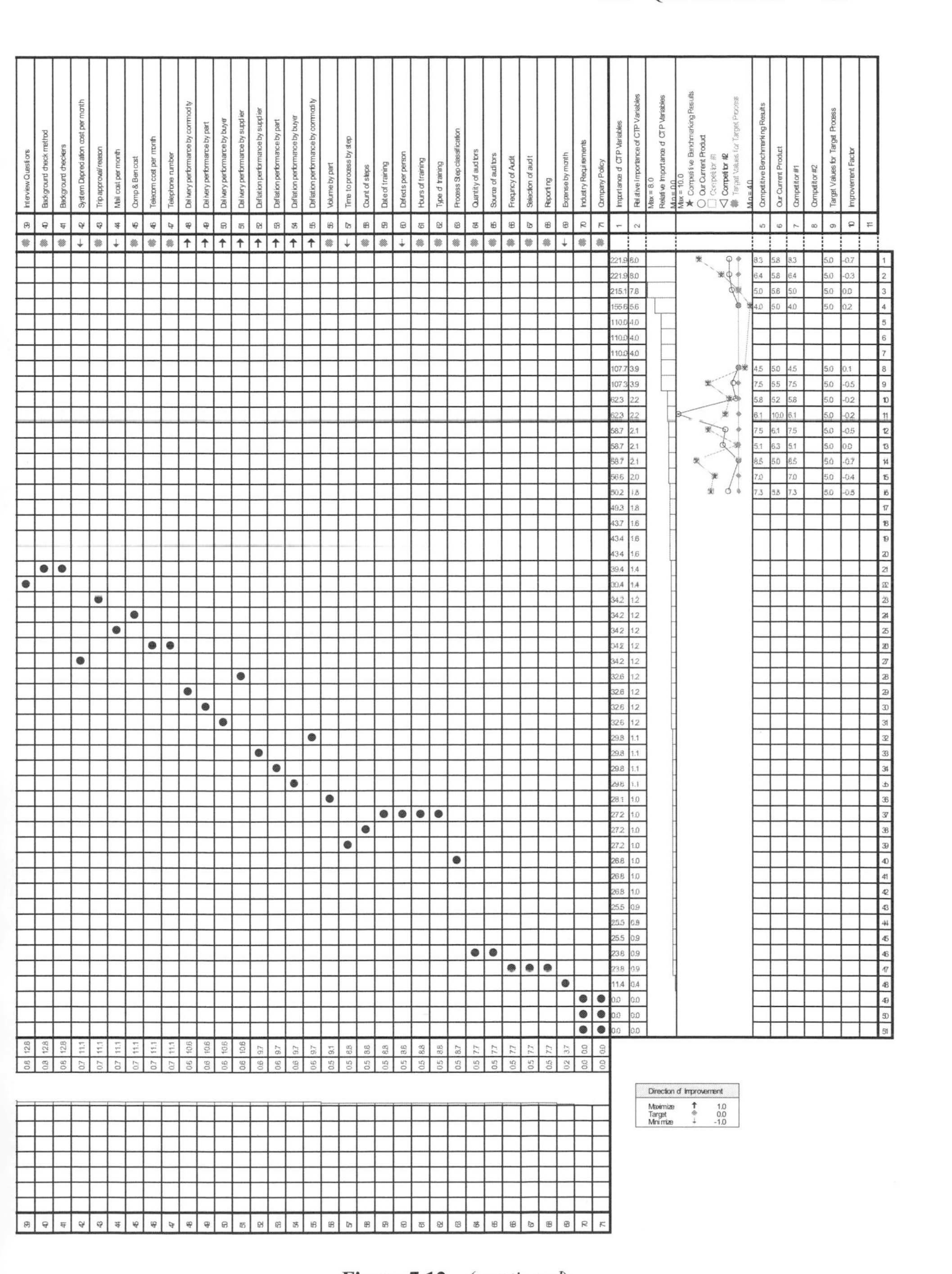

Figure 7.12 *(continued)*

Frequency of audit
Selection of audit target
Reporting

- Office expense
  - Expense by month
  - Expense category
- Number of policies
  - Industry requirements
  - Company policy
- Type of policies
  - Industry requirements
  - Company policy
- Defined policies
  - Industry requirements
  - Company policy

We place these in the matrix, define the direction of improvement, and then define the relationship between each DP and PV. We calculate the matrix and sort on the prioritized PVs. There is no need to define the trade-offs since these measures are independent of each other (or uncoupled in the context of Chapter 8). Last, we evaluate the HOQ 3 for completeness and logic.

7.12 SUMMARY

QFD is a planning tool used to translate customer needs and wants into focused design actions. This tool is best accomplished with cross-functional teams and is key in preventing problems from occurring once the design is operationalized. The structured linkage allows for rapid design cycles and effective utilization of resources while achieving Six Sigma levels of performance.

To be successful with the QFD, the team needs to avoid "jumping" right to solutions, and should process HOQ1 and HOQ2 thoroughly and properly before performing a detailed design. The team will also be challenged to keep the functional requirements solution neutral in HOQ2.

It is important to have the correct voice of customer and the appropriate benchmark information. Also, a strong cross-functional team willing to think outside the box is required in order to truly obtain Six-Sigma-capable products or processes. From this point on, the QFD is process driven but it is not the charts that we are trying to complete, it is the total concept of listening to the voice of the customer throughout the design effort.

8

PROCESS MODELING AND PROCESS MANAGEMENT

8.1 INTRODUCTION[1]

In this chapter, we will cover process modeling from the simplest SIPOC (see Chapter 2) to functional modeling and demonstrate the available evaluation techniques. By the end of this chapter, the DFSS green belt or black belt should be able to prepare a detailed process map. Process mapping and functional modeling techniques are key for service DFSS projects in the context of the road map depicted in Figure 5.1. The objective of this chapter is to present the case for process mapping, review the different types of process maps, and discuss techniques and tools used to map a process.

Process mapping can be used when designing or improving processes or parts of processes within the DFSS project scope. By breaking the process down into its smaller elements and making its value and nonvalue portions apparent, process mapping helps teams to visualize the process. In a DFSS project, processes are usually identified via tools such as SIPOC, flow planning studies, and simulation analysis (if any). Once the processes are identified, tools such as process mapping are recommended to enable the identification of non-value-added steps of a process for further treatment, transfer function detailing, defect locations, and pinpointing where improvement can be sought. In addition to documenting the process, process maps can be used to communicate information and train employees in the process aspects.

Design teams usually think they have a pretty good understanding of how their project processes work. But, oftentimes, the actual process works somewhat or significantly differently. Within the process, different steps may be taken or performed in different order. Some of these variations may happen for legitimate or illegitimate reasons. To complete a process map, the team should walk through the actual

[1]It is recommended that the reader familiarize him or herself with the concepts of Chapter 2 prior to reading this chapter, in particular Sections 2.2, 2.5, and 2.6. These sections discuss the concepts of value in service, lean processes, yield, and service metrics.

process and talk with all those involved in the process to understand it. This will help highlight opportunities for improving the process using quick-hit, actionable items and pave the way for the redesign effort leading to the map of "how the team would like it to be." When working with multilocation or cross-functional process flows, the process map enables the graphical representation that allows everyone in the process to see the entire flow rather than their isolated portions.

In this chapter, we will cover several mapping tools. Some are graphical in nature, whereas the rest are logical and prescriptive. The premier attribute of the graphical method is simplicity and ease of communication to different audiences with different levels of involvement. The tools are handy and easy to use in redesign projects. On the other hand, the major advantage of logical and prescriptive tools is their logical enforcing functions when applied to new and creative design situations. The penetration of these tools is very immature in the service industry today. However, their power is really remarkable. This book is an attempt to expose them to service Six Sigma populations. In addition, this chapter features the business process management system (BPMS, see Chapter 2), which directs the ongoing accountability and performance of a process.

8.2 PROCESS MODELING TECHNIQUES

8.2.1 SIPOC

There are two approaches possible with the SIPOC tool. The first approach is to gather all of the items within each of the columns without regard to the order in which they are applied in the actual process. This is a brainstorming tool for use with teams at the macro process level. The other method is to take each subprocess step, one at a time, and apply the SIPOC at the micro level. The order of approach is the same for both methods. The design team starts at the "process "column, ask what "inputs" are required, and asks then who the "suppliers" are for these inputs. At this time, they also assess what input characteristics the process requires. These are usually in the framework of speed, quality or cost. The next step is to assess what is the "output" of the process, who is the "customer," and what characteristics are required of the output by the customer. Figure 8.1 illustrates a generic SIPOC methodology. Figure 8.2 shows an application of the tool in BPMS (see Chapter 2 and Section 8.3).

8.2.2 Process Mapping

Like most of the DFSS tools, process mapping requires a cross-functional team effort with involvement from process owners, process members, customers, and suppliers. Brainstorming, operation manuals, specifications, operator's experience, and process walkthroughs are very critical inputs to the mapping activity. A detailed process map provides input to other tools such as FMEA, transfer function DOEs, capability studies, and control plans.

Process maps can be created for many different levels, zooming in and out of the targeted process and delivering the service under consideration in the DFSS project.

Suppliers	Inputs	Input Characteristics	Process	Outputs	Output Characteristics	Customers
7. Who are the suppliers of the inputs?	6. What are the inputs of the process?	8. What are the characteristics of the inputs?	2a. What is the start of the process? 1. What is the process? 2b. What is the end of the process?	3. What are the outputs of the process?	5. What are the characteristics of the outputs?	4. Who are the customers of the outputs?

Figure 8.1 Generic SIPOC table.

Level 1

Process Steps: (Level 2)				
1. Description and Start & Stop Points				
2. Process Output				
3. Process Owner				
4. Ultimate Customer				
5. Primary Customer				
6. Stakeholder				
7. Output CTQ (2)				
8. Process CTQ (CTP) (2 & 7)				
9. Stakeholder CTQ (6)				
10. Output Measures (7 & 2)				
11. Process Measures (8 & 3)				
12. Stakeholder Measures (9 & 6)				

Figure 8.2 BPMS input analysis.

Depending on the detail the team requires, the map should be created at that level. If more details are required, then a more detailed map of the subprocess should be completed. The team should objectively demand verification with hands-on exposure to local activities, identifying rework loops and redundancies, and seeking insight into bottlenecks, cycle times, and inventory. Rework loops and redundancies are non-value-added items costing time, effort, money, and other resources, and they often referred to as the "hidden factory" in the Six Sigma literature.

To make the most improvements in any existing process, it is necessary to understand the actual way that the process works. Within this context, it is easy to understand the reasons for the flow problems and then address the causes.

In process mapping, symbols represent process steps, measurements, queues, storage, transportation (movement), and decisions (Figure 8.3).

There may be three versions of a process map (Figure 8.4). There is "what it was designed to be," which is usually a clean flow. There is the "as-is" process map, with all the gritty variety that occurs because of varying suppliers, customers, operators, and conditions. Take the ATM machine as an example. Does everyone follow the same flow for getting money out? Some may check their balance first and then make a withdrawal, whereas others may type in the wrong PIN and have to retype it. The last version is "what we would like it to be," with only value-added steps. It is clean, intuitive, and works right every time.

A process map is a pictorial representation showing all of the steps of a process. As a first step, the team should familiarize themselves with the mapping symbols, then walk throught the process by asking questions such as: "What really happens

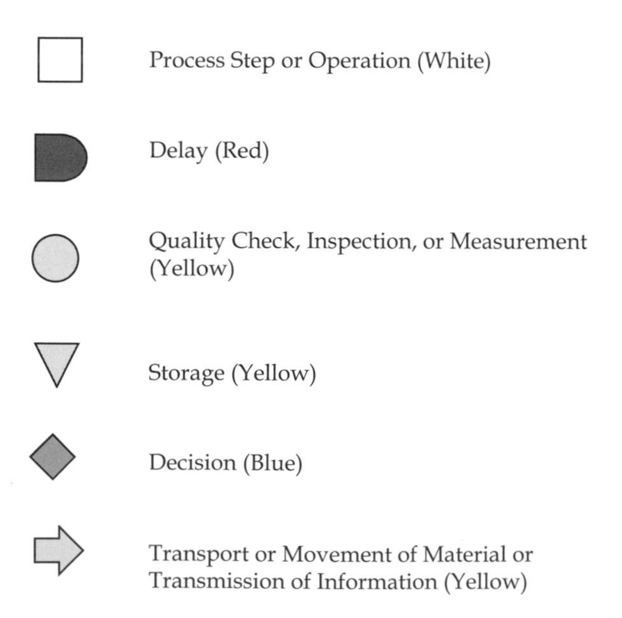

Figure 8.3 Process mapping standard symbols. Due the constraints of black-and-white printing, colors are rendered as shades of gray.

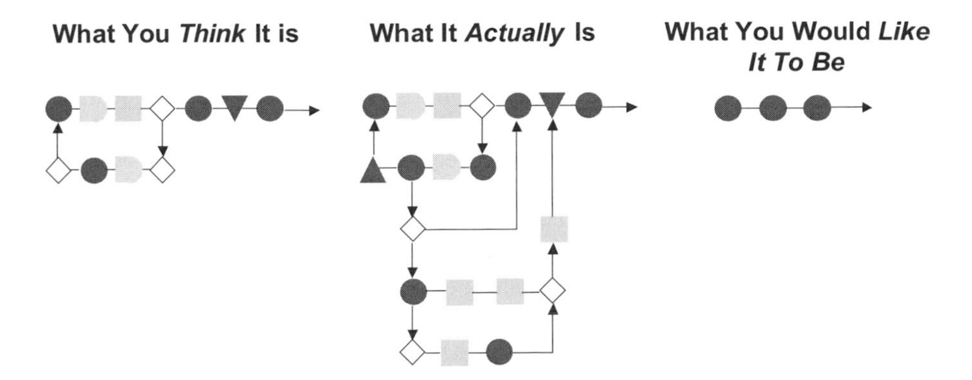

Figure 8.4 Three versions of a process map.

next in the proccss?" "Does a decision need to be made before the next step?" or "What approvals are required before moving on to the next task?" The team then draws the process using the symbols on a flip chart or overhead transparency. Every process will have a start and an end (ovals). All processes will have tasks and most will have decision points (diamonds). Upon completion, the team should analyze the map for such items as non-value-added steps, rework loops, duplication, and cycle time. A typical high-level process map is shown in Figure 8.5.

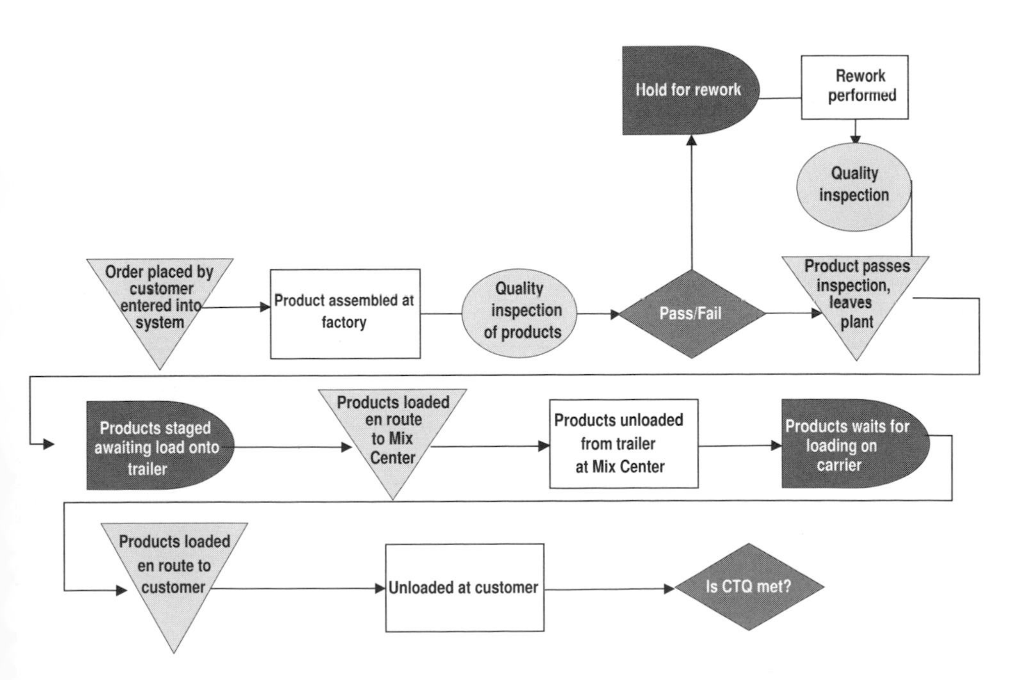

Figure 8.5 A high-level process map example.

A good process map should identify all process steps or operations as well as visible measurement, inspection, rework loops, and decision points. In addition, "swim lanes" are often used when mapping information flow for service-type transactional and business processes. We believe that swim lanes are appropriate for all types of DFSS projects. The swim lanes segregate steps by who does them or where they are done and makes hand-offs visual. The map is arranged in a table in which the rows indicate the "who" that owns or performs the process step (process owner), and the process flows that change "lanes" indicate hand-offs. Hand-off points are where lack of coordination and communication can cause process problems. An example is depicted in Figure 8.6.

Clear distinctions can be made between the warehousing and those process steps where customer interactions occur. The swim lane process map example shows a high-level portion of the order receiving process. Another level of detail in each of the process steps will require further mapping if they are within the scope of the project.

Process mapping is a methodology composed of the following steps and actions:

Step 1. Define the process

- Objective/scope of the project defined
- Revisit the objective to select and prioritize processes to be mapped
- Focus on times, inputs, functions, hand-offs, authority and responsibility boundaries, and outputs
- Appropriate level of detail discussed
- Linkages to other analyses introduced
- List of high-level process steps
- Document the overall flow of the process between the start and stop boundaries

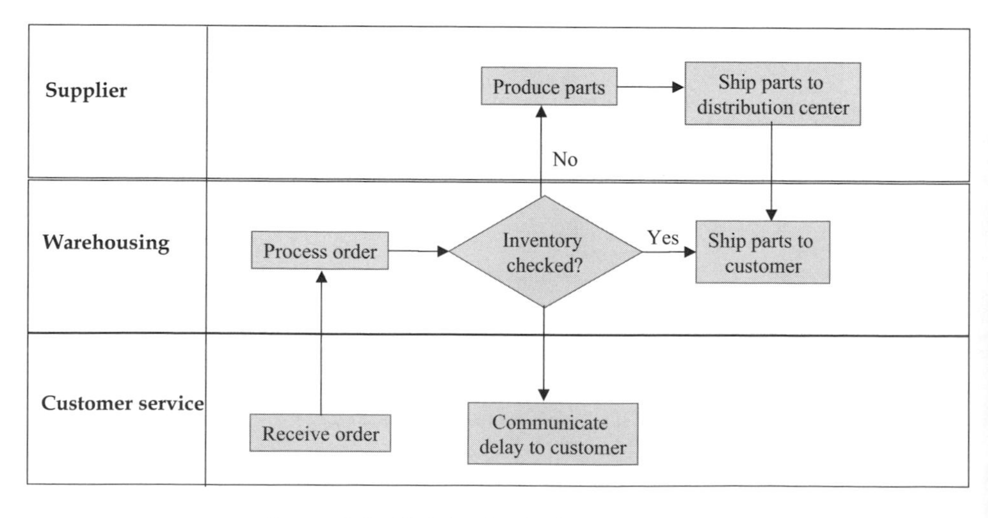

Figure 8.6 High-level swim lanes process map.

Step 2. Collect information

- Data required. Use sampling and collection techniques (focus groups, interviews, observations)
- Resources required in this mapping step and the mapping exercise in general
- Process walk-through schedule based on shifts, days, and weeks where unique work occurs
- When do we zoom in to map every process step?
- Continuous flow
- Process stops (when flow is disrupted or disconnected)
- Have process inputs and output variables (PIV, the *x*s, and POV, the *y*s) been identified. Remember that we are after a transfer function that fully describes what is happening/should be happening to confirm that the solution is really optimal.
- Consider merging of multiple steps
- Add operating specifications
- Be clear about terminology and nomenclature so that everyone is speaking the same language. Avoid biased, nontechnical language. Be consistent.
- Accumulate a list of quick-hits items and develop an action plan for their implementation.

Step 3. Validate and verify

- Ensure the use of common terms and nomenclature
- Consolidate, compile, and reconcile information
- Summarize findings
- Validate information
- Confirm controversial data

Step 4. Build the process maps

- Sequence of activities and work steps based on the previous steps
- Swim lanes (if necessary)
- Beginning and end of process
- Work steps assigned to participants
- Parcel out times for each work step
- Initial assessment of control plan
- Start an initial assessment after process map is complete
- Add measurement techniques
- Define operating specifications
- Determine targets
- Know the context of the process mapping within the project
- Does process mapping fit with previous, parallel, and proceeding analyses?
- Conduct benchmarking

- Baseline current performance
- Document comments

Step 5. Analyze and draw conclusions
- Identfy characteristics of the process steps
- Define hypotheses about how the process steps inputs link to output variables relative to targets/historical means and variances
- Plan follow-up work if any (e.g., new measurement systems, SPC charting)
- View improvement or redesign opportunities
- Apply “if–then” simulation scenarios for any layout changes (see Chapter 13)

Step 6. Communicate recommendations, findings, and conclusions
- Customize the presentation to the audience with insights into change implications
- Update specifications, control plans, procedures, training, and so on.

8.2.3 Value Stream Mapping

Value stream mapping (VSM) was initially used to document manufacturing processes that are to be improved using lean manufacturing methods. VMS is equally applicable to service processes as well. They are most useful with higher-frequency processes, even mixed-model processes. In this section, we will define value stream mapping and its value, and present the different steps for its construction. Usually, whenever there is a product/service for a customer, there is a value stream. Therefore, value stream mapping is applicable whenever the design team is trying to make waste in a process visible to eliminate non-value-added steps, actions, and activities. A value-added activity or step is any activity or thing the customer is willing to pay for. Non-value-added activities are the activities that do not add market form or function or are not necessary. They should be eliminated, simplified, reduced, or integrated.

Visibility to waste can be assured via a scorecard documenting performance of variables (metrics) such as non-value-added cycle time, inventory, rework, and defects. Such measurement should be taken within a broader scope of a value stream management process. Value stream management is the process of increasing the ratio of value to nonvalue by identifying and eliminating sources of waste. Waste and non-value-added activities exhibit themselves in several formats:

- Unnecessary movement of people
- Overproduction ahead of demand
- Unnecessary movement of materials
- Waiting for the next step (idle time)
- Overprocessing

- Production of defective products (measured in services per million, defects per million, or defects per million opportunities)
- Extra inventory

Lean manufacturing techniques remove waste and non-value-adding activities from processes to enable the production and delivery of products and services at customer demand and at lower cost. Value stream mapping utilizes the icons shown in Figure 8.7. Several flows can be entertained in this technique. The icons include material *push* (schedule-driven) flows, and material *pull* (demand-driven) flows with depicted information flow mappings that parallel the movement of material and parts.

Current-state-process maps, using value stream mapping, demand a process walkthrough by the team. It is also useful to create a baseline scorecard for process waste where necessary, assuming credible data availability. Like other techniques, the value stream mapping technique provides a common language and understanding for talking about manufacturing and transactional processes. It is a guide to developing vision (future-state map) and identify and prioritize initial opportunities and improvement actions.

In value stream mapping, highly detailed process mapping is not supported nor is it the purpose of this approach to do process mapping. Individual process steps of the flow are not usually shown. Organizational structures are shown explicitly in the identification of customer and supplier interfaces. The technique is intended to show one area of material flow that does not typically cross organizational boundaries.

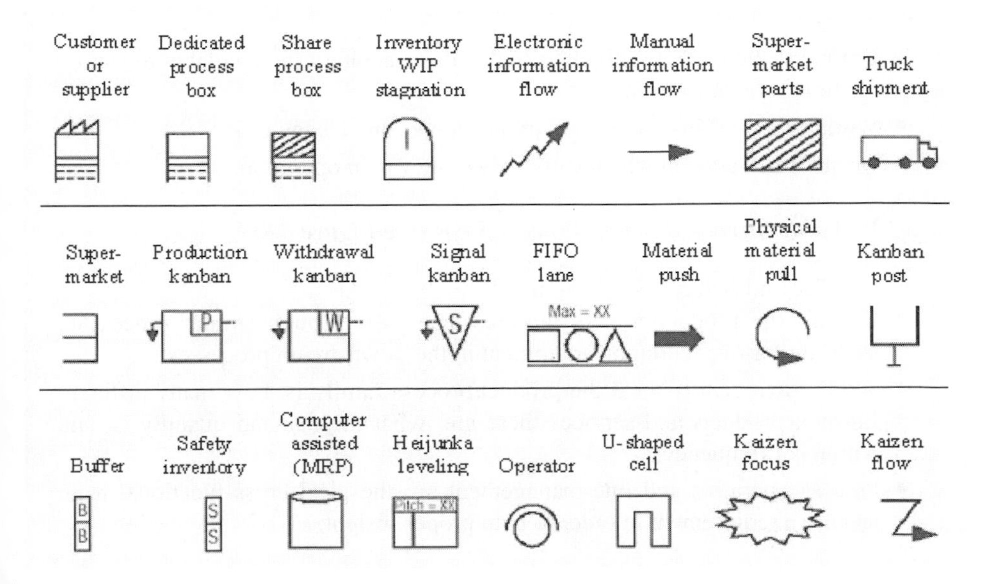

Figure 8.7 Value stream mapping icons.

Decisions loops, traditional logical operators, and if–then statements are explicitly provided. Typically, they are implied in lean concepts such as *kanban, supermarket,* and *work-in-process* inventory points. Most of the logic will occur within the processes themselves (box icons) and will not be specifically identified by the map. Value stream mapping is a great tool that helps tremendously in visualizing the service operation. It helps us visualize more than just the single-process level by exposing the flow of material, information, and sources of waste. In addition, this technique provides an organized, visual structure that includes certain data to the DFSS team and their management. It represents a gateway to institutionalize lean thinking by linking to lean concepts and techniques that avoid "cherry picking" or "Kamikaze Kaizen." It forms the basis for a lean implementation plan. In this regard, value stream mapping enables a whole door-to-door flow, a blueprint for lean implementation. The technique provides a current baseline by obtaining current operating and measurable information that will enable teams to identify waste and develop a vision (future state) with defined actions needed to reach such a state. Several time metrics are used to gauge improvement of the future state from the current state. They are:

- Lead-time. The time it takes for one unit to complete the process, including non-value-added time that the unit or product waits between operations.
- Total cycle time. Cumulative time it takes for unit or product to complete all operations, not including wait time between operations. Also known as summation of touch time but not process time.
- Throughput time. How often a unit or product is completed by the operation from start to finish.
- Changeover time. Time to switch from producing one product type to another.
- Uptime. Calculated by dividing the actual machine time available to run by the time scheduled to run.
- Working time. Time per period minus breaks and cleanup.
- Queue time. Time in which work is waiting for an operation.

8.2.3.1 Value Stream Mapping Process Steps (see Figure 8.8)

Step 1. Define and Select VSM Resources

- Identify the product/process families that pass through similar processing steps and/or over common equipment in the downstream processes.
- Write down clearly what the product/process family is, how many different finished products and services there are, what the demand quantity is, and with what frequency.
- Process operators and line management are the ideal cross-functional team and are a requirement in order to gain proper insight.

Step 2. Set up a "war room" with all logistics (e.g., paper, markers, sticky notes) and computer resources needed.

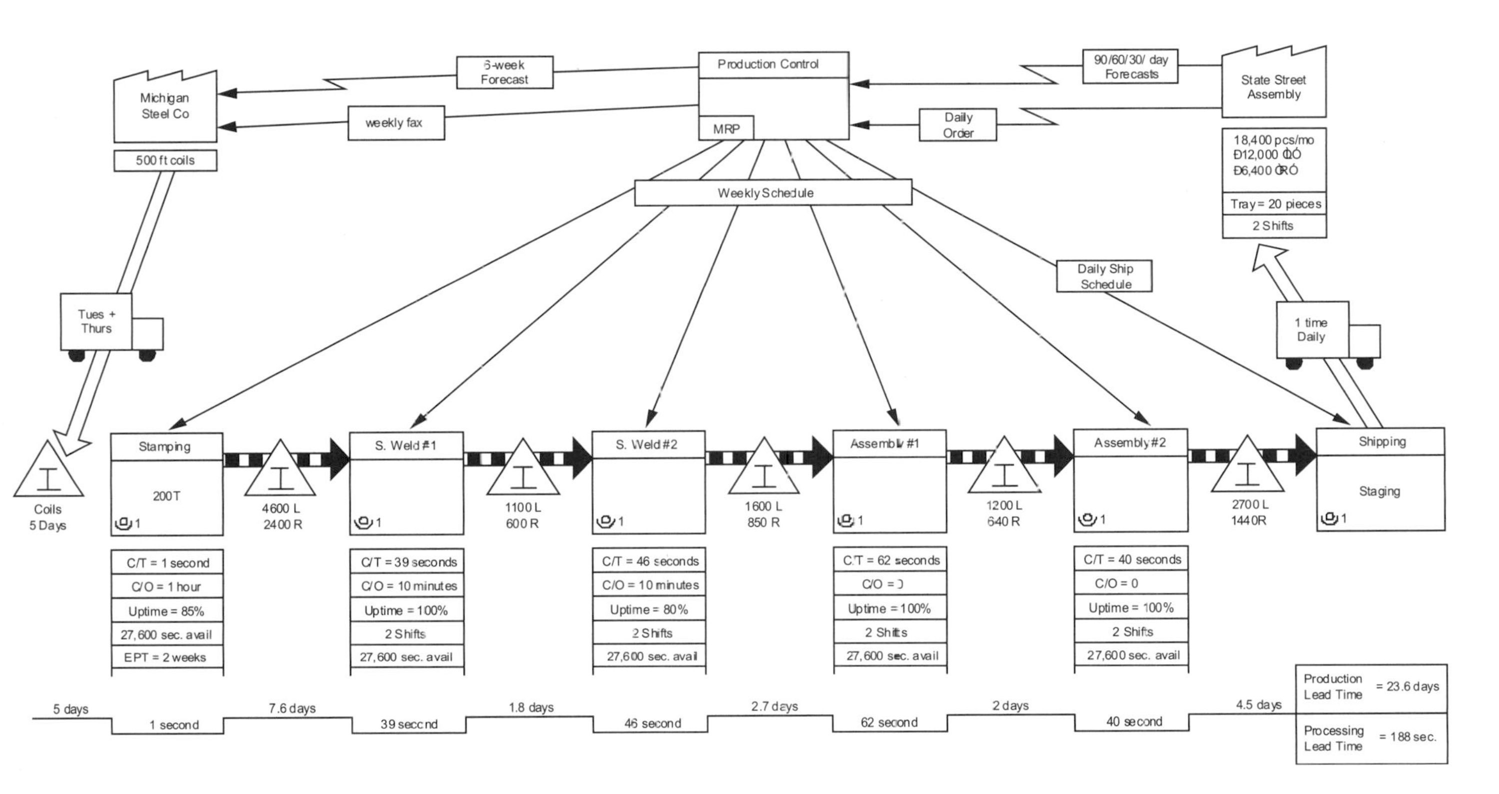

Figure 8.8 Value stream mapping example.

Step 3. Draw the current state map

- Collect current-state information while walking through the actual pathways of material/content and information flows. Begin with a quick walkthrough of the entire door-to-door value stream to get a sense of the flow sequence of the processes. Instead of starting at the receiving dock and walking downstream, begin at the shipping end and walk upstream to begin with the processes that are linked most directly to the customer. Start with the concept of what the customer values, and as you move upstream, challenge each process step for how it adds to or creates customer value.
- Every team member draws a map when walking through the process.
- Develop a specific set of questions before going to the work area.
- While walking through the process talk to the employees.
- Map the "current condition," using the standard icons to convey key information (material/content and information flow). Emphasize the importance of data collection involved in the VSM effort.
- Start with the customer on the upper-right corner, and draw a factory icon, as shown if the example in Figure 8.8. Underneath this icon, draw data boxes, with the customer name in the top box. Make a record of other information such as services ordered per day, week or month required, packaging requirements, and number of shifts the customer runs. Process boxes (rectangles), cycle times, shifts operated, and changeover time and information such as number of people, cycle time, changeover time, uptime, shifts, and available working time must be noted and mapped. Add inventory locations (triangles), including material/content type and quantity stored at each location, and material/content flow (fat arrows), including indication of whether a "push" or "pull" discipline is followed.
- Insert a truck icon and the broad arrow to indicate movement of finished goods to the customer.
- At the other end of the map, insert a factory icon and a data box for the supplier.
- Populate the supplier and the trucks data boxes with appropriate data.
- Draw the information flow lines as appropriate. Add the information flow by adding narrow lines. A small box icon is used to describe the different information-flow arrows.
- A process box indicates the production control department.
- Add "push" and pull" arrows.
- Draw triangles to indicate inventory and populate the boxes with appropriate information. Other icons and information indicate external suppliers' delivery frequency and so on. Significant problems (e.g., frequent machine breakdowns) are represented by "bad clouds" or starbursts as depicted in Figure 8.9.
- Identifying the lead-time ladder along the bottom is the next step. To get the total production lead time, the team needs to take the inventory number and

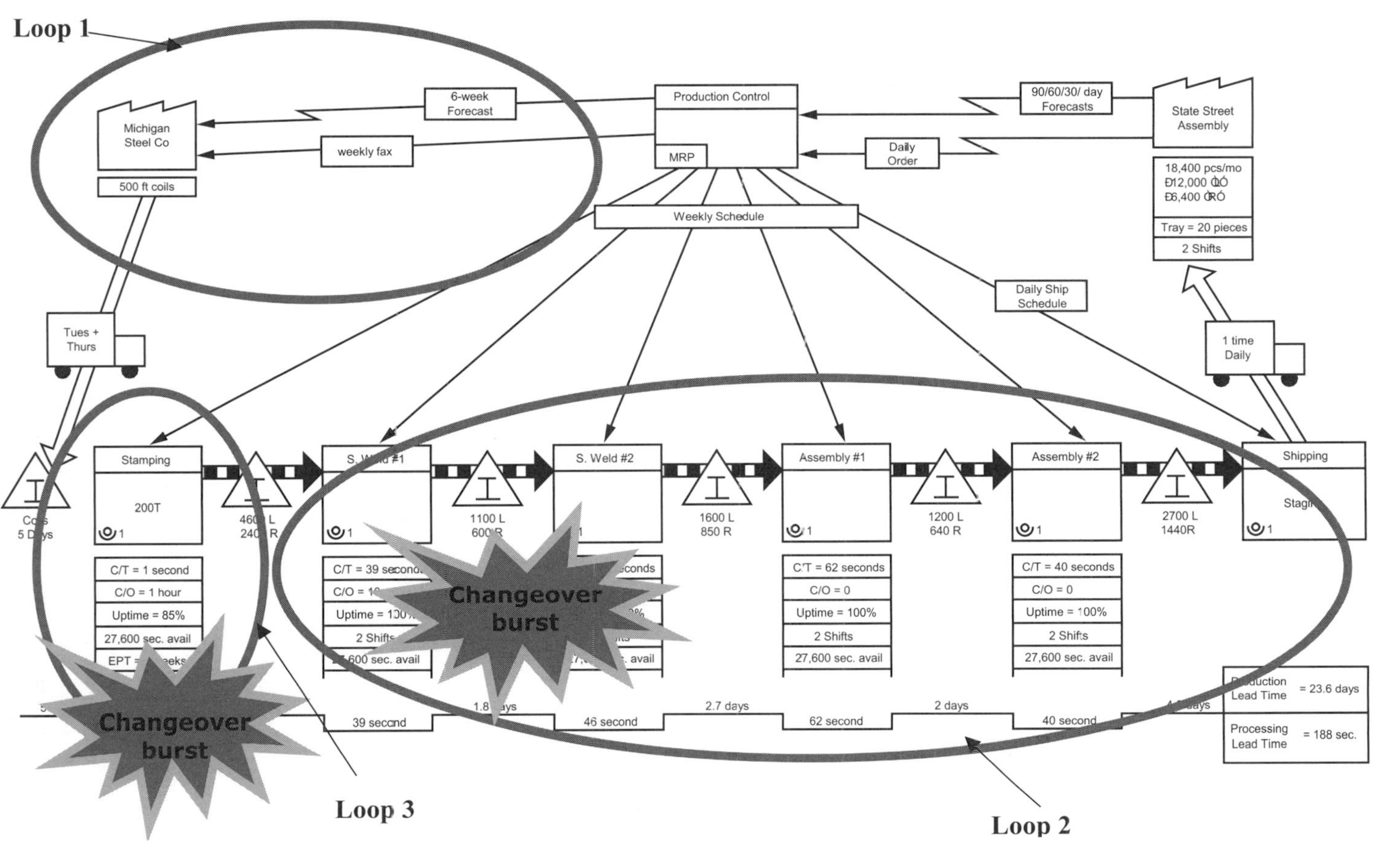

Figure 8.9 Value stream map of Figure 8.8 with loops and starbursts added.

divide it by the daily customer requirements. To get the cycle time (or value-added time), you need to add up all the cycle time numbers. For maps with multiple upstream flows, use the ratio of longest lead time to total the lead time. When it is all added up, it usually indicates astonishing numbers.

- When done walking through the process the team goes back to the "war room"[1] and, as a team, integrates and develops one current state map. The most important reason, though, for establishing a team war room is to ensure that the team gets its information right, up front, to prevent expensive changes downstream.
- Conduct formal brainstorming of all issues based on the observations made during the walkthrough and questions to the employees. The brainstorm items should be recorded.
- On the value stream map, the team needs to draw loops around parts of the map where improvements can be sought. Examples include combining processes to create a better flow and eliminate batching of inventory between those processes. Cell loops, pacemaker loops, and constraints are typical. A pull-system "supermarket" usually corresponds to a loop. Within each loop, the team should use lightening bursts to highlight improvement ideas, starting with loop that has the biggest issue and and will require the most fixing. Figure 8.9 replicates Figure 8.8 with added loops.

Step 4. Draw and implement the future-state map. The power behind value stream mapping is always the future-state map. To develop a future-state map and implementation plan, the team needs a well-defined map with loops identified (Figure 8.9). The team needs to be exposed to flow concepts, setup reduction, cellular concepts, and total preventive maintenance. The design team keeps updating the current-state map with material/content flows. They look for waste and creative ideas like moving processes closer together, combining operations, and eliminating transportation. Basically, the steps are:

- Utilizing the map with loops identified, map a future-state proposal reflecting what the situation will be after improvements are made.
- Develop implementation plans to achieve the future-state map.
- Present the future-state map and the implementation to management for their buy-in.
- Execute actions to achieve the future-state condition, then stabilize the new condition.
- Repeat current-state mapping, analysis, looping, and future-state mapping steps to achieve the next future-state target.

[1]The war room is the service DFSS team information center. The value in having a dedicated team room is in creating and maintaining facilities and environments to foster individual and collective learning of service and process knowledge, including project direction/vision and prior history. The war room could serve as a focus for team activities to foster team building/growth and inter- and intradisciplinary communication and collaboration.

Step 5: Develop and implement a communication plan. Develop a communication plan that supports the project and communicates objectives, accomplishments, and team efforts. The communication plan must be synchronized with the tollgate review schedule (making it timely) and continue after project closure, if needed. Conduct a value stream mapping overview for workshop participants

An example of a value stream map for typical manufacturing operations is shown in Figure 8.8 (see Rother and Shook, 2000). Take special note of the ruler running along the bottom of the map that shows the flow time at each step in the plant. Notice how the high-level processes, process level, and plant levels are drawn with very little detail. The value stream map is a good overview, helps the team focus on specific processes, and shows both informational and material/content flows.

After creating the current-state map in the charcaterize phase of the DFSS road map, the next step is to conduct analysis and determine the team thinking for the future-state map. The future-state map is part of the optimize phase. The future-state mapping may include future-state cell design, implementation plan (with who?, what?, and when?), and FMEA of the future-state implementation plan.

The purpose of the future-state map is to reflect the looped map in identifying waste and determining what can become a reality within a short period of time. The implementation is time bounded. After improvements in the value stream have been implemented, the future-state map becomes the current-state map, and a new future-state map is developed.

8.3 BUSINESS PROCESS MANAGEMENT SYSTEM (BPMS)

BPMS is used to establish ongoing integrated ownership, accountability, and performance of a process. Oftentimes, the results of transactions in one function are the inputs to transactions in a different functional process. A good example of this is when sales creates a sales order that determines the quantity and desired delivery date as well as the price. This information meanders through various processes and ends up with a build order and associated purchase order at a supplier. Common sense tells us that accuracy and timeliness of the information flow is paramount in order to fulfill the customer need, but how is this related to the infrastructures and procedures that exist today in non-Six Sigma processes?

8.3.1 What is BPMS?

BPMS is a system composed of teams organized around a process with a clear mission, process objectives, roles and responsibilities, and performance expectations. There are process management charts that document the procedures, measurements, monitoring, and reaction plans for the process. The teams hold formal reviews focused around the process metrics displayed in a dashboard manner. The teams share accountability and, through effective communication and feedback, they drive commitment, reenforcing practices and process improvements.

8.3.2 How Is BPMS Implemented?

The deployment of BPMS can follow a DMAIC framework (see Chapter 2). The define phase consists of defining the purpose, boundaries, and goals. The measure phase consists of process mapping the "as is" state and determining the inputs, outputs, and internal process measures, as well as the suppliers and customers of each of these. Measures of effectiveness, efficiency, and cycle time are determined. The analyze phase consists of evaluating the capability of the process to meet the process goals and determining which set of measures and control indicators are the minimum number that will appear on the process dashboard. The improve phase consists of deployment of the defined process, with clearly defined roles and responsibilities documented in procedures. In this phase, active data collection begins to feed the control phase. The final phase is the control phase, which consists of ensuring that the process is performing as expected and the structures and procedures are being adhered to. This final phase of the BPMS deployment is an ideal building block for either DMAIC or DFSS process improvements.

8.4 RELATIONSHIP OF PROCESS MODELING TO DISCRETE EVENT SIMULATION

There is a strong link between process modeling and discrete event simulation, which will be further explored in Chapter 14. Simulation employs a set of theoretical concepts and computational methods that describe, represent, and simulate the functioning of real processes. Computer simulations are becoming popular in performing research, and are expanding traditional experimental and theoretical approaches. Simulation can be regarded as a numerical experiment, but it often requires advanced theories. It can provide information that is impossible or too expensive to measure, as well as insights that are not amenable to or are too complicated for analytical methods.

Simulation models are simplified abstractions of reality representing or describing its most important and driving elements and their interactions. Simulation can be used for a multitude of purposes such as analysis and understanding of design, redesign and problem solving projects, testing of if–then hypotheses and theories, prediction of process behavior under various conditions and scenarios in support of decision making, and analysis with visualization of results.

Real processes are sometimes complex and often include nonlinear behavior as well as stochastic components and loops (e.g., rework) over spatial and time scales; therefore, models can represent the processes only at a certain level of simplification. Empirical knowledge is based on statistical analysis of observed data, and they are usually applicable only to the same conditions under which the observations were made. Process mapping is based on understanding the processes within the scope and their input–output relationships and logical linkages. Discrete event simulation models of complex systems often use a combination of empirical knowledge and process-based approaches. Process simulation software enables the following outcomes:

- Design and redesign of new processes to optimize them
- Increase service productivity
- Reduce total process cycle time
- Increase throughput
- Reduce waiting time
- Reduce activity cost
- Reduce inventory costs
- Troubleshooting of existing processes
- What–if studies

In a DFSS project, the primary emphasis is on using discrete system models to analyze administrative, decision-making, service development, manufacturing, and service delivery processes. Discrete system models characterize the system as a flow of entities that enter and move through various processes and queues according to probability functions specified by the modeler. Some processes may also exhibit continuous characteristics, in which case continuous model constructs may be applied. Continuous system models utilize the numerical integration of differential equations to simulate behavior over time. Such models are often used for studying the systems containing feedback loops, in which the outputs are fed back and compared with control inputs. In this book, discrete event simulation and its unique challenges will be covered in some detail. Discrete event simulation is best suited for business and service-type processes due to its transactional and event-driven behavior.

Business processes are often too complex and dynamic to be understood and analyzed by flowcharting and spreadsheets alone. Herein lies the opportunity for Six Sigma deployment to institutionalize simulation as the standard tool. Simulation is the only tool that can provide both accurate analysis and visualization of the process and alternatives. Process simulation is the technique that allows representation and interaction of processes, people, and technology. There are essentially four steps in doing it: (1) building a model, (2) validating and running a model, (3) analyzing the performance measures, and (4) evaluating alternative scenarios.[2]

A model mimics the operations of a business. This is accomplished by stepping through the events in compressed time while displaying an animated picture of the flow. Because simulation software keeps track of statistics about model elements, performance of a process can be evaluated by analyzing the model output data.

By modifying decision variables in the model, without the cost and risk of disrupting existing operations or building a new system, the team can accurately predict, compare, or optimize the performance of the reengineered process. Simulation models can provide the most accurate and insightful means to analyze and predict the performance measures of business processes in these scenarios. Discrete-event-based simulation is the most capable and powerful tool for business process simula-

[2]See Kerim Tumay, http://www.reengineering.com/articles/janfeb96/sptprochr.htm, Business Process Simulation: Matching Processes with Modeling Characteristics, ER SPOTLIGHT.

tion that are discrete-event-driven in nature. These tools provide modeling of entity flows, with animation capabilities that allow the user to see how flow objects are routed through the system. Some of these tools even provide object-oriented and hierarchical modeling, which simplifies development of large business process models. Simulation process modeling is aimed at improving our understanding and predicting the impact of processes and their interactions. It provides supporting tools for modeling, especially data management, analysis, and visualization. Process mapping models describe the behavior of phenomena represented by a set of individual processes with various types of value-added and non-value-added step interactions at local, regional, or global value streams. Simulation models can be fully integrated or linked through data and interfaces.

8.5 FUNCTIONAL MAPPING

A functional map can be defined as the functional elements and the set of logical and cause–effect relationships amongst them. Mathematically, design mapping, transfer functions, and simulation models can be defined as functional mappings. Graphically, functional mapping is depicted in a block diagram that is composed of nodes connected by arrows depicting the relationships. Both functional and process models should capture all design elements and ensure correct relationships between process variables (*x*s or PVs) and functional requirements (*y*s or FRs). A functional model is captured mathematically using transfer functions with matrices belonging to a design *hierarchical* level. Hierarchy is built by breaking down the design into a number of simpler functional design mappings that collectively meet the high-level functional requirements of the design. There are two recognized mappings in the service DFSS road map of Chapter 5:

- The functional mapping between the functional requirements (FRs) and the design parameters (DPs), the subject of this section
- The process mapping between the DPs and the process variables (PVs), the subject of Section 8.2

The functional mapping is usually developed first to define the design concept. Preliminary concepts are best selected using the Pugh selection method (Section 8.6). The preliminary work to verify mapping choices should help the DFSS team to get started on concept generation. The team needs to select the best solution entity in terms of design parameters (DPs) to meet or exceed functional requirements. New technologies (DPs) can enable new process mappings. The pursuit of linked technology and mapping options may reveal new opportunities for customer satisfaction (see Section 7.7). Process mappings need to be robust against customer use, misuse, and abuse; errors in requirements and specifications; unanticipated interactions with other portions of the solution entity; or process variations. The functional requirements that are derived from VOC should be verified over a range of operational parameters that exceed known requirements and specifications. This may require sensitivity studies in the form of a classical DOE or Taguchi's parameter design. Determining sensitivity of design performance due to changes in operating conditions (including local environment and solution entity interactions) over the expected operating range is an essential task for transfer function optimization within the DFSS road map.

8.5.1 Value Analysis/Engineering FAST Technique

The Society of American Value Engineering defines value engineering as "the systematic application of recognized techniques which identify the functions of a product or service, establish a monetary value for that function and provide the necessary function at the lowest overall cost." Value engineering (VE) or value analysis (VA) is the application of a wide variety of skills to reduce cost for the manufacturing or service industries. These skills range from statistics, engineering, and psychology to accounting and selling. Value engineering relates the worth of a product or service to its cost.

It is very well known that VE is obsessed with definitions. Defining the design problem is the milestone in this type of expertise. Defining a system (problem under study) in terms of its functional structure is synonymous to understanding it. A key technique used to define and decompose the functional structure is the function analysis system technique (FAST). In service design, FAST is capable of identifying the functionality of processes and steps, and isolating their basic functions together with the primary *logical* paths by using a complete set of logical questions. *Basic functions* are the principal functions that need to be delivered to satisfy customer needs. *Primary paths* are the functional paths that evolve from the basic functions. Their root, the basic functions, uniquely characterizes them. In the primary path, many other supporting functions called secondary functions will follow the basic function. Secondary functions in the primary path are there because of the logical thinking that a design team's collective experiences uses to fulfill the basic function. Their development is governed by the design team's creativity and knowledge. In a dynamic service environment, basic functions are work functions (e.g., transmit information). Secondary functions, on the other hand, are subjective in nature (e.g., create style) and constitute a high percentage of total cost. Another factor in practicing FAST is decomposition, that is, breaking down the design project into well-defined smaller and manageable processes and subprocesses. This is achieved through the set of logical questions listed below:

- HOW is the function accomplished?
- WHY is the function performed?
- WHEN is the function performed?
- WHAT mechanism (process step, hardware, software) performs it?

Value engineering lays the design problem out in a series of steps, each of which is progress toward a possible solution. A subdivision can be created so that the facts are capable of having a direct or indirect bearing on the system. A VE project must be done over six consecutive phases called the *job plan* (Park 1992).

In value engineering, the functional mapping is a starting point. The service is viewed as a set of steps and subprocesses acting in harmony to deliver desired features and analyzed from the functionality point of view. The value engineering FAST technique produces a *logical functional map* for the design. There may be many different mechanisms to provide the same process function, out of which a limited and creative subset of options deserves careful consideration. Function analysis makes it possible for us to set up aggressive cost reduction goals while satisfying customer requirements (Park, 1992). Function definition is a very critical step and may lead to unfortunate consequences if conducted arbitrarily. For example, in a simple business office, the task of filing may be defined as "store files." Such a functional definition has operational limitations and may constrain opportunities for improvement and creativity. It may be better to describe it as "store data," since there are many ways to achieve this function and "filing" is only one of them. Data may be stored in computers (software, hardware), microfiches, transparencies, notes, tags, CPU memory, and so on. Thus, defining the function in the broadest terms allows greater freedom to creatively develop alternatives (Park, 1992). This type of naming nomenclature is very similar to the QFD process when defining solution free functional requirements (see Chapter 7). In order to facilitate conducting this task, some guidelines were developed to help steer the definition activity and keep it on the right track. However, these guidelines are not rules, and their successful implementation is contingent on the design team qualities and experience. The of guidelines are:

- Define functions using the broadest possible viewpoint in order to unlock creative possibilities for improvement (make the process solution free).
- Try to limit an expression of the function to two words only, a verb and a noun. Using this constraint, we avoid hidden or multiple functions when defining a functional map of a service entity. The use of abstraction can facilitate this descriptive activity. Avoid using names in the noun part; that is, generality in the noun part (a type of fuzzy uncertainty, nonspecific) is desired in function definition in the concept stage. The functional requirements (FRs) represent the noun part of the function definition.

8.5.1.1 The Generation of FAST. FAST which was developed by Charles W. Bytheway in 1963, is a primary tool with the appealing feature of defining a functional logical map by employing logical questions. It is not a flow chart, although its appearance is similar to a flow chart. FAST is a logical diagram that utilizes four logical questions: how, why, when, and what. When a service design is physically stated, levels of hierarchy prevail through physical connections. Dynamic information and energy flow as shown in Figure 8.10. These physical relations are not reflected in FAST, which is mainly a functional diagram. The FAST steps are:

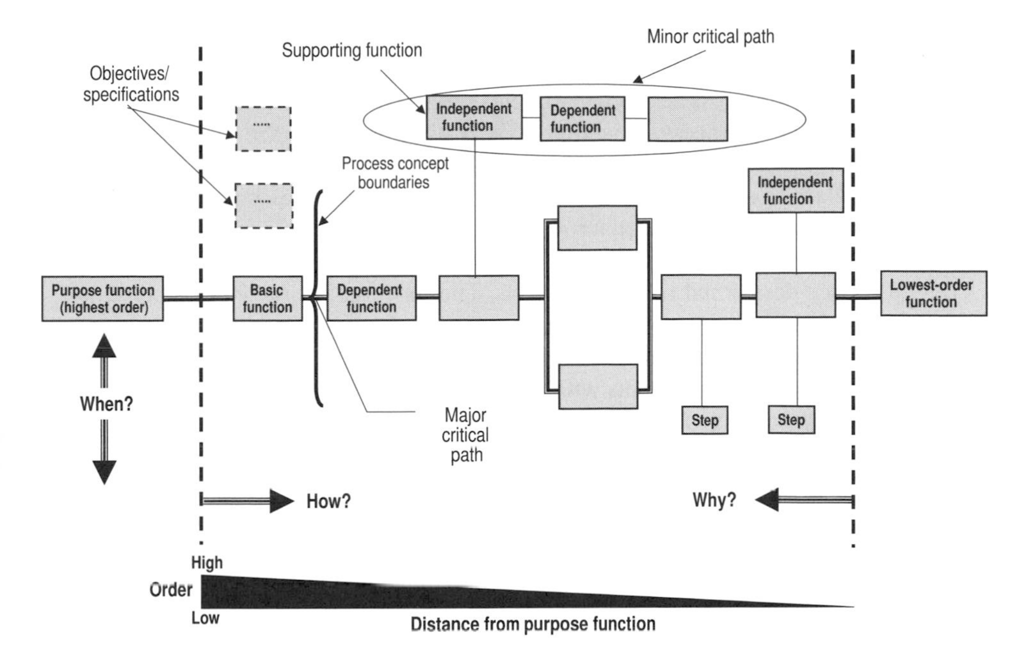

Figure 8.10 Functional analysis systems technique (FAST) logical mapping.

- Determine a scope for the conceptual process. Some functions may be discovered to be outside the desired scope of the design as identified by the purpose function, and, hence, to pursue them may be a waste of time and resources.
- Identify the basic functions of the new concept design. Basic functions are the primary reason for the concept to exist physically. If the system is complex, there may be more than one basic function. Use a numbered block diagram to represent functions. Usually, basic functions are numbered first. Secondary functions have higher numbers.
- Apply the logical questions to test the definition of the basic function:

  How is the function accomplished?

  Why is the function performed?

  When is the function performed?

  What mechanism (process step, hardware, software) performs it?

To illustrate, the following FAST example is developed for a DFSS initiative plan deployment. This provides the means of cascading leadership policy down through all levels of company. The statement of leadership policy should be in general terms with objectives or purpose functions such as "grow the business" and "increase customer satisfaction." Staff management (reporting to the leadership team) restates such purpose functions in measurable terms within their control. For example, a senior leader may have the objective "increase customer satisfaction by 15% for the next 3 years." His/her direct report manager's cascaded objective might be "deploy Six Sigma in design departments with services released at stage 7 [Figure 1.2] with high process capability." Each objective cascaded down levels of management becomes more concrete in its statement of objective relative to the deployment plan. Six Sigma deployment plan operatives such as black belts, green belts, and champions (see Chapter 4) deliver on the plan objectievs. FAST was originally developed to improve product design, but has been found to be as effective in service design and redesign projects. The function "increase customer satisfaction by 15% for the next 3 years" is identified as the purpose function for the deployment. How does one increase customer satisfaction? The answer is to deploy Six Sigma in design departments with services released at stage 7 (Figure 1.2) with high process capability. This function is identified as a basic function for the plan. By "deployment" it is meant to put in place processes, targets, standards, procedures, and resources that will result in improved satisfaction by the end of the third year of active deployment. The common practice is to develop a glossary of terms of functional meanings as FAST develops, to maintain a consensus and distinguish functions from each other. The FAST functional mapping is carried out in the following logic dimensions:

- The "why" logic dimension:

  Function = "Increase customer satisfaction by 15 %," operative = "senior leadership"

  Function = "deploy Six Sigma with services released at stage 7 (Figure 1.2) with high process capability," operative = "Champion"

  . . .

  Function = "monitor satisfaction scorecard," operative = "process owner"

  Function = "work on weak scorecard metrics projects," operative = "belt"
- The "how" logic dimension. There are, of course many departments that are involved in deploying the plan, and what is represented in Figure 8.11 is only a partial depiction. "Increase customer satisfaction" is accepted as the "why" end (the purpose function) in the FAST diagram by the "senior leadership" operative who identifies "how" (the means of attaining such purpose). The "champion" Six Sigma (see Chapter 4) operative accepts this objective and continues the cascading of deployment objectives.

How does one increase customer satisfaction? The answer is by deploying DFSS, deploying Six Sigma, and marketing new services. Why deploy DFSS? The answer is to increase customer satisfaction. The logical loop is closed since for each question there is an answer. The limit of the scope of the FAST in the "why" direction should be the customer's needs. When the logic questions are satisfied, we can go to the next function in the diagram. The newly defined functions may be added to the glossary of terms. A hypothetical portion of the FAST diagram is shown in Figure 8.11.

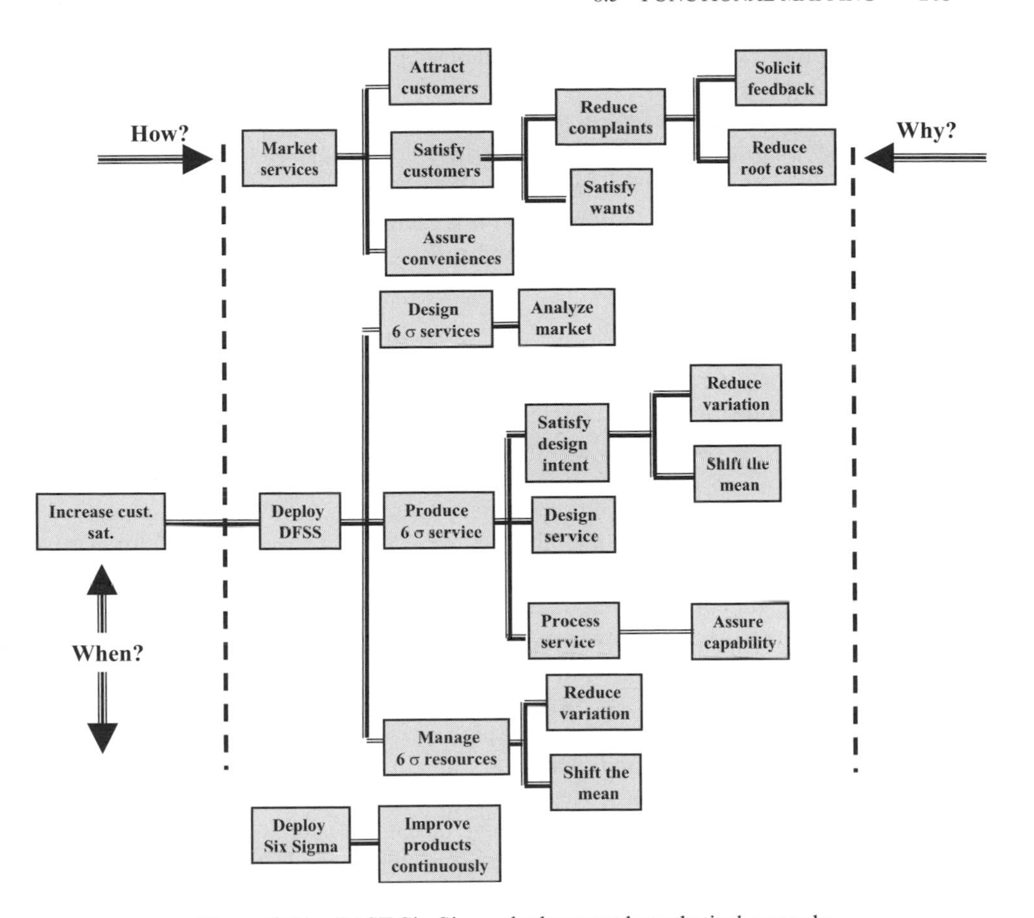

Figure 8.11 FAST Six Sigma deployment hypothetical example.

With DFSS, attention has begun to shift from the improvement of design in downstream development stages to early upstream stages. This shift is motivated by the fact that the design decisions made during the early stages of the product development cycle have the largest impact on total life cycle cost and quality of the design. The DFSS deployment is currently receiving an increased focus on addressing service industry efforts to shorten lead times, cut development and production costs, lower total life cycle cost, and improve the quality of products and services. The term "quality" in the context of this chapter can be defined as the degree to which the design vulnerabilities do *not* adversely affect service performance. This definition as well as most of the developments of this book, are equally applicable to all design stages (see Chapter 1) because the design principles, in particular those promoted to *axiom* status,[3] are universal. In the context of the DFSS road map, the major design vulnerabilities are categorized as:

[3]Fundamental knowledge that cannot be tested, yet is generally accepted as true, is treated as an axiom.

- Conceptual vulnerabilities that are established due to the violation of design principles.
- Operational vulnerabilities that are created as a result of factors beyond the control of the designer called "noise factors." These factors, in general, are responsible for causing a functional requirement or process performance to deviate from target values. Controlling noise factors is very costly or difficult, if not impossible. Operational vulnerability is usually addressed by robust design (see Chapter 13).

Conceptual vulnerabilities will always result in operational vulnerabilities. However, the reverse is not true. That is, it is possible for a healthy concept that is in full obedience to design principles to be operationally vulnerable (El-Haik, 2005). Conceptual vulnerabilities are usually overlooked during development due to the lack of understanding of the principles of design, the absence of a compatible systemic approach to find ideal solutions, the pressure of deadlines, as well as budget constraints. Prior to DFSS, these vulnerabilities are usually addressed by traditional quality methods. These methods can be characterized as after-the-fact practices since they use lagging information relative to developmental activities such as warranty and customer complaint databases. Unfortunately, these practices drive development toward endless design–test–fix–retest cycles, creating what is broadly known as the "fire fighting" operation mode in many industries. Companies who follow these practices usually suffer from high development costs, longer time to market, lower quality levels, and marginal competitive edge. In addition, the fire fighting actions to improve the conceptual vulnerabilities are not only costly but also hard to implement, as pressure to achieve design milestones builds during the development cycle. Therefore, it should be an objective to implement quality thinking in the conceptual stages of the development cycle, a need that is nicely treated by the axiomatic design method (Suh, 1990, 2001). The objective of axiomatic design is to address design conceptual vulnerabilities by providing tools and formulations for their quantification, then eliminate or reduce them.

8.5.2 Axiomatic Method

Many designs have some degree of "coupling," a design vulnerability that results in a diminished degree of controllability by both the design team and the customer in usage environments. Many designs require complicated processes with many decision points in order to fulfill the many voices of the customer. Axiomatic design provides a prescriptive methodology to assess coupling and seek to resolve it as well as reduce design complexity in the early stages of design cycle (see Chapters 1 and 3).

Axiomatic design is a prescriptive[4] engineering design method. Axiomatic design is a design theory that constitutes basic and fundamental design element

[4]Prescriptive design describes how design should be processed. Axiomatic design is an example of a prescriptive design methodology. Descriptive design methods like design for assembly are descriptive of best practices and are algorithmic in nature.

knowledge. In this context, a scientific theory is defined as a theory comprising fundamental knowledge areas in the form of perceptions and understandings of different entities, and the relationship between these fundamental areas. The theorist, to produce consequences that can be, but are not necessarily, predictions of observations, combines these perceptions and relations. Fundamental knowledge areas include mathematical expressions, mapping, categorizations of phenomena or objects, and models, and are more abstract than observations of real-world data. Such knowledge and relations between knowledge elements constitute a theoretical system. A theoretical system may be one of two types, axioms or hypotheses, depending on how the fundamental knowledge areas are treated. Fundamental knowledge that is generally accepted as true, yet cannot be tested, is treated as an axiom. If the fundamental knowledge areas are being tested, they are treated as hypotheses (Nordlund, 1996). In this regard, axiomatic design is a scientific design method with the premise of being a theoretic system based on two axioms.

Motivated by the absence of scientific design principles, Suh (1984, 1990, 1995, 1996, 1997, 2001) proposed the use of axioms as the scientific foundations of design. The following are two axioms that a design needs to satisfy:

Axiom 1: The Independence Axiom
Maintain the independence of the functional requirements

Axiom 2: The Information Axiom
Minimize the information content in a design

In the context of axiomatic deployment, the independence axiom will be used to address the conceptual design vulnerabilities, whereas the information axiom will be tasked with the operational aspects of the design vulnerabilities (El-Haik, 2005). Operational vulnerability is usually minimized and cannot be totally eliminated. Reducing the variability of the design functional requirements and adjusting their mean performance to desired targets are two steps used to achieve such minimization. Such activities will also result in reducing design information content, a measure of design complexity per Axiom 2. The customer relates information content to the probability of successfully producing the design as intended and maintaining it afterward.

The design process involves three mappings between four domains (Figure 8.12). The reader may already notice the similarity with QFD phases (see Chapter 7). The first mapping involves the mapping between customer attributes (CAs) and the functional requirements (FRs). This mapping is very important as it yields the definition of the high-level minimum set of functional requirements needed to accomplish the design intent. This definition can be accomplished by the application of quality function deployment (QFD). Once the minimum set of FRs is defined, the *physical mapping* may be started. This mapping involves the FRs domain and the design parameter codomain (DPs). It represents the product development activities and can be depicted by design matrices; hence, the term "mapping" is used. This mapping is conducted over the design hierarchy as the high-level set of FRs,

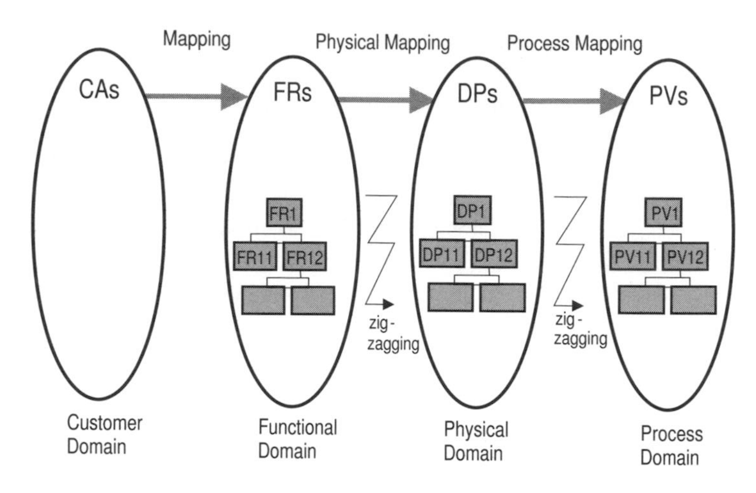

Figure 8.12 The design mapping process.

defined earlier, is cascaded down to lowest hierarchical level. Design matrices reveal coupling, a conceptual vulnerability. Matrices provide a means to track the chain of effects of design changes as they propagate across the design mapping.

Process mapping is the last mapping of axiomatic design and involves the DPs domain and the process variables (PVs) codomain. This mapping can also be represented formally by matrices and provides the process elements needed to translate the DPs to PVs in manufacturing and production domains.

The mapping equation $\boldsymbol{FR} = f(\boldsymbol{DP})$ or, in matrix notation, $\{\boldsymbol{FR}\}_{mx1} = [\boldsymbol{A}]_{mxp}\{\boldsymbol{DP}\}_{px1}$, is used to reflect the relationship between the domain, array $\{\boldsymbol{FR}\}$, and the codomain, array $\{\boldsymbol{DP}\}$, in the physical mapping, where the array $\{\boldsymbol{FR}\}_{mx1}$ is a vector with m requirements, $\{\boldsymbol{DP}\}_{px1}$ is the vector of design parameters with p characteristics, and $\boldsymbol{A}$ is the design matrix. (See Appendix 8.B for a high-level review of matrices.) Per Axiom 1, the ideal case is to have a one-to-one mapping so that a specific DP can be adjusted to satisfy its corresponding FR without affecting the other requirements. However, perfect deployment of the design axioms may not be possible due to technological and cost limitations. Under these circumstances, different degrees of conceptual vulnerabilities are established in the measures (criteria) related to the unsatisfied axiom. For example, a degree of coupling may be created because of Axiom 1 violation, and this design may function adequately for some time in the use environment; however, a conceptually weak system may have limited opportunity for continuous success even with the aggressive implementation of an operational vulnerability-improvement phase.

When matrix $\boldsymbol{A}$ is a square diagonal matrix [see the matrix in Figure 8.13(a)], the design is called *uncoupled,* that is each FR can be adjusted or changed independent of the other FRs. An uncoupled design is a one-to-one mapping. Another design that obeys Axiom 1, though with a known design sequence [see the matrix in Figure 8.13(b)], is called *decoupled.* In a decoupled design, matrix $\boldsymbol{A}$ is a lower or upper

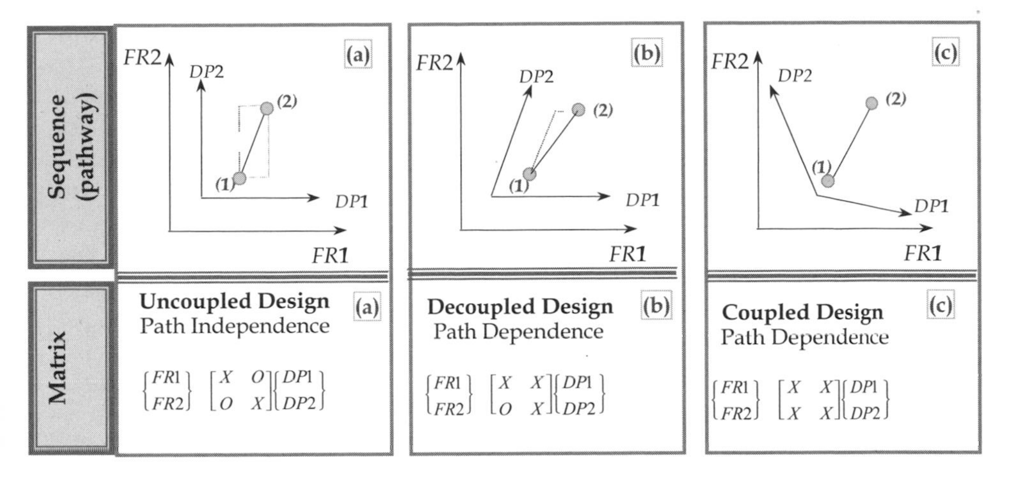

Figure 8.13 Design categories according to Axiom 1.

triangular matrix. The decoupled design may be treated as an uncoupled design when the DPs are adjusted in some sequence conveyed by the matrix. Uncoupled and decoupled design entities possess conceptual robustness (i.e., the DPs can be changed to affect specific requirements without affecting other FRs unintentionally). A coupled design definitely results in a design matrix with a number of requirements, m, greater than the number of DPs, p [see the matrix in Figure 8.13(c)]. Square design matrices ($m = p$) may be classified as coupled designs when the off-diagonal matrix elements are nonzeros. Graphically, the three design classifications are depicted in Figure 8.13 for the 2×2 design matrix case. Notice that we denote the nonzero mapping relationship in the respective design matrices by "X." On the other hand, "0" denotes absence of such a relationship.

Consider the uncoupled design in Figure 8.13(a). The uncoupled design possesses the path-independence property, that is, the design team could set the design to level (1) as a start point and move to setting (2) by changing $DP1$ first (moving east to the right of the page or parallel to $DP1$) and then changing $DP2$ (moving toward the top of the page or parallel to $DP2$). Due to the path-independence property of the uncoupled design, the team could start from setting (1) to setting (2) by changing $DP2$ first (moving toward the top of the page or parallel to DP2) and then changing $DP1$ second (moving east or parallel to $DP1$). Both paths are equivalent, that is, they accomplish the same result. Notice also that the FRs independence is depicted as orthogonal coordinates as well as perpendicular DPs axes that parallel their respective FRs in the diagonal matrix.

Path independence is characterized, mathematically, by a diagonal design matrix (uncoupled design). Path independence is a necessary property of an uncoupled design and implies full control of the design team and, ultimately, the customer (user) over the design. It also implies a high level of design quality and reliability since interaction effects between the FRs are minimized. In addition, a failure in one ($FR,$

DP) combination of the uncoupled design matrix is not reflected in the other (*FR*, *DP*) mappings within the same design hierarchical level of interest.

For the decoupled design, the path-independence property is somehow fractured. As depicted in Figure 8.13(b), decoupled design matrices have design setting sequences that need to be followed for the functional requirements to maintain their independence. This sequence is revealed by the matrix as follows. First, we need to set *FR*2 using *DP*2, then fix *DP*2. Second, we set *FR*1 by leveraging *DP*1. Starting from setting (1), we need to set *FR*2 at setting (2) by changing *DP*2, and then change *DP*1 to the desired level of *FR*1.

The above discussion is a testimony to the fact that uncoupled and decoupled designs have conceptual robustness; that is, coupling can be resolved with the proper selection of the DPs, path sequence application, and employment of design theorems (El-Haik, 2005; Suh, 2001).

The coupled design matrix in Figure 8.13(c) indicates the loss of path independence due to the off-diagonal design matrix entries (on both sides), and the design team has no easy way to improve controllability, reliability, and quality (measured by *Z*-value) of their design. The design team is left with compromise practices (e.g., optimization) amongst the FRs as the only option since a component of the individual DPs can be projected on all orthogonal directions of the FRs. The uncoupling or decoupling step of a coupled design is a conceptual activity that follows the design mapping (El-Haik, 2005).

An example of design coupling is presented in Figure 8.14, where two possible arrangements of the generic water faucet (Swenson and Nordlund, 1996) are displayed. There are two functional requirements: water flow and water temperature. The faucet in Figure 8.14(a) has two design parameters—the water valves (knobs), one for each water line. When the hot-water valve is turned, both flow and temperature are affected. The same would happen if the cold-water valve were turned. That is, the functional requirements are not independent and a coupled design matrix below the schematic reflects this. From the consumer perspective, optimization of the temperature will require reoptimization of the flow rate until a satisfactory compromise amongst the FRs, as a function of the DP settings, is obtained over several iterations.

Figure 8.14(b) exhibits an alternative design with a one-handle system delivering the FRs with, however, a new set of design parameters. In this design, flow is adjusted by lifting the handle, whereas moving it sideways will adjust the temperature. In this alternative, adjusting the flow does not affect temperature and vice versa. This design is better since the functional requirements maintain their independence per Axiom 1. The uncoupled design will give the customer path independence to set either requirement without affecting the other. Note also that in the uncoupled design case, design changes to improve an FR can be done independently as well, a valuable design attribute.

The importance of the design mapping has many perspectives. Chief among them is the identification of coupling among the functional requirements, due to the physical mapping of the design parameters, in the codomain. Knowledge of coupling is important because it provides the design team clues with which to find solu-

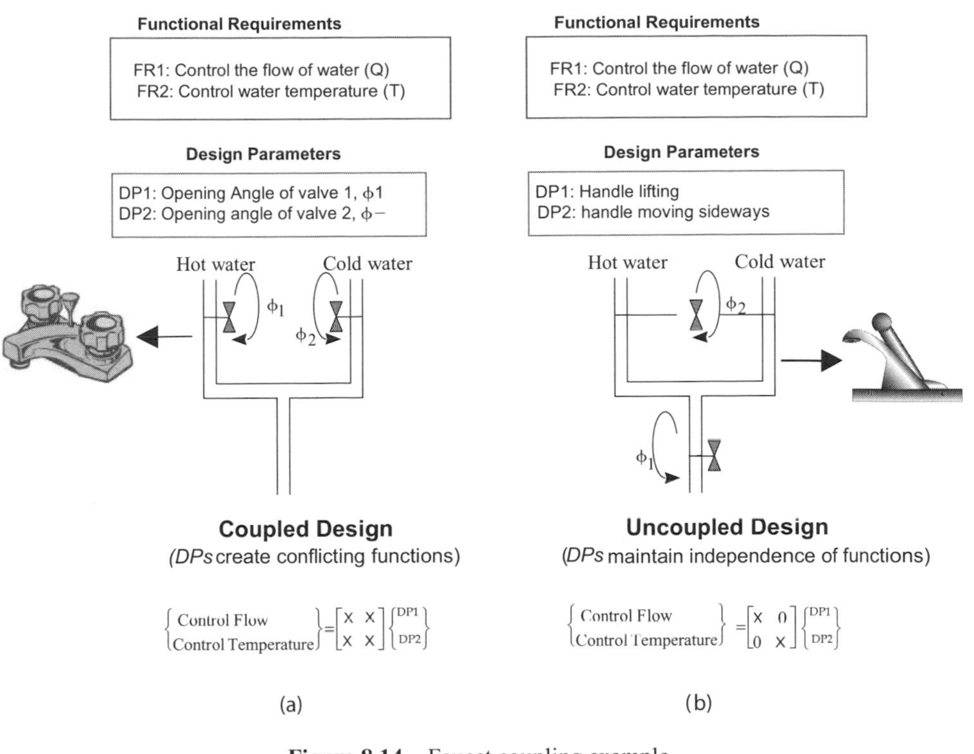

Figure 8.14 Faucet coupling example.

tions, make adjustments, or design changes in proper sequence, and maintain their effects over the long term with minimal negative consequences.

The design matrices are obtained in a hierarchy, and result from employment of the *zigzagging* method of mapping, as depicted in Figure 8.15 (Suh, 1990). The zigzagging process requires a solution-neutral environment in which the DPs are chosen after the FRs are defined and not vice versa. When the FRs are defined, we have to *zig* to the physical domain, and, after proper DP selection, we have to *zag* back to the functional domain for further decomposition or cascading, though at a lower hierarchical level. This process is in contrast to the traditional cascading processes that utilize only one domain at a time, treating the design as the sum of functions or the sum of processes.

At lower levels of hierarchy, entries of design matrices can be obtained mathematically from basic process engineering quantities, enabling the definition and detailing of transfer functions, a DFSS optimization vehicle. In some cases, these relationships are not readily available and some effort needs to be paid to obtain them empirically or via modeling. Lower levels represent the roots of the hierarchical structure at which Six Sigma concepts can be applied with some degree of ease.

Similar to CTSs , FRs, and DPs, the design specifications also need to be cascaded as well. The specifications describe the limits that are acceptable to the customer

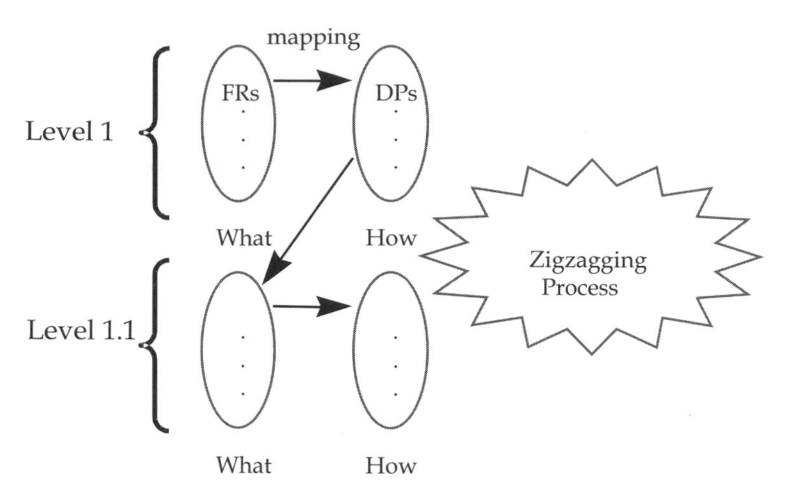

Figure 8.15 The zigzagging process.

to accomplish a design solution from their own perspective. The specification determines each limit required of a CTS and then, by mapping, to FRs and DPs, and then to PVs. That is, the specification cascading starts as a CTS specification and then is flowed down to functional requirements, design parameters, and process variables by methods such as quality function deployment (see Chapter 7) and axiomatic design. For example, in a service software design, a CTS could be "easy to use," and when cascaded (by mapping) it morphs into the specification for application programs, file layouts, data elements, reports, tables, screens, or technical communication protocols that are required to accomplish the originating CTS. The specification is written to either modify the existing FR, DP, or PV, or to introduce a new FR, DP, or PV. The technical specification in a design mapping should also describe the concept and reasons for the change, or any new functionality, as well as provide the detailed effort necessary to achieve the desired result.

8.5.2.1 Axiomatic Design Software. The application of axiomatic design processes can be nicely enhanced by using computer software packages. This book is accompanied by a copy of Acclaro DFSS Light®,[5] a limited version of Acclaro Designer® software provided by Axiomatic Design Solutions, Inc. (ADSI) via free download from the Wiley ftp site (ftp://ftp.wiley.com/public/sci_tech_med/six_sigma). ADSI is the only company dedicated to supporting the axiomatic design methodology with a portfolio of services and products. They provide a series of software solutions that can be deployed for individual, groupware or enterprise software solutions. The Acclaro Designer® software supports systems and general de-

[5]Acclaro DFSS Light® is one of the software products of Axiomatic Design Solutions, Inc. of Brighton, MA. It is protected under both copyright and pending patents. Acclaro is a registered trademark of Axiomatic Design Solutions, Inc. Browse their website at http://www.axiomaticdesign.com/default.asp.

sign problems. The Acclaro Scheduler® software supports software design optimization working with IBM's Rational Rose software development toolsets. The Acclaro Sync® software is a package that runs with Microsoft Project, permitting the effective project scheduling and synchronization of design projects using Acclaro Designer® and Acclaro Scheduler®.

Acclaro DFSS Light® is a JAVA-based software package that implements axiomatic design processes. It is about a 30 megabyte download installation file. It requires Windows 95, 2000, NT, XP, or later revisions with 256K of memory. Acclaro® software suite was *Industry Week*'s Technology of the Year and a winner of the International Institution for Production Research (CIRP) award.

Acclaro DFSS Light® enables the design team to meet the following needs in managing the DFSS ICOV process:

- Creates a process for requirements cascading (hierarchy) and analysis by implement zigzagging and decomposition trees to capture and track design logic.
- Permits design teams to collaborate at the earliest possible time with visualization and graphical representations. Geographically supports separate teams.
- Introduces a systematic process for design synthesis before document–build–test cycles.
- Analyzes design quality before committing to any cost by applying axioms.
- Enables design team to completely track the process from customer needs to requirements to design mappings to final designs.
- Visualizes and quantifies impact assessments when requirements change.
- Provides features to capture and track the design and development processes changes and instantly assess the ripple effect through the design. It is also equipped with model flow diagrams, providing capability to understand optimal design task sequencing.
- Keeps a running list of changes to the design at FR, DP, PV, or CN levels.
- Contains features to plan, estimate, and control conceptual design with coupling analysis to identify suboptimal designs with single- and full-level design matrix dependency analysis.

8.5.2.2 Axiomatic Design Mapping Example. Here we examine an implementation methodology for transition from traditional manufacturing to cellular manufacturing using axiomatic design (Bulent et al., 2002; El-Haik, 2005). In this case study, a framework to transform a traditional production system from process orientation to cellular orientation based on axiomatic design principles is introduced. A feedback mechanism for continuous improvement is also suggested for evaluating and improving the cellular design against preselected performance criteria (see Figure 8.16). The criteria, which are developed based on the independence axiom, provide the necessary steps in transforming an existing process-oriented system into a cellular manufacturing system.

Transition to cellular manufacturing follows after all cellular manufacturing

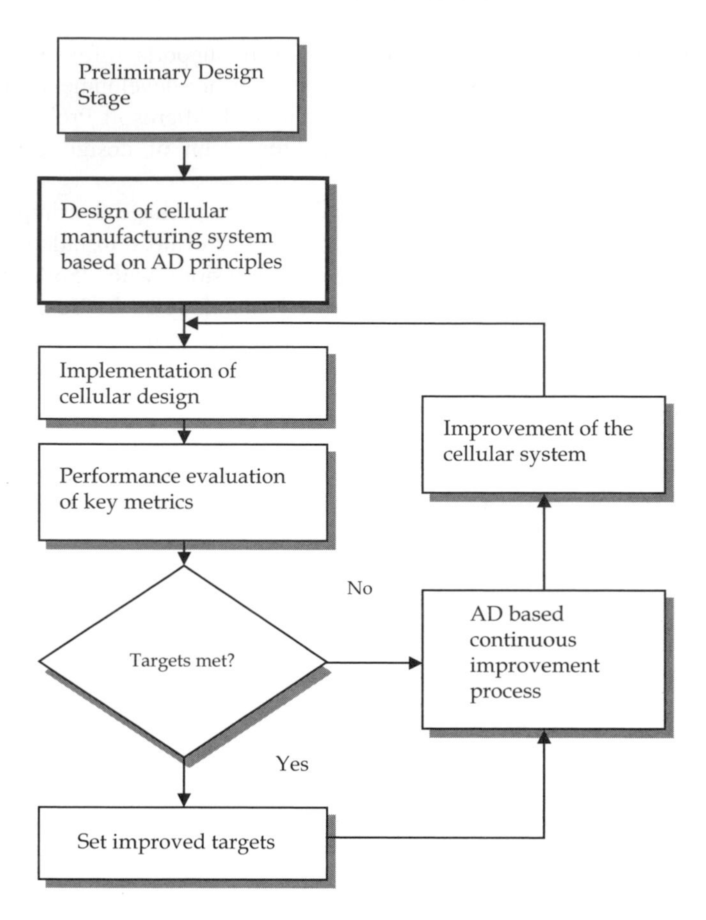

Figure 8.16 Cellular manufacturing design process with feedback mechanism.

steps are successfully completed. At this stage, the production is achieved through a cellular manufacturing system. Databases and information for comparing the system performance need to be generated with set target goals for some business metrics. Based on target values and achievements, new target values are established and appropriate system modifications and changes are effected through cellular manufacturing system improvement principles provided in the proposed procedure. These principles are also based on axiomatic design concepts. These continuous feedback and improvement principles are also in agreement with the spirit of lean manufacturing and Kaizen activities. A complete functional requirement to design parameter mappings and hierarchy for the design of a cellular manufacturing system through axiomatic design principles is provided.

The first step is to define the functional requirements (FRs) of the system at the highest level of hierarchy in the functional domain. At this stage, many functional requirements may be established. Depending on the functional definition, each functional requirement established at this stage may lead to a completely different cellular manufacturing design. The authors have selected the following as the highest functional requirements:

Hierarchical Level 1 Analysis

FR = Provide flexible production in line with customer needs

Customer needs are summarized as more product variety, smaller batch sizes with higher quality, and more frequent deliveries at lower costs. These requirements are forcing companies to reevaluate their classical production systems to obtain more flexibility in response to these customer needs. The flexibility of a production system is measured by its speed and agility to respond to rapidly changing customer needs.

In step 2, the mapping to the design parameters domain is accomplished using the zigzagging approach. At this step, design parameters (DPs), which satisfy the FRs established in the previous step, are defined by zigging. In order to make the correct DP selection, the DPs corresponding to the FRs established before must be exhaustively generated. The following DP has been selected to satisfy the FR provided above:

DP = Cellular production system design

The production system, which can meet customer's needs in an efficient way through elimination of waste, reduction of lead time, and improved quality, is a cellular production system designed with lean principles in mind.

Hierarchical Level 2 Analysis. In step 3, the zigzagging process is continued for further hierarchical level prescription of the design and to obtain further clarification. If the DPs proposed for satisfying the FRs defined earlier cannot be implemented without further clarification, the axiomatic design principles recommend returning to the functional domain (in a zagging step) for decomposing the FRs into their lower functional requirement set as follows:

FR1 = Classify and group products/components for simple material flow
FR2 = Define production strategy based on product specifications
FR3 = Rearrange resources to minimize waste
FR4 = Provide means to control production based on customer demand

In step 4, we need to find the corresponding DPs by mapping FRs in the physical domain (a zigging step). In satisfying the four FRs defined above, we move to the physical domain from the functional domain (a zigging step) and obtain:

DP1 = Procedure for defining product families
DP2 = Procedure for selecting production strategy

DP3 = Product oriented layout

DP4 = Pull production control system

The next logical step (step 5) is to determine the design matrix (DM), which provides the relationships between the FR and DP mapping elements. It is critical to insure that the DM, as established, satisfies the independence axiom. The DM is

$$\begin{Bmatrix} FR1 \\ FR2 \\ FR3 \\ FR4 \end{Bmatrix} = \begin{bmatrix} X & 0 & 0 & 0 \\ X & X & 0 & 0 \\ X & X & X & 0 \\ X & X & X & X \end{bmatrix} \begin{Bmatrix} DP1 \\ DP2 \\ DP3 \\ DP4 \end{Bmatrix} \tag{8.1}$$

A quick look reveals that the design is decoupled and, thus, satisfies Axiom 1. In the DM above, X represents a strong relationship between the corresponding FR–DP pair. The 0 indicates the absence of such relationship.

Hierarchical Level 3 Analysis. In step 6, the zigzagging process continues with FR1, FR2, FR3, and FR4 by going from the physical to the functional domain again and determining the corresponding DPs.

Step 6a. FR1—Products/Components Branch. The functional requirement FR1 as defined above may be broken down with DP1 (procedure for defining product families) in mind as:

FR11 = Determine high-volume products/components to be grouped

FR12 = Determine operations and machine types for producing each product family

FR13 = Form product families

FR14 = Determine final number of machine groups

The corresponding DPs may be stated as:

DP11 = Product–quantity Pareto analysis

DP12 = Machine–component incidence matrix

DP13 = Product grouping techniques

DP14 = Cost analysis and economic justification techniques

In this step, product families that will be economically manufactured through cellular manufacturing and their corresponding machine groups are determined by using Pareto analysis, followed by product family assignment techniques. The DM for the above vectors of FRs and DPs is

$$\begin{Bmatrix} FR11 \\ FR12 \\ FR13 \\ FR14 \end{Bmatrix} = \begin{bmatrix} X & 0 & 0 & 0 \\ X & X & 0 & 0 \\ 0 & X & X & 0 \\ 0 & X & X & X \end{bmatrix} \begin{Bmatrix} DP11 \\ DP12 \\ DP13 \\ DP14 \end{Bmatrix} \tag{8.2}$$

Once again, this is a decoupled design satisfying the independence axiom.

Step 6b. FR2—Production Strategy Branch. The functional requirement FR2 as defined above (define production strategy based on product specifications) may be decomposed with DP2 (procedure for selecting production strategy) in mind as:

FR21 = Determine the master process
FR22 = Select most appropriate process elements
FR23 = Determine required training/education needs
FR24 = Motivate labor participation

The corresponding DPs may be stated as:

DP21 = Master Process selection
DP22 = Production resources selection procedure
DP23 = Multipurpose labor training programs
DP24 = Profit sharing program

At this stage, production resources are determined following the establishment of the master process based on product specifications. Once the resource selection is complete, the education and training requirements of the workers can be established. For ensuring the full participation of workers, appropriate profit sharing programs must be established and announced to the workers to seek their dedication and involvement. The design-decoupled DM for this requirement is

$$\begin{Bmatrix} FR21 \\ FR22 \\ FR23 \\ FR24 \end{Bmatrix} = \begin{bmatrix} X & 0 & 0 & 0 \\ X & X & 0 & 0 \\ 0 & X & X & 0 \\ 0 & X & X & X \end{bmatrix} \begin{Bmatrix} DP21 \\ DP22 \\ DP23 \\ DP24 \end{Bmatrix} \tag{8.3}$$

Step 6c. FR3—Resource Rearrangement Branch. The functional requirement FR3 as defined above as "rearrange resources to minimize waste" may be broken down with DP3, defined as "product-oriented layout," in mind as:

FR31 = Minimize material handling
FR32 = Eliminate wasted motion of operators
FR33 = Minimize waste due to imbalance in the system

The corresponding DPs may be stated as:

DP31 = Material-flow-oriented layout
DP32 = Arrangement of stations to facilitate operator tasks
DP33 = Balanced resources in response to Takt time (Takt time = available time/ demand)

At this stage, lean manufacturing principles are the guiding principles of this design step. In this step, the focus is on waste elimination. Therefore, in rearranging the resources, waste due to motion, material handling, and imbalances between resources is minimized. Without this step, the designed cell will not provide the expected performance. Once again, the decoupled-design DM is

$$\begin{Bmatrix} FR31 \\ FR32 \\ FR33 \end{Bmatrix} = \begin{bmatrix} X & 0 & 0 \\ X & X & 0 \\ X & X & X \end{bmatrix} \begin{Bmatrix} DP31 \\ DP32 \\ DP33 \end{Bmatrix} \tag{8.4}$$

Step 6d. FR4—Production Control Branch. The functional requirement FR4 as defined above is "provide means to control production based on customer demand," which may be broken down with DP4 (pull production control system) in mind as:

FR41 = Ensure smooth and steady production in assembly line
FR42 = Provide material/information flow
FR43 = Provide continuous feedback information flow

The corresponding DPs may be stated as:

DP41 = Leveled/mixed production
DP42 = Card system (Kanban)
DP43 = Information/report system and visual management tools

Satisfying customers with the right amount of products produced just in time can only be accomplished through the pull system. However, just-in-time systems require a steady pull on all products in the family. To insure a steady pull, a leveled/mixed production schedule must be established. This leads us into developing the appropriate Heijunka schedule and the necessary visual management tools, including the Kanban system, for successful implementation. The DM is

$$\begin{Bmatrix} FR41 \\ FR42 \\ FR43 \end{Bmatrix} = \begin{bmatrix} X & 0 & 0 \\ X & X & 0 \\ X & X & X \end{bmatrix} \begin{Bmatrix} DP41 \\ DP42 \\ DP43 \end{Bmatrix} \tag{8.5}$$

Axiomatic design is used to develop the four design mappings described in this section. Design mappings prescribe the design blueprint and provide a conceptual means for design vulnerability elimination or reduction. Two major vulnerabilities that are treated by axiomatic design are coupling and complexity. Coupling means lack of controllability. It is corrected by setting design parameters and process variables to optimize the functional requirements and then to satisfy the customer by following design axioms. Coupling is produced as a result of violating Axiom 1, the Independence axiom.

Complexity, on the other hand, is a result of Axiom 2 violation, the information axiom. It is always wise to *uncouple* or *decouple* the design mappings (at all hierarchical levels) before the DFSS team proceeds to the next stages of service development. The theory of inventive problem solving (TRIZ) provides a road map for coupling elimination or reduction by following certain principles such as Altshuler's trade-off or contradiction matrix (see Chapter 9). Complexity can be reduced by variability reduction and means shifting of the functional requirements to deliver the CTSs. This is the essence of DFSS optimization practices. Chapters 12 and 13 are dedicated to such design treatments.

Axiomatic design can be used to design improvement and to select concepts of conceptual robustness (in obedience to design axioms) while probing the necessary sequence of design development within a prescribed design hierarchy. It can be morphed to a tool to assess technical risk and opportunities while complementing in-house current design best practices (El-Haik, 2005). As a conceptual approach, it provides the team greater flexibility to conduct trade-offs (if–then scenarios) at any hierarchical level by substituting different design parameters or process variables in a given mapping. In doing so, it highlights necessary organization communication channels amongst the different team satellite members, stakeholders, and the rest of DFSS project core team.

8.6 PUGH CONCEPT SELECTION

The key mechanisms for selecting the best possible design or process solution is the method of "controlled convergence," which was developed by Stuart Pugh (Pugh, 1991) as part of his concept selection process. Controlled convergence is a solution-iterative selection process that allows alternate convergent (analytic) and divergent (synthetic) thinking to be experienced by the team. The method alternates between *generation* and *selection* activities (Figure 8.17). El-Haik and Yang (2003) suggested the following enhancements to the controlled convergence method:

1. The "generation" activity can be enriched by the deployment of design Axiom 1 and its entire derived theoretical framework, which calls for functional requirements independence. This deployment will be further enhanced by many TRIZ methodology concepts to resolve design vulnerabilities where applicable.
2. The "selection" activity can be enhanced by the deployment of Axiom 2, which calls for design simplicity.

The controlled convergence uses comparison of each alternative solution entity to a reference datum. Evaluation of a concept is more subjective than objective. However, the method discourages promotion of ideas based upon opinions and, thus, promotes objectivity. The controlled convergence method allows for elimination of bad features and weak concepts and, thereby, facilitates the emergence of new concepts. It illuminates the best concept as the one most likely to meet the constraints and requirements of the customer (CTSs) as expressed by the specification, and the one least vulnerable to immediate project competition.

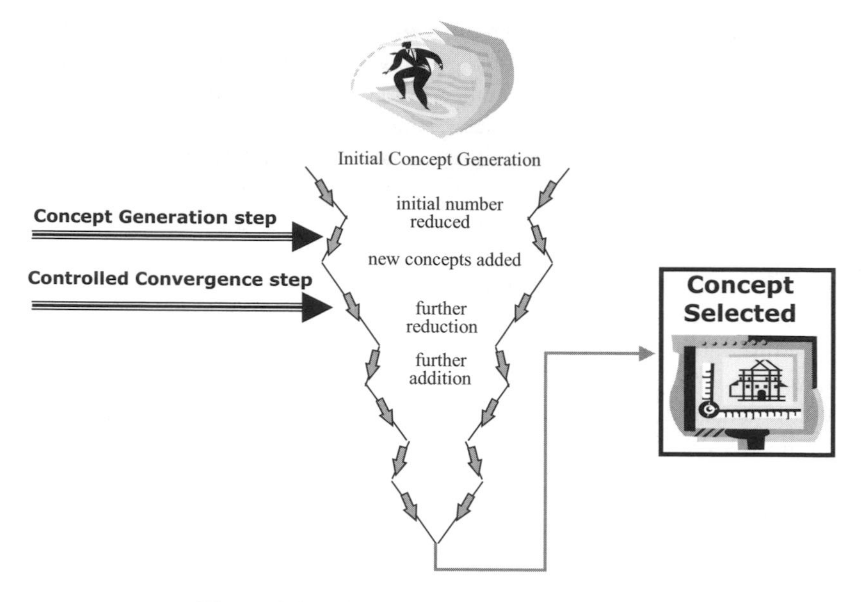

Figure 8.17 Controlled convergence method.

The development of the concepts through the combination of solution alternatives and functional requirement can be identified by a matrix called the morphological or synthesis matrix. In this matrix, the functional requirements (FRs) are listed in the rows and the solution alternatives (the design parameters and/or process variables) are laid down in the columns. At this point, process steps (subprocesses) are usually defined as the design parameters at a hierarchal level if the design mapping was conducted similar to the example in Subsection 8.5.2.1. However, this knowledge is not detailed at this stage. The functional requirements need to be listed in the order of their hierarchy by the team, to the best of their knowledge at this step, and should be grouped according to their type of DP/PV (procedure type, policy type, training type, communication type, etc.).

The concepts are synthesized and generated from all possible feasible combinations of all possible design parameters (DPs) per functional requirement (FR) in the matrix. Connecting all possible solutions using arrows between the design parameters identifies a feasible design concept. The arrows can only be connected when the team is technically confident about the functional and production feasibility (El-Haik, 2003, 2005).

In conducting this exercise, the team will identify all possible feasible design solutions. In the next step, guided by their knowledge and the DFSS road map, the team should concentrate only on promising solutions. The challenge here is to ensure that the physical and functional compatibility and other constraints are met and the appropriate flow of DPs such as procedures, policies, communication, and so on is properly identified. Normally, each functional requirement can be delivered by several possible DPs in a given hierarchical level within a concept. Therefore, the synthesis matrix exercise should be conducted at all levels of design structure. The identification of all possible alternative solutions (DPs) per functional requirement may be facilitated by the use of the morphological approaches of Zwicky (1984) and TRIZ methodology (see Chapter 9).

Several feasible high-level and undetailed concepts are usually generated using the synthesis matrix. This generation of multiple concepts poses a selection problem. Which concept should be selected for further detailing in the DFSS roadmap? The DFSS team must select the *best* concept using the Pugh concept selection method.[6]

In this step, the DFSS team produces the convergence to the best concept in iterative steps that are performed with DFSS discipline and rigor. The following sequence may be used to facilitate the convergence to the best concept by the DFSS team:

1. Develop a set of criteria based on the customer's wants and needs, the CTSs. Determine CTS importance weights. Do not forget regulatory and legal requirements in design mapping. These criteria should be measurable, defined, and understood by all members of the team.
2. Develop a way to sketch concepts developed so far.
3. Choose a datum (baseline) with which all other concepts are to be compared. The datum could be an existing baseline, as in the case of redesign situations. In new design situations, the datum could be any concept that the team may generate from the synthesis matrix.
4. Develop a group of design concepts that are aimed at satisfying the CTSs and FRs.
5. Using a simple matrix, list the criteria from step 1 above on the left and the concepts across the top. List concepts in the columns of the Pugh matrix as obtained from the synthesis matrix.
6. Compare the new concepts with the datum. Evaluate concepts versus the defined criteria. Use the traditional evaluation of plus (+), minus, or the same (s). The datum will be the neutral element(s) of the numbering system chosen. Comparing each solution entity against the datum, rate them either as plus (better than), minus (worse than), or same as the datum. (See Figure 8.18 for a Pugh matrix of a European vacation.)
7. Identify the top best new concepts, having scored them relative to the datum, and sum the ranks across all criteria to get the scores. These scores must not be treated as absolute as they are for guidance only and, as such, must not be summed algebraically. Certain concepts will exhibit relative strengths, whereas others will demonstrate relative weakness. Select the best concept with maximum number of pluses and the minimum number of minuses.
8. Make hybrids. Combine best concepts to make a new datum. Incorporate strong ideas from other concepts. Perform trade-off studies to generate alternatives using design axioms and TRIZ. Look at the negatives. What is needed in the design to reverse the negative (relative to the datum)? Will the

[6]The concept selection problem was formulated by El-Haik (2005) as an optimization problem.

CTSs \ Concepts	 ITALY	 LONDON	 PARIS	...	 SPAIN
CTS1	+				
CTS2	+				
:	-				
:	S				
CTSm					
Total (-)					
Total (+)					
Total (S)					

Figure 8.18 Pugh matrix of a European trip.

improvement reverse one or more of the existing positives due to design coupling? If possible, introduce the modified concept into the matrix and retain the original solution entity in the matrix for reference purposes. Eliminate truly weak concepts from the matrix. This will reduce the matrix size. See if strong concepts begin to emerge from the matrix. If it appears that there is an overall uniformity of strength of the concepts, this will be an indication of one of two things (or a mixture of both). (1) The criteria are ambiguous and, hence, subject to mixed interpretation by the DFSS team. (2) Uniformity of one or more of the concepts suggests that they are subsets of the others (i.e., they are not distinct). In this case, the matrix cannot make a distinction because none exists.

9. If the new datum is unacceptable, loop to Step 1, otherwise go to step 10.
10. If the new datum is not acceptable and/or close to Ideal Final Result (see Chapter 9), loop to Step 4.

Suppose you are planning a European vacation. Your concepts are the countries, such as Italy, UK, Spain, and France. What will be your CTSs (sightseeing, restaurants, shopping, countryside, history, sun and sea, etc.)? What about the cost constraints? You may want to fill out a concept selection table such as the one in Figure 8.18 and apply the above ten steps. It is a good exercise to make sure you maximize satisfaction, your return on investment. Do not forget that you can morph the concepts into multicountry trip in Step 8.

8.7 SUMMARY

In the service DFSS project road map, there are two different mappings that are required—the functional mapping and the process mapping. A functional mapping is a logical model depicting the logical and cause–effect relationships between design functional elements. There are several techniques used to produce functional mappings, such as the axiomatic design zigzagging method, value engineering, the FAST approach, and IDEF methodologies.

A process map is a schematic model for a process. A process map is considered to be a visual aid for picturing work processes, which show how inputs, outputs, and tasks are linked. There are a number of different methods of process mapping, such as graphical process mapping and value stream mapping. A process is the basic unit for service Six Sigma design projects. The business process management system (BPMS) is an effective tool for improving overall performance. Simulation is a tool that can be used within the DFSS map for redesign of projects at Six Sigma levels.

Pugh concept selection method is used after design mapping to select a winning concept that will be detailed, optimized, piloted, and validated in the following design stages.

APPENDIX 8.A: IDEF MAPPING TECHNIQUE

During the 1970s, the U.S. Air Force Program for Integrated Computer Aided Manufacturing (ICAM) sought to increase manufacturing productivity through systematic application of computer technology. The ICAM program identified the need for better analysis and communication techniques for people involved in improving manufacturing productivity. As a result, the ICAM program developed a series of techniques known as the Integration Definition For Function Modeling (IDEF) techniques, which include those discussed in this section.

In 1983, the U.S. Air Force Integrated Information Support System program enhanced the IDEF1 information modeling technique to produce IDEF1x (IDEF1 Extended), a semantic data-modeling technique. The IDEF family of techniques is widely used in the government, industrial, and commercial sectors, supporting modeling efforts for a wide range of enterprises and application domains. In 1991, the National Institute of Standards and Technology (NIST) received support from the U.S. Department of Defense, Office of Corporate Information Management (DoD/CIM), to develop one or more Federal Information Processing Standards (FIPS) for modeling techniques. The techniques selected were IDEF0 for function modeling and IDEF1x for information modeling. These FIPS documents are based on the IDEF manuals published by the U.S. Air Force in the early 1980s.

Today, IDEF is a family of mapping techniques with different objectives. For example, IDEF0 is a technique used to model a wide variety of systems that use hardware, software, and people to perform activities. IDEF1x semantically models the relationships between various pieces of data, and IDEF2 captures the dynamic behavior of a system. IDEF3 was created specifically to model the sequence of activities in manufacturing systems. IDEF5 models domain ontologies, and IDEF6 defines the motives that drive the decision-making process. In this section, we will only examine IDEF0 and IDEF3 due to their application to service mapping.

In IDEF0 and IDEF3, multiple flows are shown by how the arrows enter process boxes (Figure 8A.1). Each side of the function box has a standard meaning in terms of box–arrow relationships. The side of the box with which an arrow interfaces reflects the arrow's role. Arrows entering the left side of the box are inputs (I). Inputs are transformed or consumed by the function to produce outputs. Arrows entering the box on the top are controls (C). Controls specify the conditions required for the function to produce correct outputs, usually interpreted as information. Arrows leaving a box on the right side are outputs (O). Outputs are the data or objects produced by the function, as shown in Figure 8A.1. Mechanisms are the means (design parameters or process variables) by which functions are delivered, and they enter the box from the bottom.

Function boxes are linked together to map the process. Some boxes are complex enough that the task within the box can have its own *child* process map diagram. These *parent–child* diagrams are linked together in a context diagram, which can be represented as a tree or as an indented list, as shown in Figure 8A.2.

An IDEF3 map typically contains four to six process boxes per map. Additional detail will force the mapping to continue onto another map at a lower level in the hi-

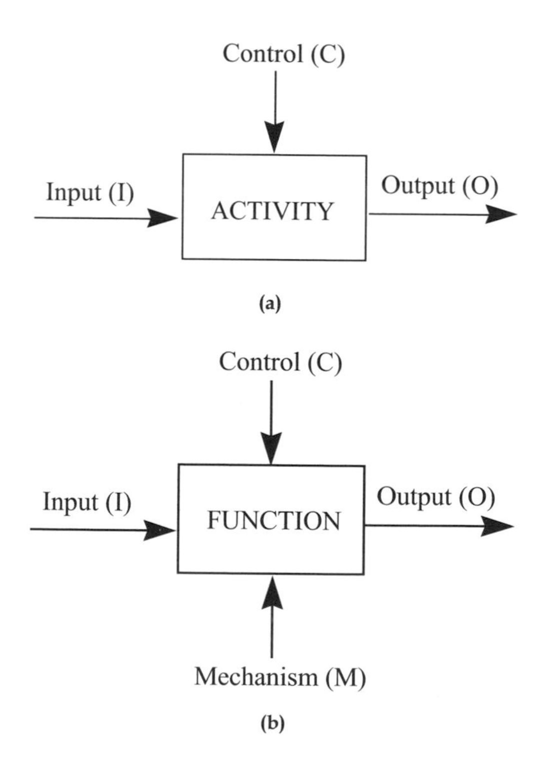

Figure 8A.1 (a) IDEF3 function box and interface arrows. (b) IDEF0 function box and interface arrows.

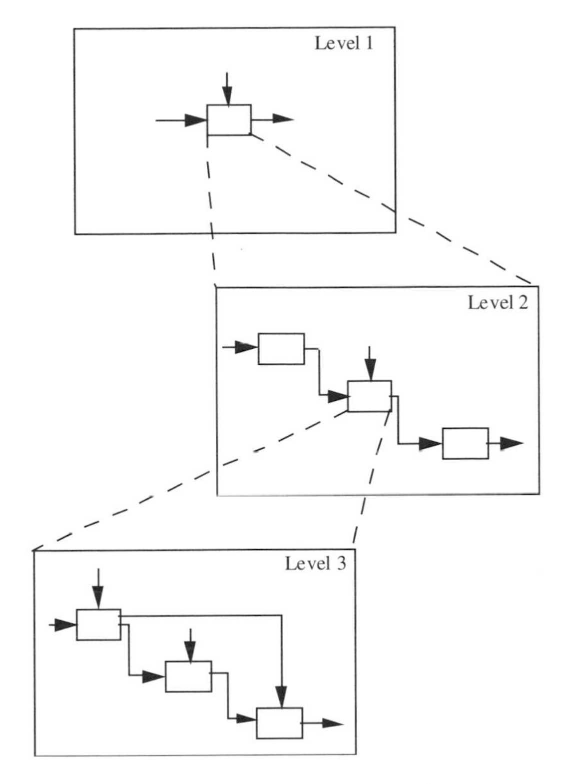

Figure 8A.2 IDEF0 and IDEF3 decomposition structure.

erarchy with limited numbering schemes. The limit to the number of boxes per map does not impede the documentation and presentation of detailed process information. Utilizing multiple levels of decomposition provides the capability to specify at whatever level of detail is needed. Moreover, those needing less detail are not forced to absorb all the details. Logical operators that control "if–then" or iterative loops are provided by the IDEF0 and IDEF3 family using Figure 8A.3.

8.A.1 IDEF0 Technique

The standard "Draft Federal Information Processing Standards Publication 183" of IDEF0 was published in 1993. The standard is composed of normative and informative sections. Compliance with the normative sections (Sections 1 through 3) is required. The informative sections (Annexes A through D) provide additional suggestions and guidance.

IDEF0 (Integration DEFinition language 0) is based on SADTä (Structured Analysis and Design Techniqueä), developed by Douglas T. Ross and SofTech, Inc. In its original form, IDEF0 includes both a definition of a graphical modeling language (syntax and semantics) and a description of a comprehensive methodology for developing models.

Symbol	Meaning	Definition
&	Asynchronous AND	INPUTS: All preceding activities must complete OUTPUTS: All following activities will start
&	Synchronous AND	INPUTS: All preceding activities must complete simultaneously OUTPUTS: All following activities will start simultaneously
O	Asynchronous OR	INPUTS: One or more of the preceding activities must complete OUTPUTS: One or more of the following activities will start
O	Synchronous OR	INPUTS: One or more of the preceding activities must complete simultaneously OUTPUTS: One or more of the following activities will start simultaneously
X	Exclusive OR	INPUTS: Exactly one of the preceding activities must complete OUTPUTS: Exactly one of the following activities will start

Figure 8A.3 IDEF family logical operators.

IDEF0 may be used to model a wide variety of automated and nonautomated systems. For new systems, IDEF0 may be used first to define the requirements and specify the functions, and then to design an implementation that meets the requirements and performs the functions. For existing systems, IDEF0 can be used to analyze the functions the system performs and to record the mechanisms (means) by which these are done.

The result of applying IDEF0 to a system is a model that consists of a hierarchical series of diagrams, text, and glossary cross-referenced to each other. The two primary modeling components are functions (represented on a diagram by boxes) and the data and objects that interrelate those functions (represented by arrows). As a function modeling language, IDEF0 has the following characteristics:

- It is comprehensive and expressive, capable of graphically representing a wide variety of business, manufacturing, and other types of enterprise operations to any level of detail.
- It is a coherent and simple language, providing for rigorous and precise expression, and promoting consistency of usage and interpretation.
- It enhances communication between systems analysts, developers, and users through ease of learning and its emphasis on hierarchical exposition of detail. It is well tested and proven through many years of use in Air Force and other government development projects, and by private industry.

- It can be generated by a variety of computer graphics tools; numerous commercial products specifically support development and analysis of IDEF0 diagrams and models.

In addition to definition of the IDEF0 language, the IDEF0 methodology also prescribes procedures and techniques for developing and interpreting models, including ones for data gathering, diagram construction, review cycles, and documentation. Materials related solely to modeling procedures are presented in the informative annexes of the standard. In this section, we will limit the exposure to the IDEF0 algorithmic steps and follow with an illustrative example.

Step 1. Define the mapping scope. IDEF0 process experts are identified and added to the DFSS team. Team members are to be provided with basic training in the methodology. Team meets regularly to perform tasks related to the process modeling effort, that is, information collection and model construction.

Step 2. Define appropriate activities

Step 3. Arrange activities in phased sequence

Step 4. Identify and define input and output objects in all stages of the process life cycle stages

Step 5. Determine decision points and flow junctions

Step 6. Identify and define activity controls and mechanisms

Step 7. Define notifications and messages

Example: Buying a house (Figure 8.A.4)

Glossary of activities at level 1:

Obtain Loan Amount—Determine maximum amount that can be loaned
Select a house—Find a house that suits the buyer's needs
Negotiate Price—Determine a mutually agreed upon price
Complete Paperwork—Sign loan and legal closing documents

Glossary of inputs and outputs at level 1:

Loan Application—The bank's loan application
Loan Amount—The amount the bank will loan
Selected house—The house the buyer chose
Agreed-Upon Price—Mutually agreed-upon price
New house—The newly purchased house

Glossary of controls and mechanisms at level 1):

Credit History—Past history of the buyer's ability to pay back loans
Loan officer—Person who determines the buyer's fitness for a loan
Bank regulations—Rules regarding bank applications
Market analysis—Analysis providing purchasing information for house with similar attributes in the area of interest
Real estate agent—An individual who assists in house purchase

Area attractions—Buyer's assessment of the area the house is in
Experience—Buyer's background in the purchasing process
Insurance agent—Person handling insurance policy for the house

Glossary of activities at level 2:

Apply For Loan—Fill out paperwork to initiate loan process
Discuss Acceptance—Discuss conditions of loan approval
Get Maximum Loan Amount—Find out the maximum loan amount

Glossary of inputs and outputs for the get loan amount process at level 2:

Loan Application—The bank's loan application
Scheduled Meeting—Meeting with loan officer to discuss loan application
Approved—Met all requirements
Rejected—Did not meet all requirements
Loan Amount—Maximum loan amount

Glossary of controls and mechanisms for the get loan amount process at level 2:

Income—The buyer's annual income
Credit History—Past history of the buyer's ability to pay back loans
Loan Officer—Person who determines the buyer's fitness for a loan
Bank Regulations—Rules regarding bank applications
Paperwork—The papers requiring completion for a loan

The reader is encouraged to follow these steps to finish all Level 2 functions by mapping the balance of the Level 1 map, namely, "Select a House," "Negotiate Price," and "Completer Paperwork." The process will continue for levels three onward.

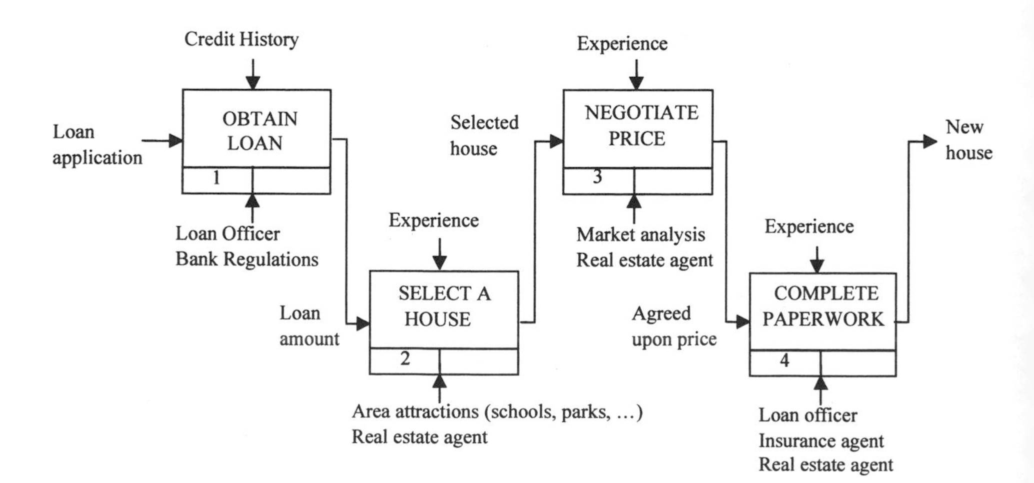

Figure 8A.4 "Buying house" IDEF0 process Level 1 map.

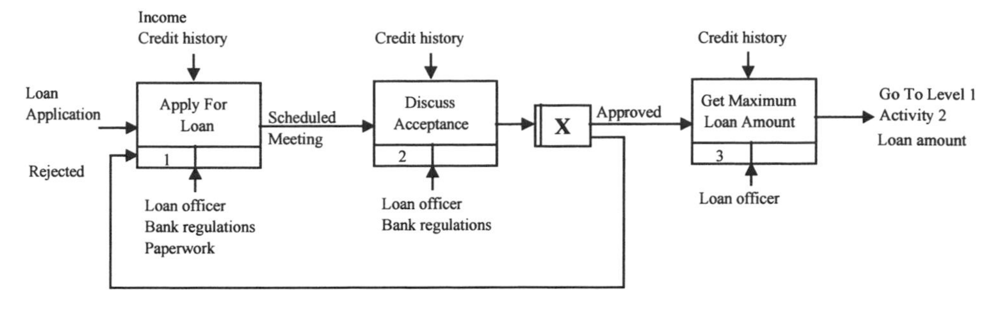

Figure 8A.5 "Obtain loan" IDEF0 process map, Level 2.

APPENDIX 8.B: MATRIX REVIEW[7]

A matrix can be considered a two-dimensional array of numbers. They take the form

$$\boldsymbol{A} = \begin{bmatrix} a_{11} & a_{13} & a_{13} \\ a_{21} & a_{23} & a_{23} \\ a_{31} & a_{33} & a_{33} \end{bmatrix}$$

Matrices are very powerful and form the basis of all modern computer graphics because they are so fast. We define a matrix with an upper-case bold italic letter. Look at the above example. The dimension of a matrix is its height followed by its width, so the above example has dimension 3 × 3. Matrices can be of any dimension but, in terms of computer graphics, they are usually kept to 3 × 3 or 4 × 4. There are a few types of special matrices; these are the column matrix, row matrix, square matrix, identity matrix, and zero matrix. A column matrix is one that has a width of 1 and a height greater than 1. A row matrix is a matrix that has a width greater than 1 and a height of 1. A square matrix is one in which both dimensions are the same. For instance, the above example is a square matrix because the width equals the height. The identity matrix is a special type of matrix that has values of 1 in the diagonal from top left to bottom right and the remaining values are 0. The identity matrix is indicated the letter ***I***, where

$$\boldsymbol{I} = \begin{bmatrix} 1 & 0 & 0 \\ 0 & 1 & 0 \\ 0 & 0 & 1 \end{bmatrix}$$

The identity matrix can be any dimension, as long as it is also a square matrix. The zero matrix is a matrix that has all its elements set to 0. The elements of a matrix are all the numbers in it. They are numbered by the row/column position so that a_{13} means the matrix element in row # 1 and column # 3.

[7]Adopted from Phil Dadd, http://www.gamedev.net/reference/articles/article1832.asp.

9

THEORY OF INVENTIVE PROBLEM SOLVING (TRIZ) FOR SERVICE

9.1 INTRODUCTION

As we design services, there are many contradictions or trade-offs that must be resolved. Often we discover, when the design is operationalized, that the design will not achieve the desired level of performance because of the unresolved or suboptimized resolution of contradictions or because of coupling (see Chapter 8). What if a tool existed that allowed for effective resolution of these contradictions, expanded the design teams knowledge, and allowed for safe and effective innovation? Wouldn't you be interested in such a tool?

Well, there is such a tool, the theory of inventive problem solving (TRIZ, also know as TIPS) that was introduced in the United States circa 1991 and is the culmination of efforts by Genrich S. Altshuller of Russia. The purpose of this chapter is to familiarize the reader with the basics of TRIZ, and provide some practical applications and references for further reading. TRIZ is a tool that can be used heavily in the DFSS road map conceptualize phase (see Figure 5.1). Opportunities for TRIZ applications are vital in other phases moving forward.

9.2 TRIZ HISTORY

The history of TRIZ is interesting and mirrors the life of its inventor, Genrich S. Altshuller. Altshuller was born on October 15, 1926 in Tashkent, Uzbekistan. While in the ninth grade, he obtained his first patent (Author's Certificate) for an underwater diving apparatus. In the tenth grade, he developed a boat propelled by a carbide-fueled rocket engine. In 1946, he invented a method for escaping from an immobilized submarine, and this led to employment in the patent office of the Caspian Sea Military Navy. The head of this department challenged Altshuller to solve a difficult problem and he was successful, but a series of events, meetings, and letters led to imprisonment four years later. He was driven to invent and assist others to invent. He was troubled by the common misconception that invention was the result of accidental enlightenment or genealogy, so he sought to discover a methodology for inventing.

He called upon a former schoolmate, Rafael Shapiro, also a passionate inventor, to work with him to discover the methodology of invention. By now, Altshuller believed that invention was no more than the removal of technical contradictions with the help of certain principles. Shapiro was excited about the discovery and they researched all the existing patents and took part in inventing competitions in their search for new methods of invention. They received a National Competition Award for the design of a flame- and heat-resistant suit. In recognition of that, they were invited to Tbilisi, a town in Georgia, where upon arrival they were arrested, interrogated and sentenced to 25 years in Siberia. It turns out that all of the letters that Alshuller had written to Stalin had him pegged as an intellectual and there was only one place for intellectuals in the USSR.

Altshuller used his TRIZ model to survive, minimizing harmful function and optimizing useful function. In Siberia, he worked 12 hours per day logging and decided it was better to be put in solitary confinement. Later, he was transferred to the Varkuta coal mines where he had to toil for 8–10 hours per day. In 1954, a year and a half after Stalin's death, Altshuller was released from prison.

In 1956, the first paper by Altshuller and Shapiro "Psychology of Inventive Creativity," was published. In 1961, his first book, *How to Learn to Invent,* was published, and 50,000 readers paid around 25 cents to read about the First 20 Inventive Principles. In 1959, he wrote his first letter to the highest patent organization in the former Soviet Union, VOIR, requesting a chance to prove his theory. After writing hundreds of letters, 9 years later they promised a meeting no later than December 1968. When the meeting was held, he met for the first time many people who considered themselves his students. In 1969, he wrote a new book, *Algorithm for Inventing,* in which he delivered the mature 40 Principles and the first algorithm to solve complex inventive problems.

For a period from 1970 through 1980, many TRIZ schools were opened throughout the USSR and hundreds of students were trained. During this period, Altshuller traveled to conduct seminars. This stage ended in 1980 when the first TRIZ specialist conference took place in Petrozavodsk, Russia. During the next period, from 1980 through 1985, TRIZ received much publicity in the USSR. Many people became devotees of TRIZ and of Altshuller, and the first TRIZ professionals and semiprofessionals appeared. Altshuller was highly efficient at developing his TRIZ model due to the large number of seminars he conducted, the various TRIZ schools that were established, and the individual followers who joined the ranks, allowing for the rapid testing of ideas and tools. TRIZ schools in St. Petersburg, Kishinev, Minsk, Novosibirsk, and other cities became very active under Altshuller's leadership.

In 1986, the situation changed dramatically. Altshuller's illness limited his ability to work on TRIZ and control its development. Also, for the first time in the history of TRIZ, *perestroika* allowed it to be applied commercially. The Russian TRIZ Association was founded in 1989 with Altshuller as president. The period from 1991 on saw the rapid deterioration of economic conditions in the former USSR, and many capable TRIZ specialists, most of whom who had established their own businesses, had to move abroad. Many of the TRIZ specialists immigrated to the United States and started promoting TRIZ individually. Others found international partners and established TRIZ companies. Today, there are many consultants and firms offering training, consultation, and software tools. Trizjournal.com has become a force in disseminating knowledge and information on the development and application of TRIZ worldwide.

9.3 TRIZ FOUNDATIONS

Contradictions and trade-offs are constraints that often create design issues for which neither QFD nor other tools in the DFSS toolkit provide a means for resolution. Contradiction is a conflict, the result of opposing interests. For this reason, TRIZ is a welcome improvement on past innovation tools.

9.3.1 Overview

When Genrich Altshuller completed his research of the world patent base, he made four key observations:

1. There are five levels of invention:
   Level 5: Discovery of new phenomena
   Level 4: Invention outside a design paradigm requiring new technology from a different field of science
   Level 3: Invention inside a design paradigm that requires the resolution of a physical contradiction
   Level 2: Improvement by invention that requires the resolution of a technical contradiction
   Level 1: Apparent solution (no innovation) results in simple improvement
2. Inventive problems contain at least one contradiction. Altshuller recognized that the same design problem that includes contradiction had been addressed by a number of inventions in different areas of industries. He also observed the repetition of using the same fundamental solutions, often separated by several years. Altshuller concluded that if the later designer had the knowledge of the earlier solution, his/her task would have been simpler. He sought to extract, compile, and organize such information, leading him to observation (3) and observation (4) below.
3. The same principles are used in many inventive designs and can therefore be considered solution patterns. An inventive principle is a best practice that has been used in many applications and has been extracted from several industries. For example, the "nested doll principle" offers the most efficient use of internal cavities of objects like holes and dents to store material. These hollow spaces may enclose objects of interest to the design, a packaging benefit.

Helen Richare, footwear designer from Great Britain, invented a method to help women following a youthful fashion (platform shoes) eliminate carrying a purse. Richard designed a number of models in which the cavity in the shoe is used as a depository for necessary little things (Figure 9.1). The cavity is accessed through a special door in the shoe's heel.

4. There are standard patterns of evolution. To create a product or service, it is necessary to forecast and make analogies to future situations for similar concepts in terms of design functionality. The past evolution of a design is examined and then an analogy is applied to predict the future of the design of interest. For example, when searching for variants of attaching a detachable soap dish to a wall, one may use an analogy with a load-handling crane, as in Figure 9.2.

Exhaustive study of the world's patents reveals that the same principles have been used in innovative solutions to problems in different industries and fields, sometimes with many years elapsing between applications. Access to this information is one of the contributions of TRIZ. For example, Figure 9.3 shows the multiple applications of radio transmission. In 1897, Marconi obtained a patent for radio transmission. In 1901, he transmitted a message across the Atlantic. In 1922, the British Broadcasting Corporation (BBC) was founded. In 1989, 200 patents were issued with radio in their abstract, increasing to 1194 in 2002. Radio waves have been applied to television broadcasting and to the remote controls for televisions (noninfrared types). Wireless telephones, wireless speakers, wireless internet connections, magnetic resonance imaging, and RF ID tags on products for inventory control all demonstrate how one invention can be applied to multiple industries to enable unique solutions.

Defining function is the very basis of the TRIZ methodology. In fact, TRIZ is obsessed with functional definition. For example, the main useful function of a vehicle is "to transport people or load." The vehicle is a technical system that con-

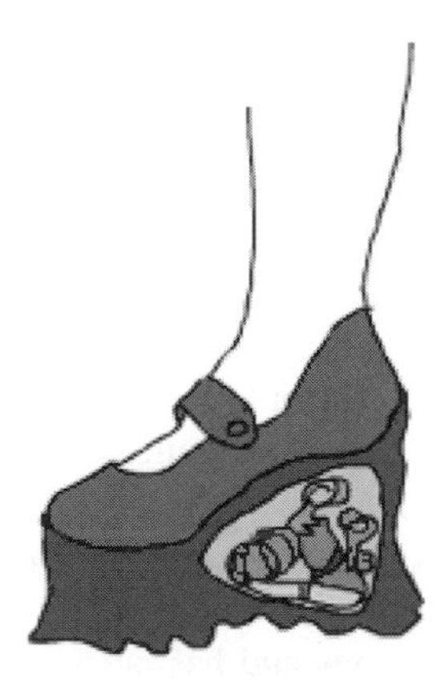

Figure 9.1 Nested doll TRIZ principle applied to shoe design. (From Shpakovsky, N. and Novitskaya, E., 2002, "Multifunctional Heel," downloaded from Generator Website at http://www.gnrtr.com/solutions/en/s040.html.)

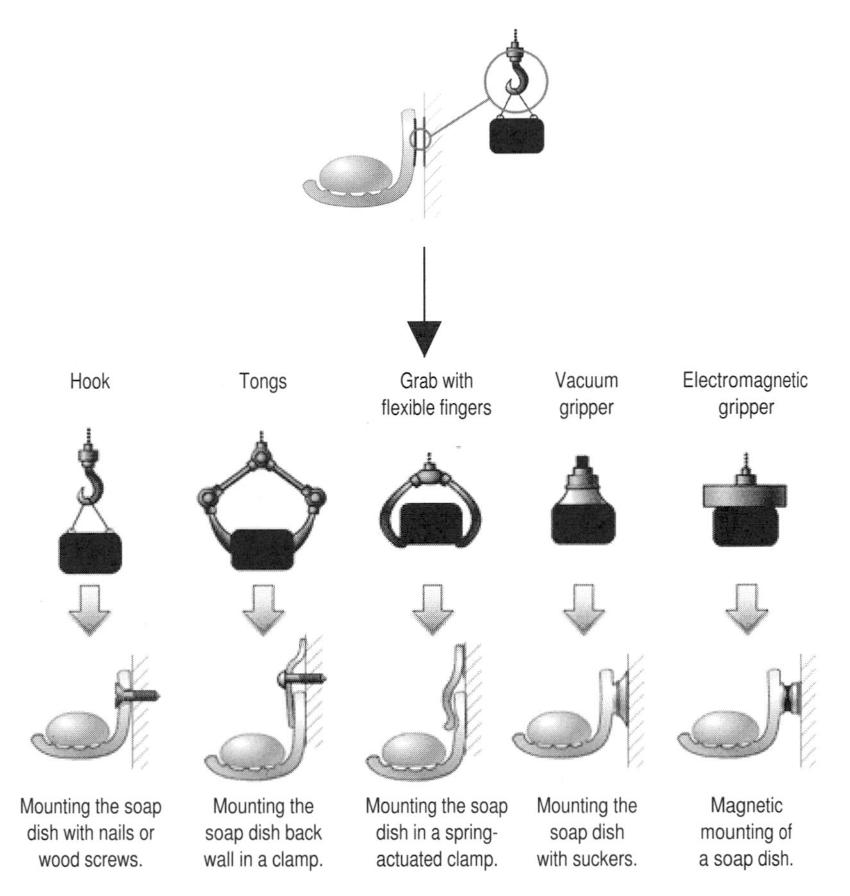

Figure 9.2 Design evolution by analogy. (From Novitskaya, Elena, 2002, "Transformation of structurally similar elements of technical system," downloaded from Generator Website @ http://www.gnrtr.com/tools/en/a09.html.)

tains, among other thousands of parts, two elements: a fuel tank and the fuel itself. The function of the fuel tank is "to enclose liquid." A coffee vending machine performs it's main useful function, "to make and sell coffee." Part of this technical system is a foam cup, which contains the coffee. A cup and coffee are two elements of the design parameters (structure) delivering the function "to enclose liquid." A cup with coffee and a tank with fuel have a similar function and to some degree similar design parameters, "shell and filler" (Figure 9.4). With such a function definition approach, it is possible to analyze a selected system and search for solutions.

A quick survey of our environment reveals that the same patterns of evolution exist in very diverse products and services. One such pattern is the introduction of modified substances. Look at the variety of soft drinks in the store today: regular, diet, cherry, vanilla, and so on. Another pattern is mono-bi-poly, in which a system moves from a mono system such as a fast food restaurant and then moves to a bi

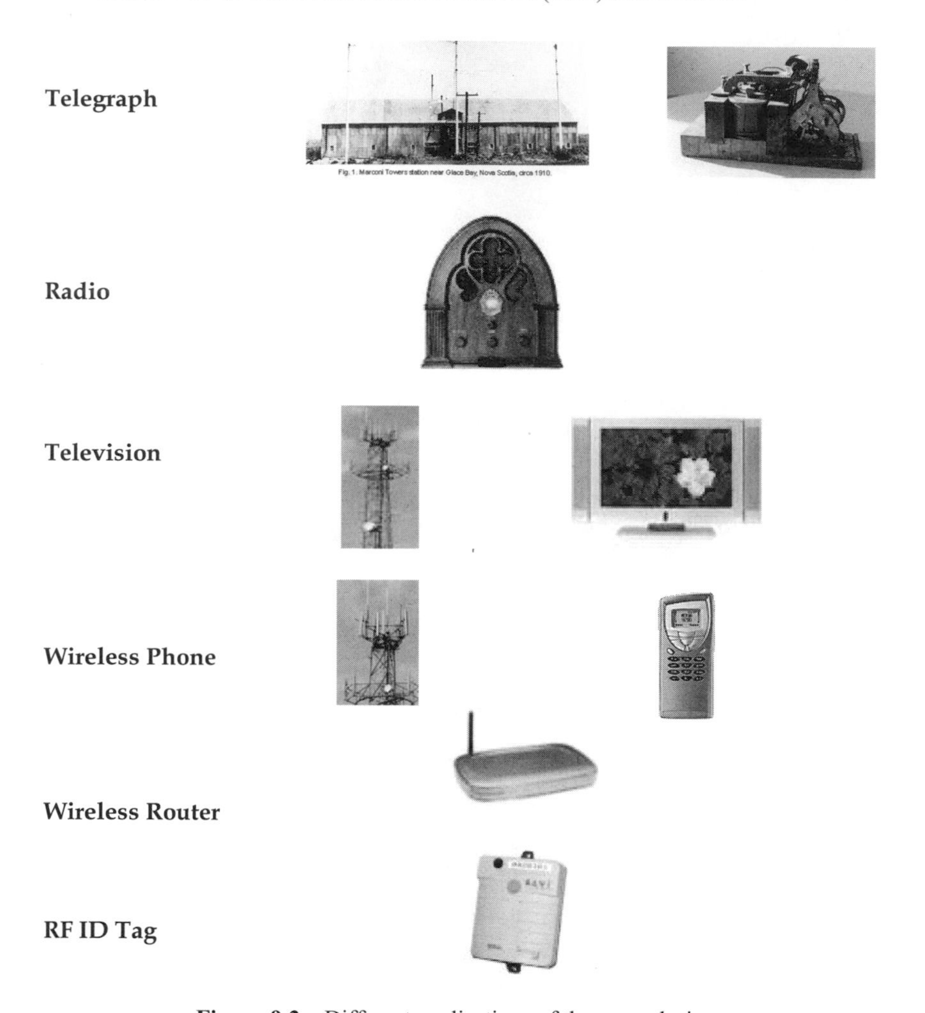

Figure 9.3 Different applications of the same design.

system such as Baskin-Robbins™ and Dunkin Donuts™ colocated in the same building, and then to a poly system and combined poly-system. In the business world, we have observed the evolution from mainframe computers with "batch" processing to distributed personal computers that have the same functionality. We have also observed the need for centralized specialty services such as forms providers and marketing brochures to evolve to distribution via desktop publishing. TRIZ offers eight patterns of evolution containing 280 lines of evolution from which tomorrow's products and processes can be designed today.

Using the TRIZ methodology, it is possible to generate concepts for reducing negative effects and improving the performance of existing designs. TRIZ includes four analytical tools used to structure the innovative problem and six knowledge-based tools used to point in the direction of solution concepts.

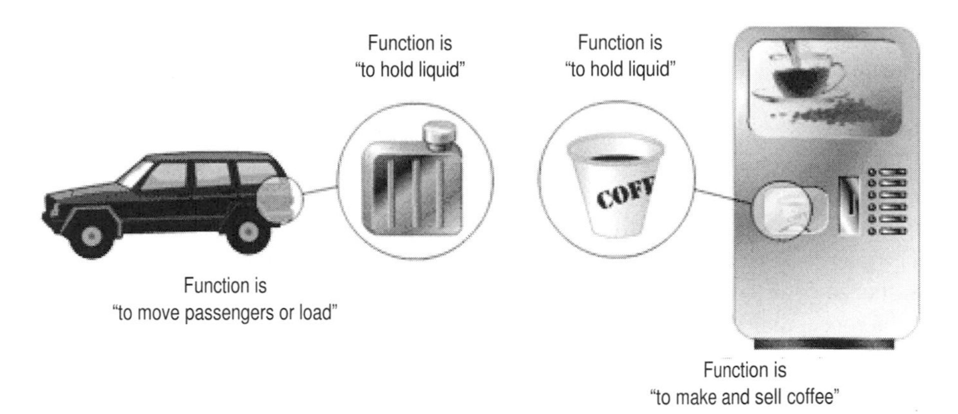

Figure 9.4 Different application with similar functions. (From Novitskaya, Elena, 2002, "Transformation of structurally similar elements of technical system," downloaded from Generator Website @ http://www.gnrtr.com/tools/en/a09.html.)

9.3.2 Analytical Tools

Within TRIZ, there is a set of analytical steps forming a methodology to focus on the correct problem or opportunity. These are listed below:

A1. Clarify the Opportunity—Gathers all the relevant data for analysis and ensures focus on the correct opportunity, not just symptoms.

A2. Function Analysis—Takes a single problem statement and, through the use of linked cause-and-effect statements, generates an exhaustive list of more explicit problems. The objective of functional analysis is simplification, a trend in design development in which the number of the design parameters decreases but can not be less than the number of functional requirements (see Chapter 8). In this case, the cost of the design decreases while its functionality remains within permissible specifications. Trimming, a functional analysis role-based technique, is used to simplify the design mapping.

A3. Algorithm for Inventive Problem Solving (ARIZ)—An alternative way to structure problem definitions for more difficult problems. This is used by experienced TRIZ practioners and requires over 80 hours of training to properly use this process.

A4. Substance-Field Analysis (Su-Field)—Models a problem into three components for breakthrough thinking with regard to system structure and energy sources.

9.3.3 Knowledge-Based Tools

The knowledge-based tools listed here represent the key observations that Altshuller made about improving the efficiency and effectiveness of resolving contradictions and generating inventive breakthrough concepts.

K1. Patterns/Predictions of Evolution. These are descriptions of the sequence of designs possible for a current design. One prediction, for example, describes the evolution of a system from a macro level to a micro level. Examples of this are a full-service bank to an ATM machine to at-home banking, or a hospital to a walk-in clinic to home diagnosis and treatment (e.g., WebMD™). To create a competitive product or service, it is necessary to forecast the future situation for similar functional designs. This is usually done through the method of analogy and exploration. The past evolution of a design functional requirement is examined and plotted on a S-shaped evolution curve, as in Figure 9.5. Then, a conclusion is made about probable conceptual alternatives of its evolution, with proper consideration giving to evolutionary trends and the design parameters differences. TRIZ evolution studies are specific to certain design hierarchies (components, subsystems, or systems) within a design concept. It should be noted that some S-curves describe the evolution of a total system. To predict the evolution of a current design, use an analogy with a specific design element. A specific solution depends on the structure of the design mappings to be transformed, as well as on the parameters to be improved and the available resources.

K2. Inventive Principles and Contradiction Table. Design contradictions between two performance parameters (drawn from a total of 39 parameters; see Table 9.1) may be resolved by using one or more of 40 innovation principles. Successfully used principles for 1201 contradictions are presented in a matrix (see Appendix A). This can be very useful but it is difficult to apply to processes/services due to the strong database of product-related examples. As an example let us look at improving the productivity (parameter 39) of an organization by creating specialists. This

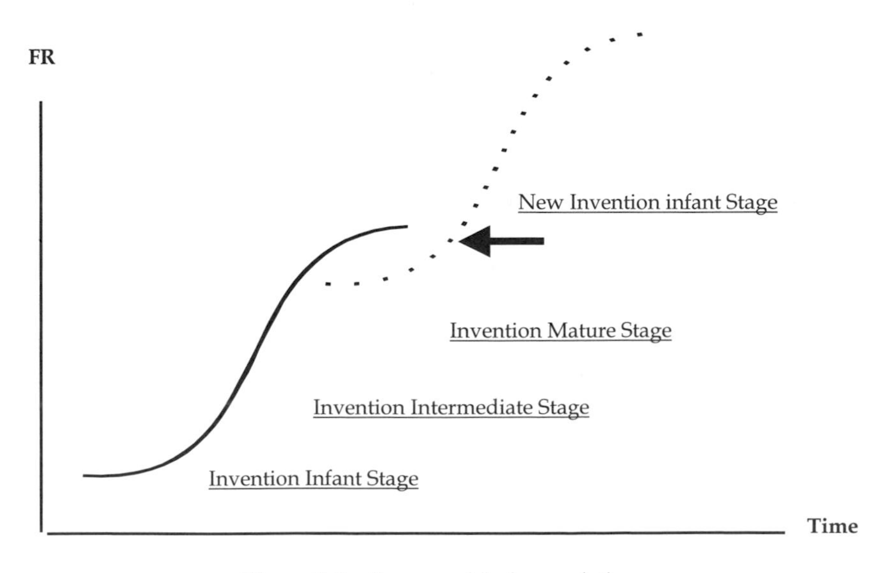

Figure 9.5 S-curve of design evolution.

Table 9.1 TRIZ engineering parameters*

1. Weight of moving object	21. Power
2. Weight of stationary object	22. Energy loss
3. Length of moving object	23. Substance loss
4. Length of stationary object	24. Information loss
5. Area of moving object	25. Waste of time
6. Area of stationary object	26. Quantity of a substance
7. Volume of moving object	27. Reliability
8. Volume of stationary object	28. Accuracy of measurement
9. Velocity	29. Manufacturing precision
10. Force	30. Harmful actions affecting the design object
11. Stress or pressure	31. Harmful actions generated by the design object
12. Shape	32. Manufacturability
13. Stability of object's composition	33. User friendliness
14. Strength	34. Repairability
15. Duration of action generalized by moving object	35. Flexibility
16. Duration of action generalized by stationary object	36. Complexity of design object
17. Temperature	37. Difficulty to control or measure
18. Brightness	38. Level of automation
19. Energy consumed by moving object	39. Productivity
20. Energy consumed by stationary object	

*From Nordlund, M., *An Information Framework for Engineering Design based on Axiomatic Design,* The Royal Institute of Technology, Sweden, 1996; and Ivanov G. I., *Creative Formulas, or How to Invent* (in Russian), M. Prosvenia Publishing, Russia, 1994.

creates a contradiction of loss of strength (parameter 14), in that if one of the specialists is out sick how do you cover for him? If we use the contradiction matrix (Appendix A), and the 40 inventive principles (see Section 9.8) we get possible solutions of

- 10. Preliminary actions
- 18. Mechanical vibration
- 28. Mechanical interaction substitution
- 29. Pneumatics and hydraulics

Using TechOptimizer™ (Figure 9.6) as a tool, we also get several examples for each suggested solution. It is easy to translate these TRIZ suggestions into practical suggestions. Translation of these four follows:

- 10. Preliminary action—ensure that incoming information is accurate and complete, have customers fill in information, have tools and forms available.
- 18. Mechanical vibration—have specialists rotate jobs to maintain breadth of experience.

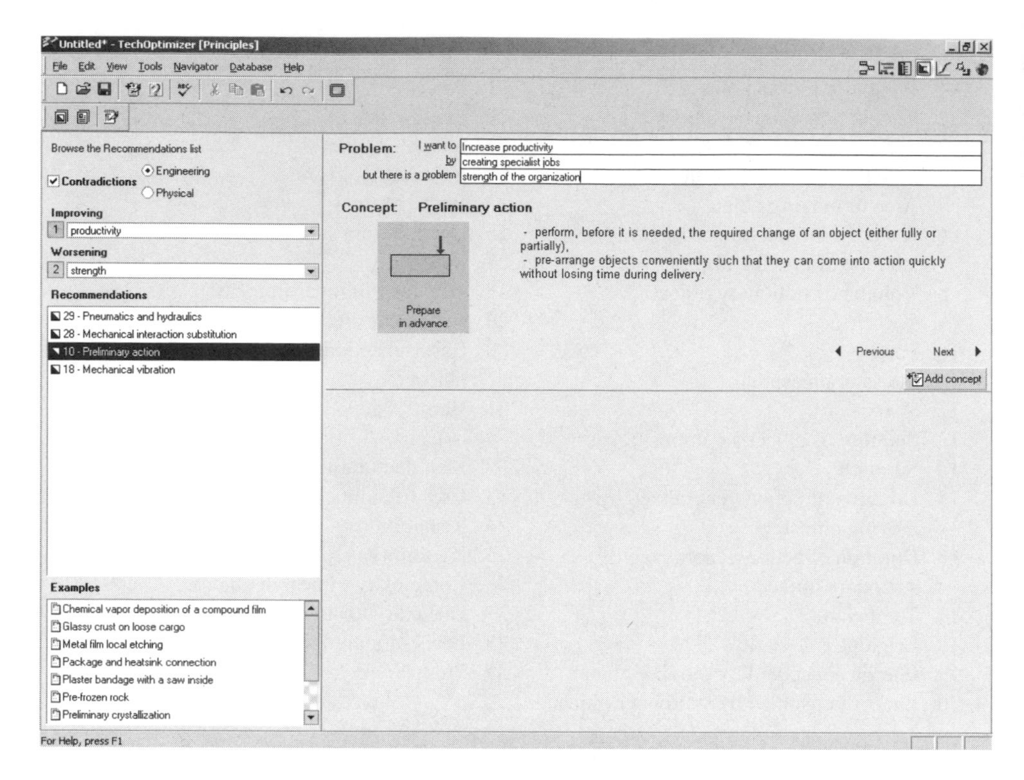

Figure 9.6 TechOptimizer™ contradiction definition screen.

- 28. Mechanical interactions substitution—Use sensory means of detection instead of manual effort. Auto check for completeness of fields, validate Zip Codes, etc.
- 29. Pneumatics and hydraulics—Use voice or data transmission instead of hard copy, use central database (hydraulic reservoir).

K3. Separation Principles. Inherent contradictions, sometimes called physical contradictions, are the simultaneous occurrence of two mutually exclusive conditions that can be resolved using the separation principles. For example, the frying pan has a very simple design with the following design parameters: a bottom metal disc, sides, a handle, and a (possibly) cover. Historically, the pan has been improved by using better materials, for instance, nonstick coatings, or by changing the handle shape as well as the cover design. While frying, food, say, a cutlet, must be turned from time to time. Over the years, an improvement was made by replacing the conventional conical shape of the sides with a toric one. The cutlet is pushed up against the sides and then turns over following the geometry. However, there is a contradiction. It is very difficult to extract the cutlet from the frying pan because of the shape of the sides, but easy to turn the cutlet in such a frying pan. This contradiction can be resolved by the space separation principle. For this purpose, the sides are made traditionally (conical) on one side of the frying pan and toric on the other side, as in Figure 9.7.

K4. 76 Standard Solutions. These are generic system modifications for the model developed using Su-Field Analysis. These solutions can be grouped into five major categories:

1. Improving the systems with no or little change—13 standard solutions
2. Improving the system by changing the system—23 standard solutions
3. System transitions—6 standard solutions
4. Detection and measurement—17 standard solutions
5. Strategies for simplification and improvements—17 standard solutions

For example, in recycling household wastes we used to place all recyclable waste (paper, metal, plastic, and other waste) into a single trash container. Today, many municipalities require separation of waste into specialized categories. This is one of the standard solutions suggested—segmentation of flow into many parts.

K5. Effects. An effect is a physical action of an object that produces a field or another action as a consequence. As a general rule, these are phenomena related to product design. Physical, chemical, geometric, and other effects offer "free" resources commonly forgotten and sometimes even incompatible with the system as designed:

- Material resources
  - System elements
  - Inexpensive materials
  - Modified materials
  - Waste
  - Raw materials

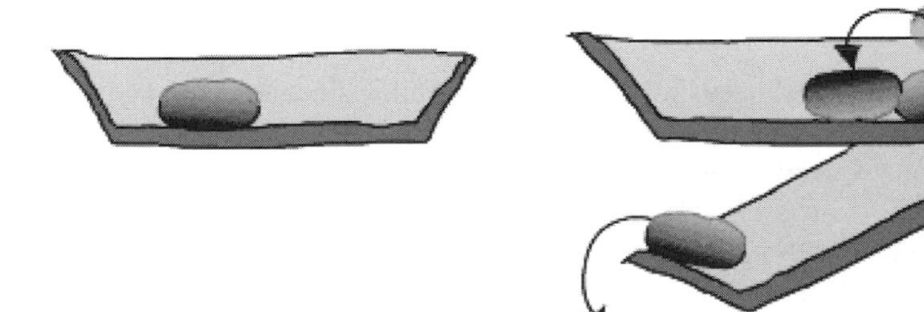

Figure 9.7 TRIZ separation example for frying pan. Left, traditional pan. Right, pan with both conical and toroidal sides. (From Shpakovsky, N. and Novitskaya, E., 2002, "Frying Pan," downloaded from Generator Website at http://www.gnrtr.com/solutions/en/s054.html.)

- Time resources
  - Parallel operations
  - Pre/post work
- Information resources
- Field resources
  - Energy in system
  - Energy in environment
- Space resources
  - Empty space
  - Another dimension
  - Nesting
- Function resources
  - Harmful functions that can be converted to good ones
  - Enhance secondary effects of functions

K6. System of Operators. When analyzing a function model after the clarify the opportunity step, we can approach the model from three perspectives or strategies. These strategies are (1) eliminate harmful effects, (2) eliminate excessive actions, and (3) enhance useful actions. Universal operators are recommendations that are potentially applicable to any situation, such as excessive action (e.g., e-mailing every employee or customer when the message is only relevant to a small subgroup). General operators are recommendations applicable to improving functionality and eliminating undesired effects, such as elimination of a harmful action (e.g., separation or transfer to a subsystem). Specialized operators are used to improve specific parameters or features of a product or process, that is, they enhance useful actions (examples include those for increasing speed, accuracy, and reliability). All of the TRIZ knowledge-based tools yield concepts that require conversion to practical solutions in order to satisfy the needs of the current problem.

9.4 TRIZ PROBLEM SOLVING FLOWCHART

The application of TRIZ needs to follow a logical flow since the methodology can solve many different problems or provide many enhancements within your design project. Figure 9.8 shows one such flow. This flow emulates the structure of the TechOptimizer™ software.

The flow begins with your practical issue and must end with a practical solution or set of solutions. Once you enter the realm of TRIZ, the first step is to clarify the opportunity. In this step, the model is created with true functions, covered in Section 9.6, and the supersystem that interacts with the model. The ideal final result should be formulated at this point to assist in the last step of determining the ideality of any solution arrived at.

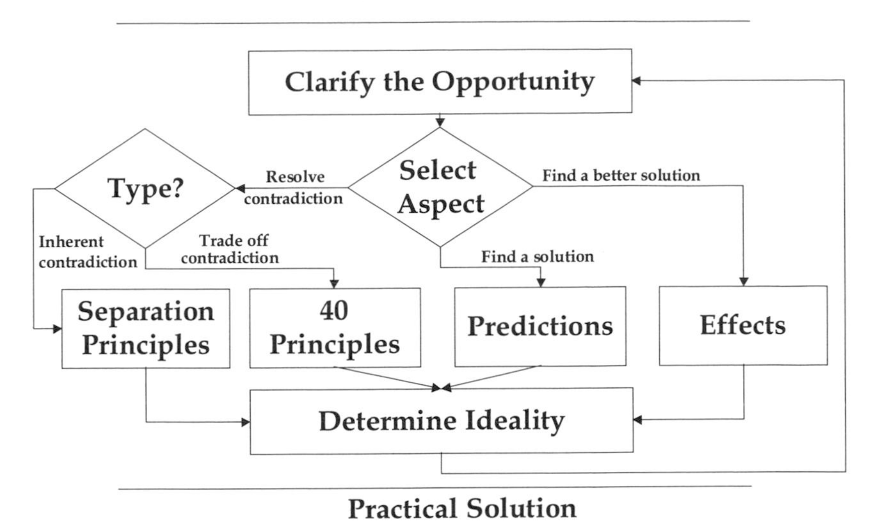

Figure 9.8 TRIZ flowchart.

Next, we look at the model and decide which aspect to resolve. There are three options at this point. We either have contradictions that need to be resolved or we need a solution or a better solution.

If we know we have a contradiction, we need to determine the type of contradiction. It is either a trade-off contradiction or an inherent or design contradiction.

Tradeoff contradictions, sometimes referred to as technical contradictions result when we try to improve one attribute and this leads to deterioration in another attribute. We can often see these contradictions in the roof of the HOQ when we have both a plus sign and a minus sign.

Inherent contradictions, sometimes referred to as physical contradictions, are the more difficult ones to resolve. These contradictions result when the same object is required to have two opposite properties.

If we are just in search of a solution, then we can use the predictions element of the TRIZ patterns/predictions of evolution covered in Section 9.11, but if we need a better solution we can use the TRIZ effects and standard solutions generator.

Each time we apply this methodology and discover a solution to our problem, we need to determine how good the solution is against our standard of the ideal final result, explained in Section 9.5.

9.5 IDEAL FINAL RESULT

The ideal final result (IFR) is the situation that suits us most of all, when the required action is performed by design objects themselves without any additional cost or effort. In solving an inventive problem, the notion of IFR creates a solution ideality that the design team should target. An IFR describes (defines) an ideal system that delivers benefit without harm. It should

- Occupy no space
- Have no weight
- Require no labor
- Take no time
- Require no maintenance, and so on

There are three methods for defining the IFR: itself, ideality checklist, and the ideality equation.

9.5.1 Itself

Using the itself method, teams can look at functions and define them in an ideal state for use in developing concept solutions. In order to use this method, express the IFR as itself, then reexpress it in terms of actual circumstances of the problem. For example, an ideal computer is one that performs its functions without actually existing. There is no such computer, or course, but we can try to approach the ideal. This concept is explored in the following examples.

Example 1. "Grass mows itself" can be reexpressed as "grass keeps itself at an attractive height." How can grass do that? How do other biological systems do it? In the southern parts of the United States, landscapers use grasses such as Zoysia, Bermuda, and St. Augistine, which are slow-growing, creeping-type grasses. They may also spray the grass with a growth inhibitor that slows down the growth rate. So there are two elements used to achieve the close-to-ideal result of "grass keeps itself at an attractive height."

Example 2. "Data enters itself" can be reexpressed as "data stays accurate, current, and in correct form." This example is the basis of many Web-enabled, self-service solutions.

9.5.2 Ideality Checklist

When any concept is created, the design team can test the concept against the following checklist:

1. Eliminates deficiencies of original system
2. Preserves advantages of original system
3. Does not make system more complicated (uses free or available resources)
4. Does not introduce new disadvantages

Anytime there is a "no" answer to the checklist, the concept should be reevaluated for usefulness.

9.5.3 Ideality Equation

With the ideality equation, we define the benefits in terms of useful functions and look at the associated costs as well as any harmful effect or functions that may result from the delivery of the useful function:

$$Ideality = \frac{\Sigma\ Benefits}{\Sigma\ Cost + \Sigma\ Harm} \tag{9.1}$$

The objective is to maintain or maximize the benefits while minimizing or eliminating the cost and harmful effects. The IFR is obtained when the denominator in Equation (9.1) is zero. TRIZ software packages such as TechOptimizer™ automatically create the ideality equation through interactive modeling and dialogue boxes. This method allows for baselining improvement toward the goal of the ideal.

9.6 BUILD SUFFICIENT FUNCTIONS

One of the key requirements of using TRIZ is that we must focus on the correct problem or opportunity. In order to do this, we have to be careful in the choice of words we use to describe the functions that we refer to. Too often, we use high-level descriptions that cover several specific functions. The key to determining if we have a sufficient function is that a function requires an "Object A" that does something to "Object B" that changes Object B somehow. In service design, objects can be design parameters, process variables, or resources in the design environment. The change can be in space, time, or some intrinsic entity such as temperature or knowledge. Take, for example, the statement "the thermometer measures temperature." This in itself is not a true function but a contraction of several functions. One set of functions may be "the body heats the thermometer" or "the thermometer informs the patient." Likewise, consider the statement "find a restaurant." What are the true functions in this statement? If we are performing an Internet search, we enter the search information into an application, the application searches databases, and then informs us of possible restaurants.

9.7 HARMFUL FUNCTION ELIMINATION

In a design, there are two types of functions: useful and harmful or undesirable. Harmful functions lead to bad things like damages and injuries. Consider the air bag system in vehicles. The primary function of an air bag is to protect passengers. Recently, however, air bags have been seriously injuring passengers. Consider the functional diagram in Figure 9.9. In this diagram, the solid line between the two ob-

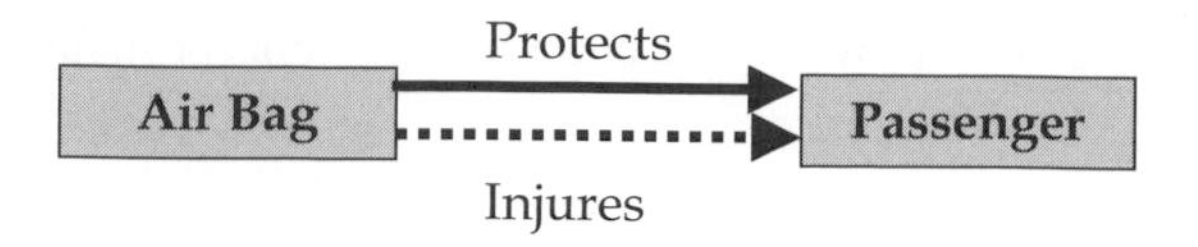

Figure 9.9 Functional diagram example.

jects indicates the useful function called "protects," and the dashed line indicates the harmful function called "injures." The air bag system is a pneumatic device producing a protective force acting between the air bag and the passenger. Useful and harmful functions coexist between the airbag and passenger. Functional analysis can be used to model this design problem. The design problem definition can be reformulated as: an air bag is needed that eliminates air-bag-caused injuries without reducing the level of protection, or adding any harmful functions.

Another example that is more closely related to services happened recently in the labor negotiations between the National Hockey League Owners and the Players Union. The players made a proposal to the owners, who had a subcommittee review the details and produce a report. The report had a useful function but, as it turned out, also had a harmful function in that it made its way to the players through a leak. This functional transaction is illustrated in Figure 9.10.

Altshuller understood the importance of a function approach in design development from the earliest days of TRIZ. For example, his concept of the ideal final result of a design says that the ideal design performs its function but does not exist, which means that it performs its function for free and with no harm. However, the need for integration of function analysis into TRIZ was recognized after developing methods to solve generic problems in innovation. Function analysis plays the key role in problem formulation.

In Figure 9.11, we see a typical flow: purchasing issues a purchase order to a supplier who produces products or services that are handed off to a logistics company for delivery to the plant. The supplier also issues an advanced shipping notice that informs purchasing and the plant as to the delivery. The supplier's hand-off to the logistics provider is an "insufficient function" and could be the focus of an improvement project. Also, directly or indirectly, a supplier could inform the competition, which is a "harmful function." The supplier may tell the competitor directly in casual conversation or indirectly by extending lead times for subsequent orders.

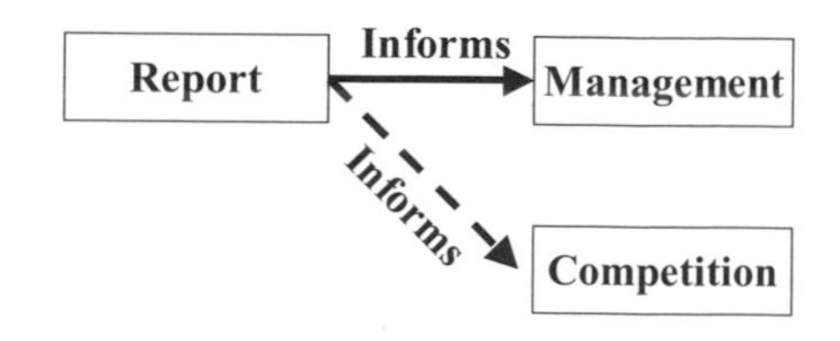

Figure 9.10 Functional transactional diagram example.

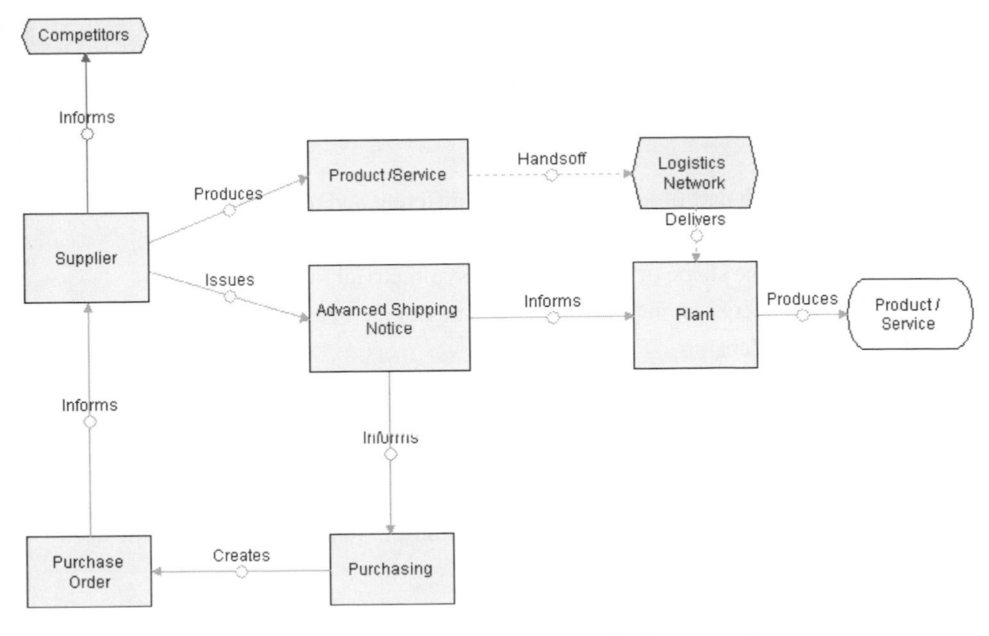

Figure 9.11 Functional diagram purchasing example.

This harmful function is also a potential focus area. By formulating the function model and using the definitions of useful function and harmful function coupled with insufficient or excessive, it is easy to see how this model could help formulate which problems to work on.

9.8 INVENTIVE PRINCIPLES

The 40 inventive principles of TRIZ provide innovators with a systematic and significant means of breaking out of current paradigms into often exciting and beneficial new ones. Although these principles were developed based on analysis of product innovations, the same psychological inertia breakthrough may also occur when the principles are applied in a business application rather than a purely engineering context. The 40 inventive principles are:

Principle 1. Segmentation

A. Divide an object into independent parts.

B. Make an object easy to disassemble.

C. Increase the degree of fragmentation or segmentation.

Principle 2. Taking Out

Separate an interfering part or property from an object, or single out the only a necessary part (or property) of an object.

Principle 3. Local Quality

A. Change an object's structure from uniform to nonuniform; change an external environment (or external influence) from uniform to nonuniform.

B. Make each part of an object function in conditions most suitable for its operation.

C. Make each part of an object fulfill a different and useful function.

Principle 4. Asymmetry

A. Change the shape of an object from symmetrical to asymmetrical.

B. If an object is asymmetrical, change its degree of asymmetry.

Principle 5. Merging

A. Bring closer together (or merge) identical or similar objects; assemble identical or similar parts to perform parallel operations.

B. Make operations contiguous or parallel; bring them together in time.

Principle 6. Universality

A. Make an object or structure perform multiple functions; eliminate the need for other parts.

Principle 7. "Nested Doll"

A. Place one object inside another; place each object, in turn, inside the other.

B. Make one part pass through a cavity in the other.

Principle 8. Anti-Weight

A. To compensate for the weight (downward tendency) of an object, merge it with other objects that provide lift.

B. To compensate for the weight (downward tendency) of an object, make it interact with the environment (e.g., use global lift forces).

Principle 9. Preliminary Anti-Action

A. If it will be necessary to perform an action with both harmful and useful effects, this action should be replaced with anti-actions to control harmful effects.

B. Create beforehand stresses in an object that will oppose known undesirable working stresses later on.

Principle 10. Preliminary Action

A. Perform, before it is needed, the required change of an object (either fully or partially).

B. Prearrange objects such that they can come into action at the most convenient place and without losing time for their delivery.

Principle 11. Beforehand Cushioning

Prepare emergency means beforehand to compensate for the relatively low reliability of an object.

Principle 12. Equipotentiality

In a potential field, limit position changes (e.g., change operating conditions to eliminate the need to raise or lower objects in a gravity field).

Principle 13. "The Other Way Round"

A. Invert the action(s) used to solve the problem (e.g., instead of cooling an object, heat it).

B. Make movable parts (or the external environment) fixed, and fixed parts movable.

C. Turn the object (or process) upside down.

Principle 14. Spheroidality—Curvature

A. Instead of using rectilinear parts, surfaces, or forms, use curvilinear ones; move from flat surfaces to spherical ones; from parts shaped as cubes (parallelepiped) to ball-shaped structures.

B. Use rollers, balls, spirals, domes.

C. Go from linear to rotary motion; use centrifugal forces.

Principle 15. Dynamics

A. Allow (or design) the characteristics of an object, external environment, or process to change to optimal or to find an optimal operating condition.

B. Divide an object into parts capable of movement relative to each other.

C. If an object (or process) is rigid or inflexible, make it movable or adaptive.

Principle 16. Partial or Excessive Actions

A. If 100% of an objective is hard to achieve using a given solution method, then, by using "slightly less" or "slightly more" of the same method, the problem may be made considerably easier to solve.

Principle 17. Another Dimension

A. Move an object in two- or three-dimensional space.

B. Use a multistory arrangement of objects instead of a single-story arrangement.

C. Tilt or reorient the object; lay it on its side.

D. Use another side of a given area.

Principle 18. Mechanical Vibration

A. Cause an object to oscillate or vibrate.

B. Increase its frequency (even up to the ultrasonic).

C. Use an object's resonant frequency.

D. Use piezoelectric vibrators instead of mechanical ones.

E. Use combined ultrasonic and electromagnetic field oscillations. (Use external elements to create oscillation/vibration).

Principle 19. Periodic Action

A. Instead of continuous action, use periodic or pulsating actions.

B. If an action is already periodic, change the periodic magnitude or frequency.

C. Use pauses between impulses to perform a different action.

Principle 20. Continuity of Useful Action

A. Carry on work continuously; make all parts of an object work at full load, all the time.

B. Eliminate all idle or intermittent actions or work.

Principle 21. Skipping

Conduct a process or certain stages (e.g., destructive, harmful, or hazardous operations) at high speed.

Principle 22. "Blessing in Disguise" or "Turn Lemons into Lemonade"

A. Use harmful factors (particularly, harmful effects of the environment or surroundings) to achieve a positive effect.

B. Eliminate the primary harmful action by adding it to another harmful action to resolve the problem.

C. Amplify a harmful factor to such a degree that it is no longer harmful.

Principle 23. Feedback

A. Introduce feedback (referring back, cross-checking) to improve a process or action.

B. If feedback is already used, change its magnitude or influence.

Principle 24. "Intermediary"

A. Use an intermediary carrier article or intermediary process.

B. Merge one object temporarily with another (which can be easily removed).

Principle 25. Self-service

A. Make an object serve itself by performing auxiliary helpful functions

B. Use wasted (or lost) resources, energy, or substances.

Principle 26. Copying

A. Instead of an unavailable, expensive, fragile object, use simpler and inexpensive copies.

B. Replace an object or process with optical copies.

C. If optical copies are used, move to IR or UV (use an appropriate out-of-the-ordinary illumination and viewing situation).

Principle 27. Cheap, Short-Lived Objects

A. Replace an expensive object with a multitude of inexpensive objects, compromising certain qualities (such as service life, for instance).

Principle 28 Mechanical Substitution

A. Replace a mechanical means with a sensory (optical, acoustic, taste, or smell) means.

B. Use electric, magnetic, and electromagnetic fields to interact with the object.

C. Change from static to movable fields, from unstructured fields to those having structure.

D. Use fields in conjunction with field-activated (e.g., ferromagnetic) particles.

Principle 29. Pneumatics and Hydraulics

A. Use gas and liquid parts of an object instead of solid parts (e.g., inflatable, filled with liquids, air cushion, hydrostatic, hydroreactive).

Principle 30. Flexible Shells and Thin Films

A. Use flexible shells and thin films instead of three-dimensional structures.

B. Isolate the object from the external environment using flexible shells and thin films.

Principle 31. Porous Materials

A. Make an object porous or add porous elements (inserts, coatings, etc.).

B. If an object is already porous, use the pores to introduce a useful substance or function.

Principle 32. Color Changes

A. Change the color of an object or its external environment.

B. Change the transparency of an object or its external environment.

Principle 33. Homogeneity

A. Make objects interact with a given object of the same material (or material with identical properties).

Principle 34. Discarding and Recovering

A. Make portions of an object that have fulfilled their functions go away (discard by dissolving, evaporating, etc.) or modify them directly during an operation.

B. Conversely, restore consumable parts of an object directly in an operation.

Principle 35. Parameter Changes

A. Change an object's physical state (e.g., to a gas, liquid, or solid).

B. Change the concentration or consistency.

C. Change the degree of flexibility.

D. Change the temperature.

Principle 36. Phase Transitions

Use phenomena occurring during phase transitions (awareness of macroscale business phenomena).

Principle 37. Thermal Expansion

A. Use thermal expansion (or contraction) of materials.

B. If thermal expansion is being used, use multiple materials with different coefficients of thermal expansion.

Principle 38. Strong Oxidants ("Boosted Interactions")

A. Replace common air with oxygen-enriched air (enriched atmosphere).

B. Replace enriched air with pure oxygen (highly enriched atmosphere).

C. Expose air or oxygen to ionizing radiation.

D. Use ionized oxygen.

E. Replace ozonized (or ionized) oxygen with ozone (atmosphere enriched by "unstable" elements).

Principle 39. Inert Atmosphere

A. Replace a normal environment with an inert one.

B. Add neutral parts or inert additives to an object.

Principle 40. Composite Structures

Change from uniform to composite (multiple) structures (awareness and utilization of combinations of different skills and capabilities).

Now, after reading this list you are probably asking “How does this apply to a services environment?” It is really the same as taking these concepts and thinking about what the analogous situation would be in an organization or service process. There are excellent references at TRIZ_Journal.com, one of which is *40 Inventive (Business) Principles with Examples* provided by Darrell Mann and Dr. Ellen Domb. For example, with Principle 1, Segmentation, Make an object easy to disassemble, we have

- Flexible pensions
- Use of temporary workers on short-term projects
- Flexible manufacturing systems
- Modular furniture/offices
- Container shipment

We think you can see that the extension of these principles will provide a rich set of concepts for your particular issue.

Another great reference (Ruchti & Livotov, 2001) provides us with a subset of translated principles, 12 innovation principles for business and management. The 12 double principles for business and management, assist the user in resolving organizational contradictions and conflicts (Table 9.2). They broaden the individual experiences and intuition of the manager and, in addition, help to quickly formulate several different approaches to difficult situations.

Each principle represents two contradictory lines of action that have to be taken into consideration when searching for solutions. There is no recommendation as to which action is the more suitable. The user is thus stimulated to think in a dialectic and creative way.

Table 9.2 Twelve principles for resolving organizational tasks in business and management

1. Combination–Separation	7. Standardization–Specialization
2. Symmetry–Asymmetry	8. Action—Reaction
3. Homogeneity–Diversity	9. Continuous action–Interrupted action
4. Expansion–Reduction	10. Partial action–Excessive action
5. Mobility–Immovability	11. Direct action–Indirect action
6. Consumption–Regeneration	12. Preliminary action–Preliminary counteraction

9.9 DETECTION AND MEASUREMENT

Although the IFR in a Six Sigma context is automatic control without detection or measurement, many contradictions require enhanced measurement or detection before evolving to this state. Detection will be defined as observing whether something occurred or not, pass or fail, or other binary conditions. Measurement is the ability to measure along a continuous scale with precision and accuracy. The concepts in TRIZ around measurement and detection are divided into three main categories:

1. Introduce Marks

   Into the object. Try to detect the feature parameter of the object by introducing a mark into the object.

   Onto the object. Try to detect the feature parameter of the object by introducing a mark onto the object.

   Into the environment. Try to detect feature parameter of the object by introducing a mark into the surroundings of the object.

2. Introduce Modified Marks

   Into the object. Try to detect the feature parameter of the object by introducing a modified mark into the object.

   Onto the object. Try to detect the feature parameter of the object by introducing a modified mark onto the object.

   Into the environment. Try to detect the feature parameter of the object by introducing modified marks into the surroundings of the object.

3. Roundabout

   Roundabout. Instead of detecting the feature parameter of the object, remove the need for detection.

   Discrete detecting. Instead of continuously measuring the parameter of the object, detect discrete changes(s). Use phase changes, instabilities, or other changes that occur under specific conditions.

   Indirect measuring. Instead of directly measuring the parameter of the object, measure some other feature(s) connected with the parameter.

   Measuring derivatives. Instead of measuring absolute values of the parameter of the object, measure relative changes of the parameter (speed of changes, acceleration, etc.).

Use these concepts to improve or resolve the need to measure or detect. If required, there are numerous sources of examples of applying these concepts, either on the Web or in TRIZ Software Packages.

9.10 TRIZ ROOT CAUSE ANALYSIS

TRIZ also relies on the traditional root cause analysis tools to ensure that the team is working on the root cause and not addressing the symptoms or conducting "science projects." One of the popular methods is to ask "Why?" until you do not know the answer; this is usually the point of impact to focus on. The other method is to ask "Who?," "What?," "When?," "Where?," "Why?," "How many?," or "How large?" This is the the 5W and 2H method.

Other published methods that can help provide focus are the Koepner–Tregoe method or the Global TOPS 8D (Team Oriented Problem Solving) method. Also, the cause and effect diagram also known as the "Ishakawa Diagram," or the "fish bone chart" can help the team focus on the true root causes. Whatever method is chosen, the team should feel confident they are working on the correct opportunity.

9.11 EVOLUTIONARY TRENDS OF TECHNOLOGICAL SYSTEMS

Technological systems are driven by demand life cycles and technology life cycles. You are familiar with the saying, "necessity is the mother of invention." Well, this is a driving force in the life cycles mentioned. When there is demand for solutions, people develop new and innovative designs and the rate of invention follows an S-curve when plotted over time. Early on, only a few inventors create some novel invention. When demand increases, many people cluster around the market and rush to create innovative applications by resolving the contradictions that have not been conquered as of yet. This flurry of activity drives the number of inventions. As demand tapers off, either due to saturation of market demand or a dramatically new technological solution, the pace of invention dwindles. Along this S-curve of evolution lie 19 definite patterns of invention. These patterns are:

1. Introduction: New Substances
    - Internal
    - External
    - In the environment
    - Between objects
2. Introduction: Modified Substances
    - Internal
    - External
    - In the environment
    - Between objects
3. Introduction: Voids
    - Internal
    - External
    - In the environment
    - Between objects
4. Introduction: Fields
    - Internal
    - External

In the environment
Between objects

5. Mono–bi–poly: Similar Objects
   Introduction of similar objects
   Introduction of several similar objects
   Combining similar objects into a common system
6. Mono–bi–poly: Various Objects
   Introduction of various objects
   Introduction of several various objects
   Combining various objects into a common system
7. Segmentation: Substances
   Monolith
   Segmented
   Liquid/powder
   Gas/plasma
   Field
8. Segmentation: Space
   Introduction of a void
   Segmentation of a void
   Creating pores and capillaries
   Activating pores and capillaries
9. Segmentation: Surface
   Flat surface
   Protrusions
   Roughness
   Activating surface
10. Segmentation: Flow
   Into to parts (paths)
   Into several parts (paths)
   Into many parts (paths)
11. Coordination: Dynamism
   Immobile
   Joint (partial mobility)
   Many joints (degrees of freedom)
   Elastic (flexible)
   Liquid/gas (transition to molecular)
   Field (transition to field)
12. Coordination: Rhythm
   Continuous action

Pulsating action
Pulsation in the resonance mode
Several actions
Traveling wave

13. Coordination: Action
    None
    Partial
    Full
    Interval
14. Coordination: Control
    Manual
    Semiautomatic
    Automatic
15. Geometric Evolution: Dimensions
    Point
    Line
    Surface
    Volume
16. Geometric Evolution: Linear
    Line
    Two-dimensional curve
    Three-dimensional curve
    Compound three-dimensional
17. Geometric Evolution: Surface
    Plane
    Single curvature of plane
    Double curvature of plane
    Combined surface
18. Geometric Evolution: Volumetric
    Prism
    Cylindrical
    Spheroidal
    Combined object
19. Trimming
    Complete system
    System with eliminated objects
    Partially trimmed system
    Fully trimmed system

It would be unusual to find a modern-day invention that relied on only one of these patterns. Consider cameras. They started out monolithic, with a single lens. Later, they became hinged and had two lenses, one for near focus and one for far. Today, the compact digital camera has an LCD (gas/plasma) display and a zoom lens (combined poly system), as well as jointed flash. Organizations and processes follow similar patterns. Just think about the paper form, which has evolved to the electronic form, a journey from immobile ink on paper to a field of electrons.

The important application of this topic is that you can actually tell where the solution you select lies along the S-curve. You do not want to be relying on a declining technology, nor do you want to be at the "bleeding edge" of invention.

9.12 TRANSITIONS FROM TRIZ EXAMPLES

Many service-improvement teams trying to use TRIZ will have difficulty finding relevant examples since the TRIZ methodology has evolved from a study of product-based patents and the ratio of product patents to service and process patents is still overwhelming. In order to assist in finding useful knowledge and examples in the TRIZ databases, a tool to transition from product-based examples to other applications is required. In this section, we will introduce a transition tool that is based on the substance-field model of TRIZ. Earlier, we discussed the function model in which Object A acts on Object B. In order for this model to function, Object A has to change something in Object B, implying that energy has to be transferred between the objects. To fulfill the transfer of energy, there are three elements required: (1) source of energy (2) transmission of energy, and (3) guidance and control. A simple transition worksheet (Figure 9.12) is used to create the full transition.

Element	Example Starting	Example Improved	My Problem Starting	My Problem Improved
Object B (acted on)				
Object A (Acting)				
Source of energy				
Transmission of energy				
Guidance and control				

Figure 9.12 Transition worksheet.

Understanding what is the starting point of your problem is the first step. So you fill in the column "My Problem Starting." Next you search the TRIZ knowledge databases for innovations that appear to be similar to what you are looking for. You are searching every field (chemical, electrical, business, psychology, sociology, etc.) for physical and scientific effects. You use the research results to complete the columns "Example Starting" and then "Example Improved." Last, you use the model to complete the transition and fill in the column ""My Problem Improved."

Many times, we see an invention reapplied in a different field; this is the result of discovery and application. Sometimes, this happens because people are in the right place at the right time. What we are proposing here is a methodology that allows for discovery of these effects in order to improve the basic functions that formulate your design. The way to perform this research varies. It may be done through benchmarking or Internet searches. It can be done through patent searches or it can be performed through use of some of the TRIZ software packages available. For instance, in TechOprimizer™ you can use the Effects Module to search for effects. Some of the typical verbs that are found in TechOptimizer™'s Effect Library that allow Object A To "insert verb" into Object B are listed in Table 9.3.

As you can see in Table 9.3, the list is comprehensive but not exhaustive. Another way of saying this is that there are only a few powerful verbs that cover most of the actions that occur between objects. What we do know is that the list is missing some of the basic transactional type of verbs, most importantly, *inform* and *enter*. Figure 9.13 shows a completed example in which the old method of delivering messages goes from land mail to facsimile and then to e-mail. An invoice that was mailed to customers can now be sent electronically. This example is contrived to demonstrate the use of the transition worksheet. In most cases, you will be working a much more elementary (lower-level) function and it will be easier to see the connections.

9.13 SUMMARY

Altshuller was an inventor's inventor and what he has provided designers throughout the world is a priceless tool to help see true design opportunities and provide us

Table 9.3 TechOptimizer™ effects verbs

Absorb	Decrease	Form	Produce
Accumulate	Deposit	Hydrate	Remove
Assemble	Destroy	Increase	Rotate
Bend	Detect	Lift	Separate
Breakdown	Disassemble	Measure	Stabilize
Change	Dry	Mix	Sublimate
Clean	Eliminate	Move	Synthesize
Combine	Embed	Orient	Vibrate
Compress	Evaporate	Phase change	
Condense	Extract	Preserve	
Decompose	Flatten	Prevent	

Creating Transitions in TRIZ				
Element	Example Starting	Example Improved	My Problem Starting	My Problem Improved
Object B (Acted on)	Message	Message	Paper invoice	Electronic invoice
Object A (Acting)	Letter	Fascimile machine	Accounts payable organization	Accounts payable system
Source of energy	Mail person / Mail vehicle	Electrical	Mail delivery	Electrical
Transmission of energy	Mail person / Mail vehicle	Telecommunication system	Invoice is carried along in vehicle	Telecommunication system
Guidance and control	Mail person	Telephone company	Post office	Internet service provider

Figure 9.13 Transition worksheet example.

with principles to resolve, improve, and optimize designs. TRIZ is a useful innovation and problem-solving tool. It mirrors the philosophy of Taguchi in the ideality equation and effective problem-solving methodologies such as TOPS 8D or Kepner–Tregoe upon which it is based. It relies on a set of principles that guide inventive activity by defining contradictions and resolving them, or by removing harmful effects or using predictions to choose superior useful functions. Applying the TRIZ "thinking" tools for inventive problem solving in engineering successfully replaces the trial-and-error method in the search for solutions in the everyday lives of engineers and developers. The majority of organizational management decisions made by the executives and managers, however, are still based on their intuition and personal experience. TRIZ is just as applicable in this environment to provide systematic, innovative, and low-risk solutions. This is part of the reason for the growing demand from management people for systematic and powerful thinking tools that assist executives in processing information and making the right decisions on time.

TRIZ-based thinking for management tasks helps to identify the technology tools that come into play:

- TRIZ tools, such as innovation principles for business and management, as well as separation principles for resolving organizational contradictions and conflicts.
- Operators for revealing and utilizing system resources as a basis for effective and cost-saving decisions.
- Patterns of evolution of technical systems to support optimum selection of the most relevant solution.

APPENDIX 9.A CONTRADICTION MATRIX

The contradiction matrix or table is one of the earliest TRIZ tools to aid designers. It is a 39 × 39 matrix that deals with 1263 common technical contradictions; that is, one parameter improves while the other degrades. The 39 parameters in the rows and columns are the same. They are listed in Section 9.8. The rows represent what needs to be improved, whereas the columns represent the parameters that degrade. In the cell of any (row, column) intersection, the reference number of the applicable inventive principles are listed as given in Section 9.8. For example, in Figure 9A.1, if we want to improve the strength of an object, its weight becomes heavy. To solve this contradiction, inventive principle #40 (composite structure), inventive principle #26 (copying), inventive principle #27 (cheap, short-lived objects), and inventive principle #1 (segmentation) can be used to solve this contradiction.

Using the Altshuller's contradiction matrix is very simple and consists of following these steps:

1. Convert your design problem statement into one of a contradiction between two performance considerations, parameters to design, or requirements.
2. Match these two requirements to any two of the 39 parameters.
3. Look up solution principles to the conflict of these two requirements using the contradiction matrix in this appendix (Figures 9A.2 to 9A.5). That is, the two requirements have numbers associated with them. Look at the corresponding row and column number's cell, which will have a list of numbers in it. These numbers are inventive principle numbers given in Section 9.8.
4. Look up the applicable principles in Section 9.8.
5. Convert these general solution principles into a working solution for your design problem.

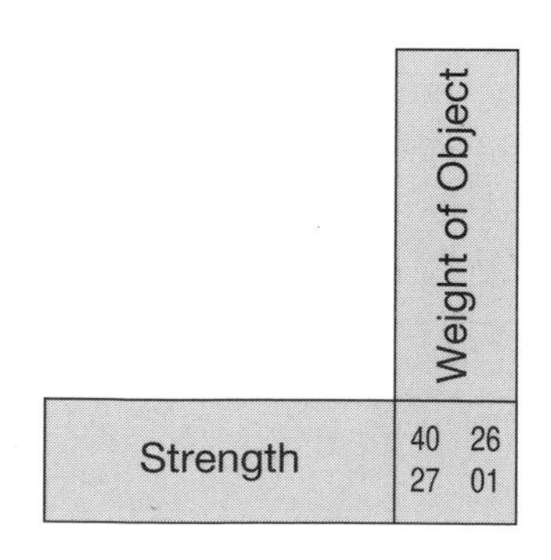

Figure 9A.1.

	1	2	3	4	5	6	7	8	9	10	11	12	13	14	15	16	17	18	19	20
1			15. 8. 29. 34		29. 17. 38. 34		29. 2. 40. 28		2. 8. 15. 38	8. 10. 18. 37	10. 36. 37. 40	10. 14. 35. 40	1. 35. 19. 39	28. 27. 18. 40	5. 34. 31. 35		6. 29. 4. 38	19. 1. 32	35. 12. 34. 31	
2				10. 1. 29. 35		35. 30. 13. 2		5. 35. 14. 2		8. 10. 19. 35	13. 29. 10. 18	13. 10. 29. 14	26. 39. 1. 40	28. 2. 10. 27		2. 27. 19. 6	28. 19. 32. 22	35. 19. 32		18. 19. 28. 1
3	15. 8. 29. 34				15. 17. 4		7. 17. 4. 35		13. 4. 8	17. 10. 4	1. 8. 35	1. 8. 10. 29	1. 8. 15. 34	8. 35. 29. 34	19		10. 15. 19	32	8. 35. 24	
4		35. 28. 40. 29				17. 7. 10. 40		35. 8. 2. 14		28. 1	1. 14. 35	13. 14. 15. 7	39. 37. 35	15. 14. 28. 26		1. 40. 35	3. 35. 38. 18	3. 25		
5	2. 17. 29. 4		14. 15. 18. 4				7. 14. 17. 4		29. 30. 4. 34	19. 30. 35. 2	10. 15. 36. 28	5. 34. 29. 4	11. 2. 13. 39	3. 15. 40. 14	6. 3		2. 15. 16	15. 32. 19. 13	19. 32	
6		30. 2. 14. 18		26. 7. 9. 39						1. 18. 35. 36	10. 15. 36. 37		2. 38	40		2. 10. 19. 30	35. 39. 38			
7	2. 26. 29. 40		1. 7. 35. 4		1. 7. 4. 17				29. 4. 38. 34	15. 35. 36. 37	6. 35. 36. 37	1. 15. 29. 4	28. 10. 1. 39	9. 14. 15. 7	6. 35. 4		34. 39. 10. 18	10. 13. 2	35	
8		35. 10. 19. 14	19. 14	35. 8. 2. 14						2. 18. 37	24. 35	7. 2. 35	34. 28. 35. 40	9. 14. 17. 15		35. 34. 38	35. 6. 4			
9	2. 28. 13. 38		13. 14. 8		29. 30. 34		7. 29. 34			13. 28. 15. 19	6. 18. 38. 40	35. 15. 18. 34	28. 33. 1. 18	8. 3. 26. 14	3. 19. 35. 5		28. 30. 36. 2	10. 13. 19	8. 15. 35. 38	
10	8. 1. 37. 18	18. 13. 1. 28	17. 19. 9. 36	28. 1	19. 10. 15	1. 18. 36. 37	15. 9. 12. 37	2. 36. 18. 37	13. 28. 15. 12		18. 21. 11	10. 35. 40. 34	35. 10. 21	35. 10. 14. 27	19. 2		35. 10. 21		19. 17. 10	1. 16. 36. 37
11	10. 36. 37. 40	13. 29. 10. 18	35. 10. 36	35. 1. 14. 16	10. 15. 36. 28	10. 15. 36. 37	6. 35. 10	35. 34	6. 35. 36	36. 35. 21		35. 4. 15. 10	35. 33. 2. 40	9. 18. 3. 40	19. 3. 27		35. 39. 19. 2		14. 24. 10. 37	
12	8. 10. 29. 40	15. 10. 26. 3	29. 34. 5. 4	13. 14. 10. 7	5. 34. 4. 10		14. 4. 15. 22	7. 2. 35	35. 15. 34. 18	35. 10. 37. 40	34. 15. 10. 14		33. 1. 18. 4	30. 14. 10. 40	14. 26. 9. 25		22. 14. 19. 32	13. 15. 32	2. 6. 34. 14	
13	21. 35. 2. 39	26. 39. 1. 40	13. 15. 1. 28	37	2. 11. 13	39	28. 10. 19. 39	34. 28. 35. 40	33. 15. 28. 18	10. 35. 21. 16	2. 35. 40	22. 1. 18. 4		17. 9. 15	13. 27. 10. 35	39. 3. 35. 23	35. 1. 32	32. 3. 27. 15	13. 19	27. 4. 29. 18
14	1. 8. 40. 15	40. 26. 27. 1	1. 15. 8. 35	15. 14. 28. 26	3. 34. 40. 29	9. 40. 28	10. 15. 14. 7	9. 14. 17. 15	8. 13. 26. 14	10. 18. 3. 14	10. 3. 18. 40	10. 30. 35. 40	13. 17. 35		27. 3. 26		30. 10. 40	35. 19	19. 35. 10	35
15	19. 5. 34. 31		2. 19. 9		3. 17. 19		10. 2. 19. 30		3. 35. 5	19. 2. 16	19. 3. 27	14. 26. 28. 25	13. 3. 35	27. 3. 10			19. 35. 39	2. 19. 4. 35	28. 6. 35. 18	
16		6. 27. 19. 16		1. 40. 35				35. 34. 38					39. 3. 35. 23				19. 18. 36. 40			
17	36. 22. 6. 38	22. 35. 32	15. 19. 9	15. 19. 9	3. 35. 39. 18	35. 38	34. 39. 40. 18	35. 6. 4	2. 28. 36. 30	35. 10. 3. 21	35. 39. 19. 2	14. 22. 19. 32	1. 35. 32	10. 30. 22. 40	19. 13. 39	19. 18. 36. 40		32. 30. 21. 16	19. 15. 3. 17	
18	19. 1. 32	2. 35. 32	19. 32. 16		19. 32. 26		2. 13. 10		10. 13. 19	26. 19. 6		32. 30	32. 3. 27	35. 19	2. 19. 6		32. 35. 19		32. 1. 19	32. 35. 1. 15
19	12. 18. 28. 31		12. 28		15. 19. 25		35. 13. 18		8. 15. 35	16. 26. 21. 2	23. 14. 25	12. 2. 29	19. 13. 17. 24	5. 19. 9. 35	28. 35. 6. 18		19. 24. 3. 14	2. 15. 19		
20		19. 9. 6. 27								36. 37			27. 4. 29. 18	35				19. 2. 35. 32		

Figure 9.A.2 Contradiction matrix 1 through 20 by 1 through 20.

	1	2	3	4	5	6	7	8	9	10	11	12	13	14	15	16	17	18	19	20
21	8. 36. 38. 31	19. 26. 17. 27	1. 10. 35. 37		19. 38	17. 32. 13. 38	35. 6. 38	30. 6. 25	15. 35. 2	26. 2. 36. 35	22. 10. 35	29. 14. 2. 40	35. 32. 15. 31	26. 10. 28	19. 35. 10. 38	16	2. 14. 17. 25	16. 6. 19	16. 6. 19. 37	
22	15. 6. 19. 28	19. 6. 18. 9	7. 2. 6. 13	6. 38. 7	15. 26. 17. 30	17. 7. 30. 18	7. 18. 23	7	16. 35. 38	36. 38			14. 2. 39. 6	26			19. 38. 7	1. 13. 32. 15		
23	35. 6. 23. 40	35. 6. 22. 32	14. 29. 10. 39	10. 28. 24	35. 2. 10. 31	10. 18. 39. 31	1. 29. 30. 36	3. 39. 18. 31	10. 13. 28. 38	14. 15. 18. 40	3. 36. 37. 10	29. 35. 3. 5	2. 14. 30. 40	35. 28. 31. 40	28. 27. 3. 18	27. 16. 18. 38	21. 36. 39. 31	1. 6. 13	35. 18. 24. 5	28. 27. 12. 31
24	10. 24. 35	10. 35. 5	1. 26	26	30. 26	30. 16		2. 22	26. 32						10	10		19		
25	10. 20. 37. 35	10. 20. 26. 5	15. 2. 29	30. 24. 14. 5	26. 4. 5. 16	10. 35. 17. 4	2. 5. 34. 10	35. 16. 32. 18		10. 37. 36. 5	37. 36. 4	4. 10. 34. 17	35. 3. 22. 5	29. 3. 28. 18	20. 10. 28. 18	28. 20. 10. 16	35. 29. 21. 18	1. 19. 21. 17	35. 38. 19. 18	1
26	35. 6. 18. 31	27. 26. 18. 35	29. 14. 35. 18		15. 14. 29	2. 18. 40. 4	15. 20. 29		35. 29. 34. 28	35. 14. 3	10. 36. 14. 3	35. 14	15. 2. 17. 40	14. 35. 34. 10	3. 35. 10. 40	3. 35. 31	3. 17. 39		34. 29. 16. 18	3. 35. 31
27	3. 8. 10. 40	3. 10. 8. 28	15. 9. 14. 4	15. 29. 28. 11	17. 10. 14. 16	32. 35. 40. 4	3. 10. 14. 24	2. 35. 24	21. 35. 11. 28	8. 28. 10. 3	10. 24. 35. 19	35. 1. 16. 11		11. 28	2. 35. 3. 25	34. 27. 6. 40	3. 35. 10	11. 32. 13	21. 11. 27. 19	36. 23
28	32. 35. 26. 28	28. 35. 25. 26	28. 26. 5. 16	32. 28. 3. 16	26. 28. 32. 3	26. 28. 32. 3	32. 13. 6		28. 13. 32. 24	32. 2	6. 28. 32	6. 28. 32	32. 35. 13	28. 6. 32	28. 6. 32	10. 26. 24	6. 19. 28. 24	6. 1. 32	3. 6. 32	
29	28. 32. 13. 18	28. 35. 27. 9	10. 28. 29. 37	2. 32. 10	28. 33. 29. 32	2. 29. 18. 36	32. 28. 2	25. 10. 35	10. 28. 32	28. 19. 34. 36	3. 35	32. 30. 40	30. 18	3. 27	3. 27. 40		19. 26	3. 32	32. 2	
30	22. 21. 27. 39	2. 22. 13. 24	17. 1. 39. 4	1. 18	22. 1. 33. 28	27. 2. 39. 35	22. 23. 37. 35	34. 39. 19. 27	21. 22. 35. 28	13. 35. 39. 18	22. 2. 37	22. 1. 3. 35	35. 24. 30. 18	18. 35. 37. 1	22. 15. 33. 28	17. 1. 40. 33	22. 33. 35. 2	1. 19. 32. 13	1. 24. 6. 27	10. 2. 22. 37
31	19. 22. 15. 39	35. 22. 1. 39	17. 15. 16. 22		17. 2. 18. 39	22. 1. 40	17. 2. 40	30. 18. 35. 4	35. 28. 3. 23	35. 28. 1. 40	2. 33. 27. 18	35. 1	35. 40. 27. 39	15. 35. 22. 2	15. 22. 33. 31	21. 39. 16. 22	22. 35. 2. 24	19. 24. 39. 32	2. 35. 6	19. 22. 18
32	28. 29. 15. 16	1. 27. 36. 13	1. 29. 13. 17	15. 17. 27	13. 1. 26. 12	16. 4	13. 29. 1. 40	35	35. 13. 8. 1	35. 12	35. 19. 1. 37	1. 28. 13. 27	11. 13. 1	1. 3. 10. 32	27. 1. 4	35. 16	27. 26. 18	28. 24. 27. 1	28. 26. 27. 1	1. 4
33	25. 2. 13. 15	6. 13. 1. 25	1. 17. 13. 12		1. 17. 13. 16	18. 16. 15. 39	1. 16. 35. 15	4. 18. 31. 39	18. 13. 34	28. 13. 35	2. 32. 12	15. 34. 29. 28	32. 35. 30	32. 40. 3. 28	29. 3. 8. 25	1. 16. 25	26327. 13	13. 17. 1. 24	1. 13. 24	
34	2. 27. 35. 11	2. 27. 35. 11	1. 28. 10. 25	3. 18. 31	15. 32. 13	16. 25	25. 2. 35. 11	1	34. 9	1. 11. 10	13	1. 13. 2. 4	2. 35	1. 11. 2. 9	11. 29. 28. 27	1	4. 10	15. 1. 13	15. 1. 28. 16	
35	1. 6. 15. 8	19. 15. 29. 16	35. 1. 29. 2	1. 35. 16	35. 30. 29. 7	15. 16	15. 35. 29		35. 10. 14	15. 17. 20	35. 16	15. 37. 1. 8	35. 30. 14	35. 3. 32. 6	13. 1. 35	2. 16	27. 2. 3. 35	6. 22. 26. 1	19. 35. 29. 13	
36	26. 30. 34. 36	2. 26. 35. 39	1. 19. 26. 24	26	14. 1. 13. 16	6. 36	34. 26. 6	1. 16	34. 10. 28	26. 16	19. 1. 35	29. 13. 28. 15	2. 22. 17. 19	2. 13. 28	10. 4. 28. 15		2. 17. 13	24. 17. 13	27. 2. 29. 28	
37	27. 26. 28. 13	6. 13. 28. 1	16. 17. 26. 24	26	2. 13. 18. 17	2. 39. 30. 16	29. 1. 4. 16	2. 18. 26. 31	3. 4. 16. 35	36. 28. 40. 19	35. 36. 37. 32	27. 13. 1. 39	11. 22. 39. 30	27. 3. 15. 28	19. 29. 25. 39	25. 34. 6. 35	3. 27. 35. 16	2. 24. 26	35. 38	19. 35. 16
38	28. 26. 18. 35	28. 26. 35. 10	14. 13. 28. 17	23	17. 14. 13		35. 13. 16		28. 10	2. 35	13. 35	15. 32. 1. 13	18. 1	25. 13	6. 9		26. 2. 19	8. 32. 19	2. 32. 13	
39	35. 26. 24. 37	28. 27. 15. 3	18. 4. 28. 38	30. 7. 14. 26	10. 26. 34. 31	10. 35. 17. 7	2. 6. 34. 10	35. 37. 10. 2		28. 15. 10. 36	10. 37. 14	14. 10. 34. 40	35. 3. 22. 39	29. 28. 10. 18	35. 10. 2. 18	20. 10. 16. 38	35. 21. 28. 10	26. 17. 19. 1	35. 10. 38. 19	1

Figure 9.A.3 Contradiction matrix 21 through 39 by 1 through 20.

	21	22	23	24	25	26	27	28	29	30	31	32	33	34	35	36	37	38	39
1	12. 36. 18. 31	6. 2. 34. 19	5. 35. 3. 31	10. 24. 35	10. 35. 20. 28	3. 26. 18. 31	3. 11. 1. 27	28. 27. 35. 26	28. 35. 26. 18	22. 21. 18. 27	22. 35. 31. 39	27. 28. 1. 36	35. 3. 2. 24	2. 27. 28. 11	29. 5. 15. 8	26. 30. 36. 34	28. 29. 26. 32	26. 35. 18. 19	35. 3. 24. 37
2	15. 19. 18. 22	18. 19. 28. 15	5. 8. 13. 30	10. 15. 35	10. 20. 35. 26	19. 6. 18. 26	10. 28. 8. 3	18. 26. 28	10. 1. 35. 17	2. 19. 22. 37	35. 22. 1. 39	28. 1. 9	6. 13. 1. 32	2. 27. 28. 11	19. 15. 29	1. 10. 26. 39	25. 28. 17. 15	2. 26. 35	1. 28. 15. 35
3	1. 35	7. 2. 35. 39	4. 29. 23. 10	1. 24	15. 2. 29	29. 35	10. 14. 29. 40	28. 32. 4	10. 28. 29. 37	1. 15. 17. 24	17. 15	1. 29. 17	15. 29. 35. 4	1. 28. 10	14. 15. 1. 16	1. 19. 26. 24	35. 1. 26. 24	17. 24. 26. 16	14. 4. 28. 29
4	12. 8	6. 28	10. 28. 24. 35	24. 26	30. 29. 14		15. 29. 28	32. 28. 3	2. 32. 10	1. 18		15. 17. 27	2. 25	3	1. 35	1. 26	26		30. 14. 7. 26
5	19. 10. 32. 18	15. 17. 30. 26	10. 35. 2. 39	30. 26	26. 4	29. 30. 6. 13	29. 9	26. 28. 32. 3	2. 32	22. 33. 28. 1	17. 2. 18. 39	13. 1. 26. 24	15. 17. 1316	15. 13. 10. 1	15. 30	14. 1. 13	2. 36. 26. 18	14. 30. 28. 23	10. 26. 34. 2
6	17. 32	17. 7. 30	10. 14. 18. 39	30. 16	10. 35. 4. 18	2. 18. 40. 4	32. 35. 40. 4	26. 28. 32. 3	2. 29. 18. 36	27. 2. 39. 35	22. 1. 40	40. 16	16. 4	16	15. 16	1. 18. 36	2. 35. 30. 18	23	10. 156. 17. 7
7	35. 6. 13. 18	7. 15. 13. 16	36. 39. 34. 10	2. 22	2. 6. 34. 10	29. 30. 7	14. 1. 40. 11	25. 26. 28	25. 28. 2. 16	22. 21. 27. 35	17. 2. 40. 1	29. 1. 40	15. 13. 30. 12	10	15. 29	26. 1	29. 26. 4	35. 34. 16. 24	10. 6. 2. 34
8	30. 6		10. 39. 35. 34		35. 16. 32. 18	35. 3	2. 35. 16		35. 10. 25	34. 39. 19. 27	30. 18. 35. 4	35		1		1. 31	2. 17. 26		35. 37. 10. 2
9	19. 35. 38. 2	14. 20. 19. 35	10. 13. 28. 38	13. 26		10. 19. 29. 38	11. 35. 27. 28	28. 32. 1. 24	10. 28. 32. 25	1. 28. 35. 23	2. 24. 32. 21	35. 13. 8. 1	32. 28. 13. 12	34. 2. 28. 27	15. 10. 26	10. 28. 4. 34	3. 34. 27. 16	10. 18	
10	19. 35. 18. 37	14. 15	8. 35. 40. 5		10. 37. 36	14. 29. 18. 36	3. 35. 13. 21	35. 10. 23. 24	28. 29. 37. 36	1. 35. 40. 18	13. 3. 36. 24	15. 37. 18. 1	1. 28. 3. 25	15. 1. 11	15. 17. 18. 20	26. 35. 10. 18	36. 37. 10. 19	2. 35	3. 28. 35. 37
11	10. 35. 14	2. 36. 25	10. 36. 37		37. 36. 4	10. 14. 36	10. 13. 19. 35	6. 28. 25	3. 35	22. 2. 37	2. 33. 27. 18	1. 35. 16	11	2	35	19. 1. 35	2. 36. 37	35. 24	10. 14. 35. 37
12	4. 6. 2	14	35. 29. 3. 5		14. 10. 34. 17	36. 22	10. 40. 16	28. 32. 1	32. 30. 40	22. 1. 2. 35	35. 1	1. 32. 17. 28	32. 15. 26	2. 13. 1	1. 15. 29	16. 29. 1. 28	15. 13. 39	15. 1. 32	17. 26. 34. 10
13	32. 35. 27. 31	14. 2. 39. 6	2. 14. 30. 40		35. 27	15. 32. 35		13	18	35. 23. 18. 30	35. 40. 27. 39	35. 19	32. 35. 30	2. 35. 10. 16	35. 30. 34. 2	2. 35. 22. 26	35. 22. 39. 23	1. 8. 35	23. 35. 40. 3
14	10. 26. 35. 28	35	35. 28. 31. 40		29. 3. 28. 10	29. 10. 27	11. 3	3. 27. 16	3. 27	18. 35. 37. 1	15. 35. 22. 2	11. 3. 10. 32	32. 40. 28. 2	27. 11. 3	15. 3. 32	2. 13. 28	27. 3. 15. 40	15	29. 35. 10. 14
15	19. 10. 35. 38		28. 27. 3. 18	10	20. 10. 28. 18	3. 35. 10. 40	11. 2. 13	3	3. 27. 16. 40	22. 15. 33. 28	21. 39. 16. 22	27. 1. 4	12. 27	29. 10. 27	1. 35. 13	10. 4. 29. 15	19. 29. 39. 35	6. 10	35. 17. 14. 19
16	16		27. 16. 18. 38	10	28. 20. 10. 16	3. 35. 31	34. 27. 6. 40	10. 26. 24		17. 1. 40. 33	22	35. 10	1	1	2		25. 34. 6. 35	1	20. 10. 16. 38
17	2. 14. 17. 25	21. 17. 35. 38	21. 36. 29. 31		35. 28. 21. 18	3. 17. 30. 39	19. 35. 3. 10	32. 19. 24	24	22. 33. 35. 2	22. 35. 2. 24	26. 27	26. 27	4. 10. 16	2. 18. 27	2. 17. 16	3. 27. 35. 31	23. 2. 19. 16	15. 28. 35
18	32	19. 16. 1. 6	13. 1	1. 6	19. 1. 26. 17	1. 19		11. 15. 32	3. 32	15. 19	35. 19. 32. 39	19. 35. 28. 26	28. 26. 19	15. 17. 13. 16	15. 1. 19	6. 32. 13	32. 15	2. 26. 10	2. 25. 16
19	6. 19. 37. 18	12. 22. 15. 24	35. 24. 18. 5		35. 38. 19. 18	34. 23. 16. 18	19. 21. 11. 27	3. 1. 32		1. 35. 6. 27	2. 35. 6	28. 26. 30	19. 35	1. 15. 17. 28	15. 17. 13. 16	2. 29. 27. 28	35. 38	32. 2	12. 28. 35
20			28. 27. 18. 31			3. 35. 31	10. 36. 23			10. 2. 22. 37	19. 22. 18	1. 4					19. 35. 16. 25		1. 6

Figure 9.A.4 Contradiction matrix 1 through 20 by 21 through 39.

	21	22	23	24	25	26	27	28	29	30	31	32	33	34	35	36	37	38	39
21		10. 35. 38	28. 27. 18. 38	10. 19	35. 20. 10. 6	4. 34. 19	19. 24. 26. 31	32. 15. 2	32. 2	19. 22. 31. 2	2. 35. 18	26. 10. 34	26. 35. 10	35. 2. 10. 34	19. 17. 34	20. 19. 30. 34	19. 35. 16	28. 2. 17	28. 35. 34
22	3. 38		35. 27. 2. 37	19. 10	10. 18. 32. 7	7. 18. 25	11. 10. 35	32		21. 22. 35. 2	21. 35. 2. 22		35. 32. 1	2. 19		7. 23	35. 3. 15. 23	2	28. 10. 29. 35
23	28. 27. 18. 38	35. 27. 2. 31			15. 18. 35. 10	6. 3. 10. 24	10. 29. 39. 35	16. 34. 31. 28	35. 10. 24. 31	33. 22. 30. 10	10. 1. 34. 29	15. 34. 33	32. 28. 2. 24	2. 35. 34. 27	15. 10. 2	35. 10. 28. 24	35. 18. 10. 13	35. 10. 18	28. 35. 10. 23
24	10. 19	19. 10			24. 26. 28. 32	24. 28. 35	10. 28. 23			22. 10. 1	10. 21. 22	32	27. 22				35. 33	35	13. 23. 15
25	35. 20. 10. 6	10. 5. 18. 32	35. 18. 10. 39	24. 26. 28. 32		35. 38. 18. 16	10. 30. 4	24. 34. 28. 32	24. 26. 28. 18	35. 18. 34	35. 22. 18. 39	35. 28. 34. 4	4. 28. 10. 34	32. 1. 10	35. 28	6. 29	18. 28. 32. 10	24. 28. 35. 30	
26	35	7. 18. 25	6. 3. 10. 24	24. 28. 35	35. 38. 18316		18. 3. 28. 40	3. 2. 28	33. 30	35. 33. 29. 31	3. 35. 40. 39	29. 1. 35. 27	35. 29. 10. 25	2. 32. 10. 25	15. 3. 29	3. 13. 27. 10	3. 27. 29. 18	8. 35	13. 29. 3. 27
27	21. 11. 26. 31	10. 11. 35	10. 35. 29. 39	10. 28	10. 30. 4	21. 28. 40. 3		32. 3. 11. 23	11. 32. 1	27. 35. 2. 40	35. 2. 40. 26		27. 17. 40	1. 11	13. 35. 8. 24	13. 35. 1	27. 40. 28	11. 13. 27	1. 35. 29. 38
28	3. 6. 32	26. 32. 27	10. 16. 31. 28		24. 34. 28. 32	2. 6. 32	5. 11. 1. 23			28. 24. 22. 26	3. 33. 39. 10	6. 35. 25. 18	1. 13. 17. 34	1. 32. 13. 11	13. 35. 2	27. 35. 10. 34	26. 24. 32. 28	28. 2. 10. 34	10. 34. 28. 32
29	32. 2	13. 32. 2	35. 31. 10. 24		32. 26. 28. 18	32. 30	11. 32. 1			26. 28. 10. 36	4. 17. 34. 26		1. 32. 35. 23	25. 10		26. 2. 18		26. 28. 18. 23	10. 18. 32. 39
30	19. 22. 31. 2	21. 22. 35. 2	33. 22. 19. 40	22. 10. 2	35. 18. 34	35. 33. 29. 31	27. 24. 2. 40	28. 33. 23. 26	26. 28. 10. 18			24. 35. 2	2. 25. 28. 39	35. 10. 2	35. 11. 22. 31	22. 19. 29. 40	22. 19. 29. 40	33. 3. 34	22. 35. 13. 24
31	2. 35. 18	21. 35. 22. 2	10. 1. 34	10. 21. 29	1. 22	3. 24. 39. 1	24. 2. 40. 39	3. 33. 26	4. 17. 34. 26							19. 1. 31	2. 21. 27. 1	2	22. 35. 18. 39
32	27. 1. 12. 24	19. 35	15. 34. 33	32. 24. 18. 16	35. 28. 34. 4	35. 23. 1. 24		1. 35. 12. 18		24. 2			2. 5. 13. 16	35. 1. 11. 9	2. 13. 15	27. 26. 1	6. 28. 11. 1	8. 28. 1	35. 1. 10. 28
33	35. 34. 2. 10	2. 19. 13	28. 32. 2. 24	4. 10. 27. 22	4. 28. 10. 34	12. 35	17. 27. 8. 40	25. 13. 2. 34	1. 32. 35. 23	2. 25. 28. 39		2. 5. 12		12. 26. 1. 32	15. 34. 1. 16	32. 25. 12. 17		1. 34. 12. 3	15. 1. 28
34	15. 10. 32. 2	15. 1. 32. 19	2. 35. 34. 27		32. 1. 10. 25	2. 28. 10. 25	11. 10. 1. 16	10. 2. 13	25. 10	35. 102. 16		1. 35. 11. 10	1. 12. 26. 15		7. 1. 4. 16	35. 1. 13. 11		34. 35. 7. 13	1. 32. 10
35	19. 1. 29	18. 15. 1	15. 10. 2. 13		35. 28	3. 35. 15	35. 13. 8. 24	35. 5. 1. 10		35. 11. 32. 31		1. 13. 31	15. 34. 1. 16	1. 16. 7. 4		15. 29. 37. 28	1	27. 34. 35	35. 28. 6. 37
36	20. 19. 30. 34	10. 35. 13. 2	35. 10. 28. 29		6. 29	13. 3. 27. 10	13. 35. 1	2. 26. 10. 34	26. 24. 32	22. 19. 29. 40	19. 1	27. 26. 1. 13	27. 9. 26. 24	1. 13	29. 15. 28. 37		15. 10. 37. 28	15. 1. 24	12. 17. 28
37	19. 1. 16. 10	35. 3. 15. 19	1. 18. 10. 24	35. 33. 27. 22	18. 28. 32. 9	3. 27. 29. 18	27. 40. 28. 8	26. 24. 32. 28		22. 19. 29. 28	2. 21	5. 28. 11. 29	2. 5	12. 26	1. 15	15. 10. 37. 28		34. 21	35. 18
38	28. 2. 27	23. 28	35. 10. 18. 5	35. 33	24. 28. 35. 30	35. 13	11. 27. 32	28. 26. 10. 34	28. 26. 18. 23	2. 33	2	1. 26. 13	1. 12. 34. 3	1. 35. 13	27. 4. 1. 35	15. 24. 10	34. 27. 25		5. 12. 35. 26
39	35. 20. 10	28. 10. 29. 35	28. 10. 35. 23	13. 15. 23		35. 38	1. 35. 10. 38	1. 10. 34. 28	32. 1. 18. 10	22. 35. 13. 24	35. 22. 18. 39	35. 28. 2. 24	1. 28. 7. 19	1. 32. 10. 25	1. 35. 28. 37	12. 17. 28. 24	35. 18. 27. 2	5. 12. 35. 26	

Figure 9.A.5 Contradiction matrix 21 through 39 by 21 through 39.

10

INTRODUCTION TO DESIGN FOR TRANSACTIONAL DFX

10.1 INTRODUCTION

As we realigned our design efforts to focus on the voice of the customer and stakeholder, and strive for innovative designs through concurrent efforts, it is inevitable that some overlaps and gaps in total design quality will occur. Without strict design standards and seamless sharing between separate subsystem design teams, there is a need to step back and think about the solutions we have chosen and test them through a set of commonsense filters. These filters ask simple questions such as, "Have we maximized commonality?", "How large a solution do I require for today? What about six months from now?", "What will the reliability/availability of the process be?"

DFX or Design for X provides answers to questions such as these. Although this concept originated in the product environment, it has several useful applications in a process/service environment. In this chapter, we will introduce to you the history of DFX and its common product solutions, and show you the relevant applications to services.

It is important to understand the subtle difference between services for products and service DFX. A car requires service such as fueling, oil changes, and tune-ups. When designing the car, these service aspects should be taken into consideration to allow for "serviceability." Consider fueling a car in which you have to open the back door and fold the back seat up to access the fuel tank. This is not a very logical design, would you not agree? Think about the things that you have to service during the course of work: disk clean-up, pen refills, change ink/toner cartridges. What tools or equipment do you use in the course of your job: telephones, computers, data storage. What would happen if your phone needed to be maintained after every 10 phone calls? Service DFX, we will see, is dependent on these aspects of product DFX as well as some service process extensions of these applications.

The majority of service processes depend on some product infrastructure to support their operation, much as a restaurant requires a building and kitchen equipment

Service Design for Six Sigma. By Basem El-Haik and David M. Roy

to enable its service provisioning. Many functional organizations require buildings and office equipment such as a personal computers and telephones as well as computer applications to provide their services. These items are classified as infrastructure in a service environment. The service itself is made up of roles and responsibilities, procedures, methods, training, and other "soft" items. It should be evident that the infrastructure elements are a direct link to the tangible product environment and all of the traditional DFX applications can be directly applied as relevant. The rest of the service elements can benefit from a services extension of the DFX concepts. For additional clarification let us look at the evolution of a purchasing department in Table 10.1. We can see that in the days of only postal delivery, the infrastructure was made up of manual elements such as typewriters and forms. As telecommunications evolved, we began to rely on the facsimile machine and there was a slight change in the infrastructure. Further evolution of enabling technologies has provided a faster, highly automated infrastructure. As the design elements change over the evolution of the environment, the type of DFX application changes also.

Another aspect of a service environment is the continuum from purely product-related services, which we are most familiar with and interact with almost daily, extending down to a purely transactional service. In between these extremes are the internal process services that are often overlooked. Table 10.2 provides some examples of each of these categories.

Design for X is a family of approaches generally denoted as Design for X or DFX[1] for short. The letter "X" in DFX is made up of life cycle processes (x). In product design, for example, one of the first members of the DFX family is DFA, or Design for Assembly. DFSS teams with service projects can greatly benefit from this chapter by drawing analogies between their processes and/or services and the topics presented here, in particular, Design for Processing and Design for Reliability (DFR).

DFX techniques are part of the conceptualize phase of the DFSS road map (see Chapter 5) and are ideal approaches to improve life cycle cost,[2] quality, design flexibility, efficiency, and productivity using concurrent design concepts (Maskell, 1991). Benefits are usually categorized as improved decision making with competitiveness measures and enhanced operational efficiency.

10.2 HISTORY AND EVOLUTION

The DFX family started with Design for Assembly (DFA) and keeps increasing in number as fostered by the need for better decision making up front, in particular those related to product processability. Early efforts started in the 1970s by a group of researchers at the University of Massachusetts and in the United Kingdom, and result-

[1]See Yang and El-Haik (2003) for more in-depth handling of DFX within the DFSS product design arena.

[2]Life cycle cost is the real cost of the design. It includes not only the original cost of manufacture, but the associated costs of defects, litigation, buybacks, distribution support, warranty, and the implementation cost of all employed DFX methods.

Table 10.1 Purchasing department service evolution

Mail	Fax	Web
Information	Information	Information
Purchasing Agent	Purchasing Agent	Purchasing Agent
Typewriter	Typewriter	Computer
PO Form	PO Form	Application Software
Envelope	Facsimile Machine	Internet Infrastructure
Stamp	Telephone Infrastructure	
Postal Service		

ed in two different commercial DFA tools: the Boothryoyd–Dewhurst (1983) and Lucas DFA methods (Miles, 1989). They employed worksheets, data and knowledge bases, and systematic procedures to overcome limitations of design guidelines, differentiating themselves from the old practices. The Boothryoyd–Dewhurst DFA method evolved out of research on automatic feeding and insertion of parts to broader industrial applications including manual assembly, in particular, locomotive engines. In Design for Assembly (DFA), the focus is placed on factors like size, symmetry, weight, orientation, form features, and other factors related to the product as well as factors like handling, gripping, insertion, and other factors related to the assembly process. In effect, DFA focuses on the assembly process as part of production by studying these factors and their relationships to ease of assembly.

The success of DFA led to the proliferation of new applications of DFX, expanding the family to Design for Manufacturability, Design for Reliability, Design for Maintainability, Design for Serviceability, Design for Inspectability, Design for Environmentalism, Design for Recycability, and so on. DFX focuses on vital business elements of concurrent engineering, maximizing the use of the limited resources available to the DFSS team. The DFX family of tools collect and present data about both the design entity and its production processes, analyze all relationships between them, and measure performance as depicted by the design mappings. They generate alternatives by combining strengths and avoiding vulnerabilities and provide a redesign recommendation for improvement through if–then scenarios.

Table 10.2 Service environment continuum

Transaction-Based Services	Internal Process Services	Customer Product-Based Services
ATM Query	Purchase Order	ATM Withdrawal
Call Center	Staffing Requisition	ATM Deposit
Travel Search	Supplier Selection	Restaurant
Web Site Search		Web Mail Order
Mileage Statement		Oil Change
		Hair Cut

The objective of this chapter is to introduce a few vital members of the DFX family. It is up to the reader to seek more in-depth material using Table 10.3. Several design-for methods are presented in this chapter that provide a mix of product support processes and pure service processes.

The DFSS team should strive to design into the existing capabilities of suppliers, internal plants, and assembly lines. The idea is to create designs sufficiently robust to achieve Six Sigma performance from within current capability. Concurrent engineering enables this kind of parallel thinking. The key "design-for" activities to be tackled by the team are:

1. Use DFX as early as possible in the DFSS road map.
2. Start with Design for Variety for service projects.
3. The findings of step 2 determine what DFX to use next. Notice the product/process/service categorization in Table 10.3. Implementation is a function of DFSS team competence. Time and resources need to be provided

Table 10.3 DFX citation table*

Product/Process/Service		
Design for	DFX Method	Reference
Assembly	Boothroyd–Dewhurst DFA	O'Grady and Oh, 1992
	Lucas DFA	Sacket and Holbrrok, 1988
	Hitachi AEM	Huang, 1996
Fabrication	Design for Dimension Control	Haung, 1996
	Hitachi MEM	
	Design For Manufacturing	Arimoto et al., 1993 Boothroyd et al., 1994
Inspection and Test	Design For Inspectability	Haung, 1996
	Design For Dimensional Control	
Material Logistics	Design For material logistics	Foo et al., 1990
Storage and Distribution	Design For Storage and Distribution	Huang, 1996
Recycling and Disposal	Design ease of recycling	Beitz, 1990
Flexibility	Variety reduction Program	Suzue and Kohdate, 1988
Environment	Design For Environmentality	Navinchandra, 1991
Repair	Design For Reliability and Maintainability	Gardner and Sheldon, 1995
Cost	Design Whole Life Costs	Shelden et al., 1990
Service	Design For Serviceability	Gershenson and Ishii, 1991
Purchasing	Design For Profit	Mughal and Osborne, 1995
Sales and Marketing	Design For Marketability QFD	Zaccai, 1994 Chapter 7, this volume
Use and Operation	Design For Safety	Wang and Ruxton, 1993
	Design For Human Factors	Tayyari, 1993

*See Yang and El-Haik (2003).

to carry out the "design-for" activities. Service-specific options from Table 10.1 include Design for Life Cycle Cost, Design for Serviceability, Design for Profit, Design for Marketability, Design for Safety, Design for Human Factors, Design for Reliability, and Design for Maintainability.

10.3 DESIGN FOR X IN TRANSACTIONAL PROCESSES

The best way to approach DFX for a transactional process is to think of the process and the content as two separate design activities that need to be optimized in unison. The content is normally the application that affects the customer or forms, as in a Web page, key elements are of data structure and layout. The process is the infrastructure elements that support the content presentation or flow.

An automated teller machine (ATM) is a bank service process. It provides the end user with transactions that rely on a set of subservices, information flow, and hardware performance. The hardware is the ATM machine itself, the telephone system, the armored security vehicle, the building housing the ATM, and the power delivery system. In fact if we revisit the IPO diagram introduced in Chapter 2, we can depict this system as shown in Figure 10.1.

The ATM machine and other hardware infrastructure should go through the formal DFX for Products process to make them affordable, reliable, and maintainable. Service maintenance for the ATM will include cleaning and restocking forms and money. What are some of the considerations for the elements of accessibility and material choices that support the services of cleaning and restocking?

10.3.1 Design for Product Service (DFPS)

The following are the DFPS guidelines, which follow the Eliminate, consolidate/standardize, and simplify hierarchy:

- Reduce service functional requirements (FRs) by minimizing the need for service. This can be easily done in companies that track their service warranties or activities. The DFSS team has the opportunity to make their DFPS procedure data driven by analyzing the possible failure rates of baseline designs and rank them using Pareto analysis to address service requirement in a prioritized sequence. Use the appropriate DFX family tool such as axiomatic design and robustness techniques to improve the reliability. For example, Design for Assembly improves reliability by reducing the number of parts, and Axiom 2 helps reduce design FRs' variation (Chapter 8), which is a major cause of failures.
- Identify customer service attributes and appropriate type of service. Type of service required by any customer segment is the determinant of the DFPS technique to be used. There are three types:

  Standard operations. Consist of normal wear and tear items like replenishing funds in an ATM machine. For standard operations, service ease should be

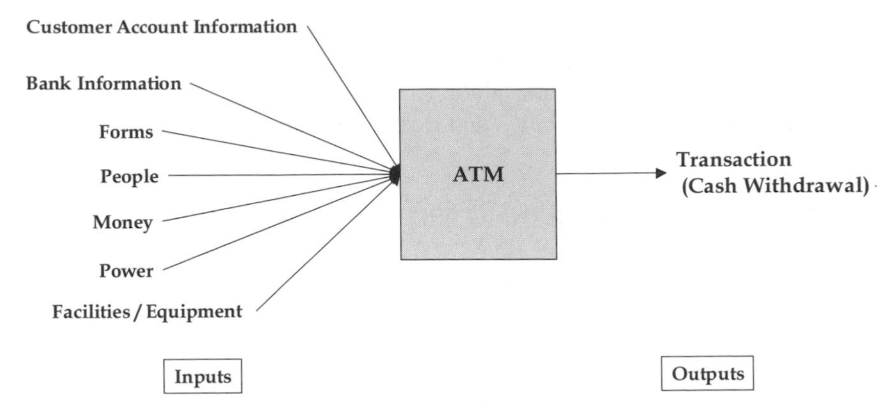

Figure 10.1 ATM IPO diagram.

optimized and applied with mistake-proofing (Poka-yoke) techniques. In many service industries, the end customer is usually the maintenance operator.

Scheduled maintenance. This is for items usually recommended in the customer manual, if any. In this category, customers expect less frequent and easier maintenance. Under the pressure of minimum life cycle cost, many companies are pushing the scheduled maintenance tasks to standard operations and "do-it-yourself" procedures. A sound scheduled maintenance procedure should call for better reliability and durability, minimum tools usage (e.g., single standard fastener size) and easy removal paths. In the ATM machine, the dispensing mechanism will require adjustments or the results may be no cash dispensed or too much.

Repair services. In a repair service, the ease of repair is key. This objective is usually challenged by limited accessibility space, time availability, and design product complexity. Repair service can be greatly enhanced by employing some sort of diagnostics system, repair kits, and modular design practices. Repair issues can have many possible causes, ranging from Type I and Type II errors in diagnostics systems, to tools and parts logistics issues, to repair technicality.

- Be vigilant about DFPS application. If the serviceability requirements did not surface by now, the DFSS team will be encouraged to use design mappings as shown in Chapter 8. Once the team identifies all serviceability process mappings, they can move to consider design alternatives. These alternatives may sometimes be inappropriate. In other cases, they may seem in conflict with each other. Nevertheless, the DFSS team should go through them if a Six Sigma capable design is to be launched in all requirements, including those related to serviceability. The serviceability set of functional requirements usually includes proper location, tools minimization, parts standardization, protection from accelerated failure, ergonomics considerations, and diagnostics functions.

The DFSS team should more or less follow the following steps to devise a sound DFPS approach:

- Review assumptions, serviceability customer requirements from the QFD, serviceability types, customer segments, and FMEA and Six Sigma targets.
- Check baseline designs and use the data available as a way to predict their design performance from the historical database. The team should also benchmark a best-in-class competition to exceed customer satisfaction.
- Identify types of services needed (e.g., standard operation, scheduled maintenance, or repair) and map them to appropriate customer segments.
- Understand all applicable service procedures including steps, sequence, and potential problems. Reference the company's lessons-learned library and controlled documentation.
- Estimate time of labor. Labor time is considered the foundation of serviceability quantification for warranty assessment purposes. Labor time is the sum of repair recognition time, diagnostic time, logistic time, and actual repair time. The team should aim to beat the best-in-class labor time.
- Minimize all problematic service areas by reviewing the customer concern tracking system (if any), determining and eliminating root causes, addressing the problem based on a prioritization scheme (e.g., Pareto analysis of warranty cost), searching for solutions in the literature (including core books), and predicting future trends.
- Determine solution approaches. The extracted information from the gathered data will lead to some formulation of a serviceability design strategy. Every separate component or critical part should be addressed for its unique serviceability requirements.
- Introduce Design for Serviceability parameters into the process mapping. These can be categorized based on the following categories:

  Orientation

    Do the components have easy removal paths (service process mapping)?

    Do the service steps require reorientation?

  Contamination

    Can the contents, if any, be contained prior to or through service?

    What is the possibility of contaminating units during service?

  Access

    Assembleability. Is it possible to group components for ease of service? Check the mapping.

      Is disassembly/sorting intuitive?

      Can asymmetrical components fit one way?

    Reachability. Can the component be reached by hand? By tool? Can the part be removed from the assembly?

Layerability. Is the part in the assembly layer correlated to frequency of service?

"Real estate." Possibility of moving or sizing parts for service space.

Efficiency. Unnecessary removal of parts that obstruct visibility or service.

Diagnostics. Can the part be accessed for diagnostics without disassembly?

Service reliability. Address potential damage of serviced or removed parts. Are all possibilities for parts minimization using DFMA exhausted?

Consider use of standard parts (e.g., fasteners).

Simplicity

Customer considerations

Tools. Design for generic tools.

Minimize special tools use.

Adjustment. Reduce customer intervention through tuning and adjustment. Use robustness techniques.

Mistake-proofing or Poka-yoke. Use color codes and very clear instructions.

For DFS, the ideal final result from a TRIZ perspective is to design the product/process for serviceability without additional cost. In reality, an incremental cost is usually needed. To decide on a budget allocation for service elements, some economic consideration is needed. If the defects per million is 10, it makes no sense to add 2¢ to a $1 part as incremental repair cost. In addition to the economic matters, the cost of losing a customer and those potential customers that the lost customer can influence must be considered.

10.4 DESIGN FOR PROCESSABILITY[3] AND ASSEMBLY

Design for Assembly (DFA) may be applied to an individual assembly or for a whole complex product like a vehicle (Sackett & Holbrook, 1990). In the latter case, a number of distinct areas can be coordinated to reflect the functional and process mappings discussed in Chapter 8, leading to the optimization of the whole product from the assembly and processability points of view. In doing so, interfaces and trade-off decisions are required if design conceptual vulnerabilities such as coupling and complexity were not resolved earlier. Therefore, linkages between teams handling several assemblies must be defined in terms of design mapping to ensure that everyone is operating in a coherent way.

Design for Processability (DFP) can be defined as designing a product to be produced in the most efficient manner possible in terms of time, money, and resources, taking into consideration how the product or service will be processed and utilizing the existing skill base to achieve the highest rolled-throughput yield possible. DFP

[3]Processability is used here in lieu of manufacturability. This section has strong linkages to product design because it assumes that designing is a process for a product (as a service design) on one hand, as well as highlighting overlaps where service is the ultimate end product.

and DFA are proven design methodologies that work for any size company. To shorten development cycle time, the early consideration of processability implications minimizes development cost and ensures a smooth launch of production. DFP promotes and ensures processability of product designs. By bridging the gap between design and process, DFP can significantly reduce critical time to market. DFSS quality themes can be designed early on with optimal parts selection and proper integration of parts for minimum vulnerabilities. By considering the cumulative effect of part quality on overall product quality, designers are encouraged to carefully specify quality.

DFP and DFA are systematic approaches within the DFX family that the DFSS team can use to carefully analyze each high-level design parameter that can be defined as a part or subassembly for manual or automated processability and assembly. The objective is to eliminate design waste. Waste, or "muda" in Japanese, may mean any of several things. It may mean features that have no function (are non-value added) and those that should have been trimmed using the mapping methods of Chapter 8. But the most leverage of DFX in the DFSS road map is in attacking the following waste sources: (1) assembly directions that need several additional operations and (2) design parameters with unnecessarily tight tolerances.

It is a golden rule of design that the DFSS team should minimize the number of setups and stages through which a part or subassembly must pass through before it becomes a physical entity within a connected whole.

As a prerequisite for DFP, the team needs to understand through experience in production how services are manufactured. Specifically, the process design should parallel the product development in a concurrent engineering setup. Developing and fostering a multifunctional team with early and active participation from purchasing, process design, finance, industrial engineering, vendors, marketing, compliance specialists, and factory workers assures a good DFP output. One of the first decisions the team has to make concerns the optimal use of off-the-shelf parts. In many cases, the product may have to literally be designed around the off-the-shelf components, but this can provide substantial benefits to the design.

Before using the DFA and DFP tools, the team should:

- Revisit the design mappings (Chapters 6 and 8) of the DFSS road map as well as the marketing strategy. The team should be aware that the DFSS road map as a design strategy is global and that the process strategy is usually local, depending on the already existing process facilities.
- Review market analysis, customer attributes, the critical-to-satisfaction characteristics (CTSs), and other requirements like packaging and maintenance. Where clarification is sought, the team may develop necessary prototypes, models, experiments, and simulations to minimize risks. In doing so, the team should take advantage of available specifications, testing, cost–benefit analysis, and modeling to build the design.
- Analyze existing processability and assembly functional requirements, operations, and sequences concurrently, using simulation tools to examine assembly and subassembly definitions of the product and service and find the best organization and production methods.

- Specify optimal tolerances (see Yang and El-Haik, 2003 and El-Haik, 2005) with the use of robust design. Understand tolerance step functions and specify tolerances wisely. Each process has its practical limit as to how tight a tolerance could be specified for a given skill level in a production process. If the tolerance is tighter than the limit, the next most precise process must be used. Design within current process capabilities. Avoid unnecessarily tight tolerances that are beyond the capability of the processes. Through design mappings, determine when new production process capabilities are needed early, to allow sufficient time to determine optimal process variables and establish a controlled process. For products, avoid tight tolerances on multiple stacked parts. Tolerances on such parts will "stack up," making maintenance of overall tolerance difficult.
- Use the most appropriate rather than the latest technology in the processes identified in the process mappings (see Chapter 8).
- Design for the minimum number of parts by physical integration. Simplify the design and reduce the number of parts because for each part, there are opportunities for process defects and assembly errors. The total cost of fabricating and assembling a product goes up with the increased number of parts. Automation becomes more difficult and more expensive when more parts are handled and processed. Costs related to purchasing, stocking, and servicing also go down as the number of parts are reduced. Inventory and work-in-process levels will go down with fewer parts. The team should go through the assembly part by part and evaluate whether a part can be eliminated, combined with another part, or the function can be performed in another way. To determine the theoretical minimum number of parts, ask the following: Does the part move relative to all other moving parts? Must the part absolutely be of a different material from the other parts? Must the part be different to allow possible disassembly? Avoid designing mirror image (right- or left-hand) parts. Design the product so the same part can function in both right- and left-hand modes. If identical parts cannot perform both functions, add features to both right- and left-hand parts to make them the same.
- Create "modular" designs using standard parts to build components and subassemblies. Standard and off-the-shelf parts are less expensive, considering the cost of design, documentation, prototyping, testing, overhead, and non-core-competency processabilities. Standard parts save the time that would be needed to design, document, administer, build, test, and fix prototype parts.
- Choose the appropriate materials for fabrication ease.
- Apply the layered assembly principles to parts handling and feeding, orientation, identification, positioning, allowable tolerances, and mating.
- Minimize setups. For machined parts, ensure accuracy by designing parts and fixtures so that all key dimensions are cut in the same setup. Use less expensive single-setup machining.
- Fixture design considerations. The team needs to understand the processabilities well enough to be able to design parts and dimension them for fixtures.

Machine tools, assembly stations, automatic transfers, and automatic assembly equipment need to be able to grip or fix the part in a particular position for subsequent operations. This requires registration locations on which the part will be gripped while being transferred, machined, processed, or assembled.

- Reduce tooling complexity and number. Use concurrent engineering of parts and tooling to reduce delivery lead time, minimize tooling complexity and cost, and maximize throughput and reliability. Apply axiomatic design information (Suh, 1990; El-Haik, 2005). Keep tool variety within the capability of the tool changer.
- Standardize fasteners and joining mechanisms. Threaded fasteners are time-consuming to assemble and difficult to automate. Minimize variety and use self-threading screws and captured washers. Consider the use of integral attachment methods such as snap-fit.
- Use the appropriate Design For Manufacturability (DFM) and Design For Assembly (DFA) tools shown in road map of Figure 10.2, as suggested by Huang (1996), who called the road map Design for Manufacturability and Assembly DFMA for short.

10.4.1 The DFMA Approach

With DFMA, significant improvement tends to arise from thinking simply, for example, reducing the number of stand-alone parts. The Boothroyd–Dewhurst DFA

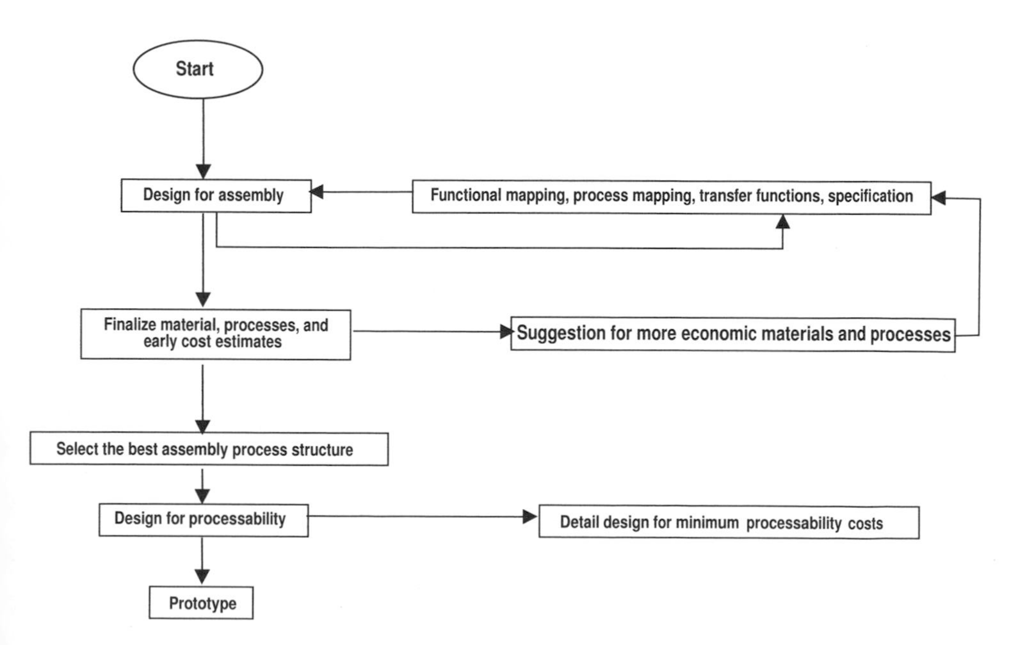

Figure 10.2 The DFMA steps.

methodology provides the following three criteria against which each part must be examined as it is added to the assembly (Huang, 1996):

1. During operation of the product, does the part move relative to all other parts already assembled?
2. Must the part be of different material than, or be isolated from, all other parts already assembled? Only fundamental reasons concerned with material properties are acceptable.
3. Must the part be separate from all other parts already assembled because the necessary assembly or disassembly of other separate parts would otherwise be impossible?

A "yes" answer to any of these questions indicates that the part must be separate or is, using DFA terminology, a *critical* part. All other parts, those which are not critical, can theoretically be removed or physically coupled with other critical parts. Therefore, theoretically, the number of critical parts is the minimum number of separate assemblies of the design. Next, the DFSS team estimates the assembly time for the design and establishes its efficiency rating in terms of assembly difficulty. This task can be done when each part is checked for how it will be grasped, oriented, and inserted into the product. From this exercise, the design is rated and from this rating standard times are determined for all operations necessary to assemble the part. The DFA time standard is a classification of design features that affect the assembly process. The total assembly time can then be assessed and, using standard labor rates, the assembly cost and efficiency can be estimated. At this stage, processability costs are not considered, but assembly time and efficiency provide benchmarks for new iterations. After all feasible simplification tasks are introduced, the next step is to analyze the manufacture of the individual parts. The objective of DFM within the DFMA structured process is to enable the DFSS team to weigh alternatives, assess processability cost, and decide between physical integration into assemblies and increased processability cost. The DFM approach provides experimental data for estimating the cost of many processes. The team should design for ease of assembly by utilizing simple patterns of movement and minimizing the axes of assembly (Crew, 1998). To reduce defective parts, features such as chamfers and tapers may be provided. In addition, parts must be designed to consistently orient themselves when fed into a process. The product's design should enable assembly to begin with a base component with a large relative mass and a low center of gravity upon which other parts are added. Assembly should proceed vertically, with other parts added on top and positioned with the aid of gravity. This will minimize the need for both reorientation and temporary fastening. Assembly that is automated will be more uniform, more reliable, and of a higher quality. Crew (1998) lists the following basic principles to facilitate parts handling and orientation:

- Product design must avoid parts that can become tangled, wedged, or disoriented. Avoid holes and tabs and "closed" parts. This type of design will allow the use of automation devices in parts handling and assembly such as vibratory bowls, tubes, magazines, etc.

- Part design should incorporate symmetry around both axes of insertion wherever possible. Where parts cannot be symmetrical, the asymmetry should be emphasized to assure correct insertion or easily identifiable features should be provided.
- With hidden features that require a particular orientation, provide an external feature or guide surface to correctly orient the part.
- Guide surfaces should be provided to facilitate insertion.
- Parts should be designed with surfaces so that they can be easily grasped, placed, and fixed. Ideally this means flat, parallel surfaces that would allow a part to be picked up by a person or a gripper with a pick-and-place robot and then easily fixed.
- Minimize thin, flat parts that are more difficult to pick up. Avoid very small parts that are difficult to pick up or require a tool such as a tweezer to be picked up, which would increase handling and orientation time.
- Avoid parts with sharp edges, burrs, or points. These parts can injure workers or customers, and they require more careful handling. They can damage product finishes and they may be more susceptible to damage themselves if the sharp edge is an intended feature.
- Avoid parts that can be easily damaged or broken.
- Avoid parts that are sticky or slippery (thin, oily plates, oily parts, adhesive-backed parts, small plastic parts with smooth surfaces, etc.).
- Avoid heavy parts that will increase worker fatigue, increase risk of worker injury, and slow the assembly process.
- Design the workstation area to minimize the distance to access and move a part.
- When purchasing components, consider acquiring materials already oriented in magazines, bands, tape, or strips.

The DFMA approach usually benefits from Poka-yoke techniques, which may be applied when components are taking form and processability and assembly issues are simultaneously considered. Poka-yoke is a technique for avoiding human error at work. The Japanese manufacturing engineer Shigeo Shingo developed the technique to achieve zero defects and came up with this term, which means, "mistake proofing." A defect exists in either one of two states: it has already occurred, calling for defect detection, or is about to occur, calling for defect prediction. The three basic functions of Poka-yoke used against defects are shutdown, control, and warning. The technique consists of steps for analyzing the process for potential problems; identifying parts by the characteristics of dimension, shape, and weight; and detecting processes deviations from nominal procedure standards. Parts should be designed so that they can be assembled by one method only. Asymmetrical holes, notches, and stops can be used to mistake-proof the assembly process.

10.5 DESIGN FOR TESTABILITY AND INSPECTABILITY

Testability is the methodology of design, execution, and optimization of the actions necessary and sufficient for test and diagnosis so that those are financially most profitable in terms of cost and accuracy. The concept of testability can imply the ability to change the means of a design functional requirement and the ability to make visible the state of such functional requirement. Test and inspection processes usually consume a significant part of the effort and cost of development, in addition to the cost and time for the acquisition of test equipment, which may be considerable. Inspectability and testability is a design requirement for continuous, rapid, and safe functionality. Testability and inspectability provide control feedback for design processes for validation and corrective actions. Early participation of testing personnel in the DFSS team will lead to design choices that can minimize the cost of testability and inspectability at the various stages of life cycle. A common understanding between team members is needed relative to the requirements for design verification and validation, acceptance after first time through production, as well as field diagnosis for defect identification. Literature on testability in electronic components exists and more common applications are now available (Drury, 1992). For example, a special color array is available for printers to ensure that all specified colors are printed on a stamp for ease of Post Office automatic inspectability. As another example, fluid containers used in cars are made from transparent materials to show the fluid levels in the container (e.g., coolant and washer fluid). Today, all computers have testing capabilities booted using the Power-On Self-Test (POST) that conducts functional inspection of several functionalities and subsystems, informs the user of any failures, and provides suggestion for corrective actions at the user's disposal. Another example comes from aerospace. Built-in Testing Equipment (BITE) is a means for evaluating functional requirements without disassembling parts, a concept similar to POST in computers. The BITE system improves accuracy and precision of the inspection function while saving time and, therefore, reducing cost.

The participation of testing personnel is required at Stage 1 of the DFSS road map to define test requirements and design the test procedures. The design, purchase or outsourcing of testing equipment and procedures can be done in parallel with design developmental activities. The understanding of design requirements will help the team in specifying the testing specification in terms of test requirements, equipment cost and utilization, and standard testing operating procedures. Testability and inspectability are usually justified by higher production rates and standardized test methods.

To deliver on testability and inspectability functions, several design rules may be listed in this context. The specification of design parameters and their tolerances should be within the natural capabilities of the production process, with process capability indices that reflect Six Sigma capability. Like Design for Serviceability (DFS), sufficient "real estate" to support test points, connections, and built-in test capabilities needs to be provided. Standard test equipment with automation compatibility to reduce setup effort, time, and cost is recommended. A built-in test and diagnosis capability to provide self-test and self-inspection in the production facility as well as in the field is a best practice to follow.

Design for Test or Testability (DFT) is basically an integral part of Design for Processability (DFM) and has become an important interface for concurrent engineering. With higher product complexity, DFT is now an increasingly important requirement for maintaining a competitive edge in the marketplace. The application of DFT in DFSS road map provides suitable test coverage while minimizing the number of required testing efforts and placing test locations to minimize test fixture complexity. The computed-based testing devices have rapidly increased in number and decreased in physical size. Many companies have been compelled to adopt DFT approaches that reduce the need for disassembly for physical access. However, the major advantages of considering testability during the DFSS project include shorter time to market due to reduced test time; lower test equipment, programming, and fixture costs; the ability to achieve adequate failure detection coverage; and the reduction or elimination of design iterations caused by test-related issues. All these benefits increase when DFT is incorporated as part of an overall DFSS project road map in new product development. If testing and inspection requirements are considered only after achieving a functional design, then one more design cycle is required to include additional testability requirements. The consideration of testability and inspectability in DFSS road map Stage 1 enforces the determination of the most practical test procedures and equipment. In addition, test fixture design is simplified and development time reduced if chosen test points are easily accessible, the test strategy is flexible, and the "real estate" is properly selected.

Defining failure modes or writing a defect list is the starting point of any Design for Inspectability process by the design team. That is why it is good idea to produce a thorough FMEA that can be used by those using this DFX method and others.

Consider the inspection process requirements and the product changes for a printed circuit board (PCB) similar to that in Figure 10.3 [see Drury & Kleiner (1984) and Anderson (1990)]. Table 10.4 lists seven defects for this example, with a description

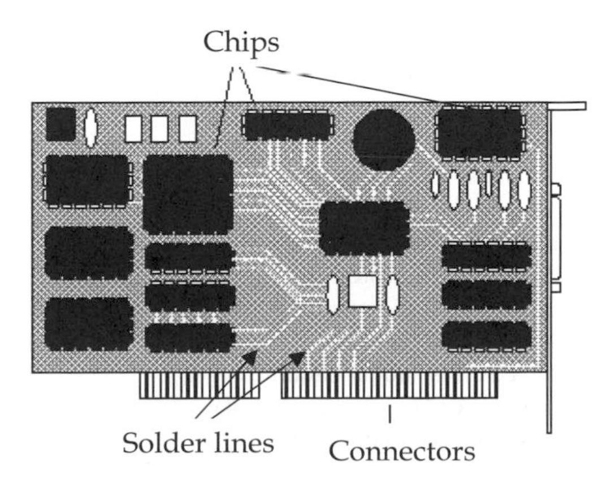

Figure 10.3 A printed circuit board.

Table 10.4 PCB example defects list

Defect Source	No.	Description
Components	1	Missing
	2	Wrong
	3	Damaged
	4	Reversed
Solder Joints	1	Missing
	2	Inadequate
	3	Excess

Table 10.5 Performance-influencing factors versus possible changes

		Changes Possible		
Activity	Influencing Factors	Product	Process	Employee
Present	Accessibility	X		
	Location of inspection areas	X	X	
	Product handling	X		X
Search	Visual			
	• Size of failure list		X	
	• Defect verification (field contrast)	X	X	X
	• Field complexity	X		
	• Failure size	X		
	• Illumination		X	
	• Peripheral acuity			
	Strategy			
	• Random vs. Systematic		X	X
	• Interfixation distance		X	X
	Timing			
	• Fixation duration		X	X
	• Time available		X	
Decision	Discrimination			
	• Failure vs. standard	X	X	
	• Standard presences		X	
	• System noise	X	X	
	• Human noise			X
	Criterion			
	• Probability of defect/defective		X	X
	(DPMO, DPO, Z-score)		X	
	• Value to cost of acceptance		X	
	• Value to cost of rejection		X	
	• Perceived costs and probabilities		X	
Action	Action complexity (e.g., defect count)	X	X	X
	Action convenience to the operator		X	X

indicating possible changes in PCB product design, process, and inspection-induced failures. Observe the number of primarily human-induced defects in Table 10.4 that can be influenced by design changes (missing, wrong insertion, damaged, etc.). This is a rich opportunity for the Design For Inspectability (DFI) application

To make use of Table 10.4 in implementing the necessary design changes, we need to assess the effect of the factors listed for each of the possible defects. The

Table 10.6 DFI design change versus defect mapping

	Components				Solder Joints		
Design Changes	Missing	Wrong	Damaged	Reversed	Missing	Inadequate	Excess
Equipment calibration	1	1	1	1	1		
Accessibility	2	2	2	2			
Location of areas of inspection	3	3					
Board handleability	4	4	4	4			
Defect/field contrast	5	6		7	10		8
Field complexity	9	9	9	10	10		
Defect size	11	12		10	10		
Defect versus standard		13	13, 14	7		13	13
System noise		15		13	10	10	16
Action complexity	17	17	17	17	17	17	17

Legend:

1. Add probe points for functional test.
2. Ensure the component is visible from a range of eye/board angle.
3. Subdivide board into visually logical areas to simplify the recognition of areas that need to be inspected.
4. Use automated insertion that crimps components to the board so that components will not fall off if the board is moved vigorously during inspection.
5. Place colored patches (of a color contrasting with the board top layer) behind components to increase discriminability of missing components.
6. If possible, code components to match identifiers on the board.
7. Use obviously asymmetric components to match identifiers on the board.
8. Use a dark-colored board undersurface to provide good contrast with excess solder.
9. Subdivide board into visually logical areas so that patterns of correct components may be easily recognized.
10. Use a regular grid of solder joints to simplify detection of missing or inadequate solder.
11. Use a regular colored patch (#4 above) to increase conspicuity of missing components.
12. Use components with lettering or identifiers in large print.
13. Provide comparison standards for all defects close to the line of sight.
14. Make colored patches same size and shape as components to simplify detection of damaged or misaligned components.
15. Reduce the number of different component types so that conspicuity of wrong components is increased.
16. Keep the design of the underface of the board as simple as possible so that bridges are not confused with legitimate connectors.
17. Reduce the number of defect types searched for one time and reported, by splitting task into populated and solder sides of the board.

PCB design team needs to produce a mapping depicting the relationship between performance-influencing factors and the list of failures and defects as given in Table 10.5. Notice that Table 10.5 includes both functional failures and potential failures found by visual inspection.

At this point, the team is ready to search for solutions to defects listed in Table 10.5 by making changes to the design requirements, including inspection. An objective is error-proof inspection (Drury and Kleiner, 1992). Product design changes are the core of DFI. Using this fact allows the design team to produce a mapping relating the design changes and the defects listed in Table 10.5. This is given in Table 10.6.

A DFI change is numbered as in Table 10.6 and defined. For example, the Design for Inspectability change No. 1 in Table 10.5 is the change referring to adding probe points for functional testing, whereas the change No. 7 refers to the use of obviously asymmetric parts for defect reversal. Notice that Table 10.6 lists inspection design changes and includes both functional failures (design and testing) and potential visual inspection defects. It also possible to add process and operator changes even in this simple example. For example, lighting changes are part of inspection process improvement (Drury, 1992), and employee selection (Thackray, 1994) , automation (Drury, 1994), and training (Kleiner and Drury, 1993) are other improvements sources. A summary of this example can be found in Huang (1996).

10.6 SUMMARY

This chapter introduces the concept of DFX as it relates to service transactions and builds from the work performed for product environments. Regardless of the environment, product or service, the concurrent design team needs to step back and assess the solution chosen from the perspective of the voice of the customer for life cycle performance. DFX for service transactions requires the process content to be evaluated much in the same light as assembly processes to minimize complexity and maximize commonality. The service process is also dependent on infrastructure support that should be processed through the DFX for products methodology. The end result will be a robust design that meets customer's needs profitably.

11

FAILURE MODE AND EFFECT ANALYSIS (FMEA)

11.1 INTRODUCTION

Failure made and effect analysis (FMEA) is a disciplined procedure that recognizes and evaluates the potential and actual effects of failure of a product or a process and identifies actions that reduce the chance of a potential failure occurring (Yang & El-Haik, 2003). It helps the DFSS team members improve their design and its delivery processes by asking "what can go wrong?" and "where can variation come from?" Service design and production, delivery, and other processes are then revised to prevent occurrence of failure modes and to reduce variation. Inputs to FMEA include past warranty or process experience, if any; customer wants, needs, and delights; performance requirements; drawings and specifications; and process mappings. For each service functional requirement and process the team needs to ask, "what can go wrong?" and determine possible design and process failure modes and sources of potential variation in all service processes under consideration, including variations in customer usage, potential causes of failure over useful life, and potential process issues such as missed tags or steps, shipping concerns, and service misdiagnosis. The team should modify the service design and processes to prevent wrong things from happening and develop strategies to deal with different situations, the redesign of processes to reduce variation, and mistake-proofing (Poka-yoke) of services and processes. Efforts to anticipate failure modes and sources of variation are iterative. This action continues as the team strives to further improve their service design and its processes.

FMEA is a team activity, with representation from project personnel involved with quality, reliability, and operations, and suppliers and customers if possible. A Six Sigma operative, typically a black or green belt, leads the team. The process owner should own the documentation.

The design team may experience several FMEA types in the DFSS road map (see Chapter 5). They are:

- Concept FMEA. Used to analyze systems and subsystems in the early concept and design stages. Focuses on potential failure modes associated with the functions of a system caused by the design.
  - Design FMEA (DFMEA). Used to analyze product/service designs before they are released to production. A DFMEA should always be completed well in advance of a prototype/pilot build. Focuses on design functional requirement. It has a hierarchy that parallels the modular structure in terms of systems, subsystems, and components.
  - Process FMEA: It is used to analyze processing, assembly, or any other processes. The focus is on process inputs. It has a hierarchy that parallels the modular structure in terms of systems, subsystems, and components.
- Project FMEA. Documents and addresses failures that could happen during a major program.
- Software FMEA. Documents and addresses failure modes associated with software functions.

We suggest using concept FMEA to analyze services and process designs in the early concept and design stages (see Figure 5.1). It focuses on potential failure modes associated with the functions of a system caused by the design. The concept FMEA helps the DFSS team to review targets for the functional requirements (FRs), to select optimum physical maps with minimum vulnerabilities, to identify preliminary testing requirements, and to determine if redundancy is required for reliability target settings. The input to DFMEA is the array of functional requirements that is obtained from the quality function deployment phases presented in Chapter 7. The outputs are (1) list of actions to prevent causes or to detect failure modes, and (2) history of actions taken and future activity. The DFMEA helps the DFSS team in:

1. Estimating the effects on all customer segments,
2. Assessing and selecting design alternatives
3. Developing an efficient validation phase within the DFSS algorithm
4. Inputting the needed information for Design for X (e.g., Design for Manufacturability, Design for Assembly, Design for Reliability, Design for Serviceability, Design for Environment)
5. Prioritizing the list of corrective action strategies that include mitigating, transferring, ignoring, or preventing the failure modes altogether,
6. Identifying the potential special design parameters from the failure standpoint, and documenting the findings for future reference

Software FMEA documents and addresses failure modes associated with software functions. On the other hand, process FMEA (PFMEA) is used to analyze service, manufacturing, assembly, or any other processes such as those encountered in transactional business processes. The focus is on process inputs. The PFMEA is a valuable tool available to the concurrent DFSS team to help them in identifying potential manufacturing/assembly or production process causes in order to place controls for either increasing detection or reducing occurrence, or both, thus prioritizing the list of corrective actions. Strategies include mitigating, transferring, ignoring, or preventing the failure modes, documenting the results of their processes, and identifying the potential special process variables that need special controls from the failure standpoint.

11.2 FMEA FUNDAMENTALS

FMEA can be described as being complementary to the process of defining what a design or process must do to satisfy the customer (AIAG, 2001). In our case, the process of defining what a design or a process must do to satisfy the customer is what we do in the service DFSS project road map discussed in Chapter 5. The DFSS team may use an existing FMEA, if applicable, for further enhancement and updating. In all cases, the FMEA should be handled as a living document.

The fundamentals of FMEA inputs, regardless of their type, are depicted in Figure 11.1 and the listed below:

1. Define scope, service functional requirements, and design parameters and process steps. For the DFSS team, this input column can be easily extracted from the functional and process mappings discussed in Chapter 8. However, we suggest doing the FMEA exercise for the revealed design hierarchy resulting from the employment of mapping techniques of their choice. At this point, it may be useful to revisit the project scope boundary as input to the FMEA of interest in terms of what is included and excluded. In DFMEA, for example, potential failure modes include the delivery of "No" FR, partial and degraded FR delivery over time, intermittent FR delivery, and unintended FR (not intended in the mapping).
2. Identify potential failure modes. Failure modes indicate the loss of at least one FR. The DFSS team should identify all potential failure modes by asking, "In what way will the design fail to deliver its FRs?" as identified in the mapping. Failure modes are generally categorized as materials, environment, people, equipment, methods, and so on, and have a hierarchy of their own. That is, a potential failure mode can be cause or effect in a higher-level subsystem causing failure in its FRs. A failure mode may occur but not necessarily must occur. Potential failure modes may be studied from the baseline of past and current data, tests, and current baseline FMEAs.
3. Potential failure effects(s). A potential effect is the consequence of the failure of other physical entities as experienced by the customer.
4. Severity. Severity is a subjective measure of "how bad" or "serious" is the effect of the failure mode. Usually, severity is rated on a discrete scale from

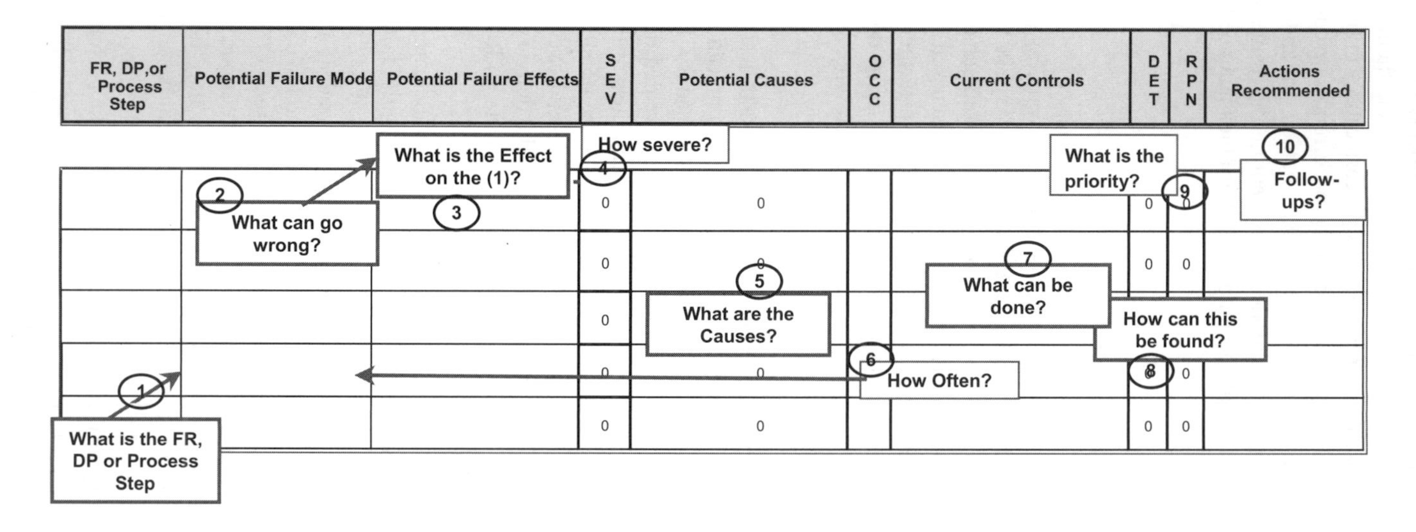

Figure 11.1 FMEA worksheet.

1 (no effect) to 10 (hazardous effect). Severity ratings of 9 or higher indicate a potential special effect that needs more attention and this typically is a safety or government regulation issue (Table 11.1). Severe effects are usually classified as "critical," "significant," or "control." A "critical" effect is usually a safety issue and requires more deep study for all causes down to the lowest level using, possibly, fault tree analysis (FTA). "Significant" effects are important for the design itself. "Control" effects are regulated by the government for any public concern. A control plan is needed to mitigate

Table 11.1 Automotive Industry Action Group (AIAG) severity ratings

Effect	Severity of Effect Defined	Rating
None	No effect.	1
Very Minor	Minor disruption to production line. A portion (less than 100%) of the product may have to be reworked on-line but in-station. Fit/finish/squeak/rattle item does not conform. Defect noticed by discriminating customers.	2
Minor	Minor disruption to production line. A portion (less than 100%) of the product may have to be reworked on-line but out-of-station. Fit/finish/squeak/rattle item does not conform. Defect noticed by average customers.	3
Very Low	Minor disruption to production line. The product may have to be sorted and a portion (less than 100%) reworked. Fit/finish/squeak/rattle item does not conform. Defect noticed by most customers.	4
Low	Minor disruption to production line. 100% of product may have to be reworked. Vehicle/item operable, but some comfort/convenience item(s) operable at reduced level of performance. Customer experiences some dissatisfaction.	5
Moderate	Minor disruption to production line. A portion (less than 100%) may have to be scrapped (no sorting). Vehicle/item operable, but some comfort/convenience item(s) inoperable. Customers experience discomfort.	6
High	Minor disruption to production line. Product may have to be sorted and a portion (less than 100%) scrapped. Vehicle operable, but at a reduced level of performance. Customer dissatisfied.	7
Very High	Major disruption to production line. 100% of product may have to be scrapped. Vehicle/item inoperable, loss of primary function. Customer very dissatisfied.	8
Hazardous: With Warning	May endanger operator. Failure mode affects safe vehicle operation and/or involves noncompliance with government regulation. Failure will occur with warning.	9
Hazardous: Without Warning	May endanger operator. Failure mode affects safe vehicle operation and/or involves noncompliance with government regulation. Failure will occur without warning.	10

the risks for the significant and critical effects. The team needs to develop proactive design recommendations. This information is carried to the PFMEA after causes have been generated in the projects.

5. Potential Causes. Generally, these are the set of noise factors and the deficiencies are designed in due to the violation of design principles, axioms, and best practices (e.g., inadequate assumptions have been made). The study of the effect of noise factors helps the DFSS team identify the mechanism of failure. The analysis conducted by the DFSS team with the help of process mapping allows for the identification of the interactions and coupling of their scoped project with the environment and with customer, and within the processes and subprocesses themselves. For each potential failure mode identified in step 2, the DFSS team needs to enter a cause in this column. See Section 11.5 for cause and effect tools linking FMEA steps 3 and 5. There are two basic reasons for these causes: (1) the design is manufactured and assembled within specifications, (2) the design may include a deficiency or vulnerability that may cause unacceptable variation (misbuilds, errors, etc.), or both.
6. Occurrence. Occurrence is the assessed cumulative subjective rating of the process entity failures that could occur over the intended life of the design, in other words, the likelihood of the cause occuring. FMEA usually assumes that if the cause occurs, so does the failure mode. Based on this assumption, occurrence is also the likelihood of the failure mode. Occurrence is rated on a scale of 1 (almost never) to 10 (almost certain) based on failure likelihood or probability, usually given in parts per million defective (PPM). See Table 11.2 for linkage to process capability. In addition to this subjective rating, a regression correlation model can be used. The occurrence rating is a ranking scale and does not reflect the actual likelihood. The

Table 11.2 FMEA Occurrence linkage to capability

Numerical Ranking	Occurrence Likelihood
1	1 in 10^6 ($C_{pk} > 1.67$)
2	1 in 20,000 ($C_{pk} = 1.33$)
3	1 in 5,000 ($C_{pk} \sim 1.00$)
4	1 in 2,000 ($C_{pk} < 1.00$)
5	1 in 500
6	1 in 100
7	1 in 50
8	1 in 20
9	1 in 10
10	1 in 2

actual likelihood or probability is based on the failure rate extracted from historical service or warranty data with the same parts or surrogates. See Table 11.3 for example.

In DFMEA, design controls help in preventing or reducing the causes of failure modes and the occurrence column will be revised accordingly.

7. Current controls. The objective of design controls is to identify and detect the design deficiencies and vulnerabilities as early as possible. Design controls are usually applied for *first-level failures*. A wide spectrum of controls is available like lab tests, project and design reviews, and design modeling (e.g., simulation). In the case of redesign of a service DFSS project, the team should review relevant (similar failure modes and detection methods experienced on surrogate designs) historical information from the corporate memory such as prototype tests, and modeling studies. In the case of white-sheet design, the DFSS team needs to brainstorm new techniques for failure detection by asking be what means they can recognize the failure mode. In addition, they should ask how they can discover its occurrence.

   Design controls span a spectrum of different actions that include physical process mappings changes (without creating vulnerabilities), special controls, design guidelines, DOEs, design verification plans, and modifications of standards, procedures and best-practice guidelines.

8. Detection. Detection is a subjective rating corresponding to the likelihood that the detection method will detect the fist-level failure of a potential failure mode. This rating is based on the effectiveness of control system through related events in the design algorithm; hence, FMEA is a living document. The DFSS team should:
   - Assess the capability of each detection method and how early in the DFSS endeavor each method will be used.
   - Review all detection methods in step 8 and achieve consensus on a detection rating.

Table 11.3 Automotive Industry Action Group (AIAG) Occurrence Rating

Probability of Failure	Occurrence	Rating
Very High: Persistent failures	$\geq$ 100 per thousand vehicles/items ($\geq$ 10%)	10
	50 per thousand vehciles/items (5%)	9
High: Frequent failures	20 per thousand vehicles/items (2%)	8
	10 per thousand vehicles/items (1%)	7
Moderate: Occasional failures	5 per thousand vehicles/items (0.5%)	6
	2 per thousand vehicles/items (0.2%)	5
	1 per thousand vehicles/items (0.1%)	4
Low: Relatively few failures	0.5 per thousand vehicles/items (0.05%)	3
	0.1 per thousand vechicles/items (0.01%)	2
Remote: Failure is unlikely	$\leq$ 0.010 per thousand vehicles/items ($\leq$ 0.001%)	1

- Rate the methods. Select the lowest detection rating in case of methods tie.

See Table 11.4 for an example.

9. Risk priority number (RPN). This is the product of the severity (step 4), occurrence (step 6), and detection (step 8) ratings. The range is between 1 and 1000. RPN numbers are used to prioritize the potential failures. The severity, occurrence, and detection ratings are industry specific and the black or green belt should use his/her own company's adopted rating system. The automotive industry converged to the ratings given in Tables 11.3 and 11.4 (compiled AIAG ratings are shown in Table 11.5). The software FMEA is given in Table 11.6.
10. Actions recommended. The DFSS team should select and manage recommended subsequent actions. This is where the risk of potential failures is high; an immediate control plan should be crafted to control the situation.

    Over the course of the design project, the DFSS team should observe, learn, and update the FMEA as a dynamic living document. FMEA is not retrospective, but a rich source of information for corporate memory. Companies should build a "corporate memory" that will record the design best practices, lessons learned, and transfer functions, and record what corrective actions were attempted and what did and did not work and why. This

Table 11.4 Automotive Industry Action Group (AIAG) detection rating

Detection	Likelihood of Detection	Rating
Almost certain	Design control will almost certainly detect a potential cause/mechanism and subsequent failure mode.	1
Very high	Very high chance the design control will detect a potential cause/mechanism and subsequent failure mode.	2
High	High chance the design control will detect a potential cause/mechanism and subsequent failure mode.	3
Moderately high	Moderately high chance the design control will detect a potential cause/mechanism and subsequent failure mode.	4
Moderate	Moderate chance the design control will detect a potential cause/mechanism and subsequent failure mode.	5
Low	Low chance the design control will detect a potential cause/mechanism and subsequent failure mode.	6
Very low	Very low chance the design control will detect a potential cause/mechanism and subsequent failure mode.	7
Remote	Remote chance the design control will detect a potential cause/mechanism and subsequent failure mode.	8
Very remote	Very remote chance the design control will detect a potential cause/mechanism and subsequent failure mode.	9
Absolute uncertainty	Design control will not and/or can not detect a potential cause/mechanism and subsequent failure mode; or there is no design control.	10

Table 11.5 AIAG compiled ratings

Rating	Severity of Effect	Likelihood of Occurrence	Ability to Detect
10	Hazardous without warning	Very high: Failure is almost inevitable	Cannot detect
9	Hazardous with warning		Very remote chance of detection
8	Loss of primary function	High: Repeated failures	Remote chance of detection
7	Reduced primary function performance		Very low chance of detection
6	Loss of secondary function	Moderate: Occasional failures	Low chance of detection
5	Reduced secondary function performance		Moderate chance of detection
4	Minor defect noticed by most customers		Moderately high chance of detection
3	Minor defect noticed by some customers	Low: Relatively few failures	High chance of detection
2	Minor defect noticed by discriminating customers		Very high chance of detection
1	No effect	Remote: Failure is unlikely	Almost certain detection

memory should include pre and post-remedy costs and conditions, including examples. This is a vital tool to apply when sustaining good growth and innovation strategies and avoiding attempted solutions that did not work. An online "corporate memory" has many benefits. It offers instant access to knowledge at every level of management and design staff. The DFSS team should document the FMEA and store it in a widely acceptable format in the company in both electronic and physical form.

11.3 SERVICE DESIGN FMEA (DFMEA)

The objective of the DFMEA is to help the design team by designing the failure modes out of their project. Ultimately, this objective will significantly improve the reliability of the design. Reliability, in this sense, can simply be defined as the quality of design (initially at the Six Sigma level) over time.

The proactive use of DFMEA is a paradigm shift as it is usually not done or looked at as a formality. This attitude is very harmful as it indicates ignorance of its significant benefits. Knowledge of the potential failure modes can be acquired from experience, discovered by the customer (field failures), or found in prototype testing, but the most leverage of the DFMEA is obtained when the failure modes are proactively identified during the early stages of the project (when it is still on paper).

Table 11.6 The software FMEA rating

Rating	Severity of Effect	Likelihood of Occurrence	Detection
1	Cosmetic error: No loss in product functionality. Includes incorrect documentation.	1 per 100 unit-years (1/50m)	Requirements/ Design reviews
2	Cosmetic error: No loss in product functionality. Includes incorrect documentation.	1 per 10 unit-years (1/5m)	Requirements/ Design reviews
3	Product performance reduction, temporary through time-out or system load the problem will "go away" after a period of time.	1 per 1 unit-year (1/525k)	Code walkthroughs/ Unit testing
4	Product performance reduction, temporary through time-out or system load the problem will "go away" after a period of time.	1 per 1 unit-month (1/43k)	Code walkthroughs/ Unit testing
5	Functional impairment/Loss: The problem will not resolve itself, but a "work around" can temporarily bypass the problem area until fixed without losing operation	1 per week (1/10k)	System integration and test
6	Functional impairment/loss: The problem will not resolve itself, but a "work around" can temporarily bypass the problem area until fixed without losing operation.	1 per day (1/1440)	System integration and test
7	Functional impairment/loss; The problem will not resolve itself and no "work around" can bypass the problem. Functionality has either been impaired or lost but the product can still be used to some extent.	1 per shift (1/480)	Installation and start-up
8	Functional impairment/loss: The problem will not resolve itself and no "work around" can bypass the problem. Functionality has either been impaired or lost but the product can still be used to some extent.	1 per hour (1/60)	Installation and start-up
9	Product halts/process taken down/ reboot required: The product is completely hung up, all functionality has been lost and system reboot is required.	1 per 10 min (1/10)	Detectable only once on line
10	Product halts/process taken down/ reboot required: The product is completely hung up, all functionality has been lost and system reboot is required.	1+ per min (1/1)	Detectable only once "on line

The DFMEA exercise within the DFSS road map is a function of the hierarchy identified in the service design mapping. First, the DFSS team will exercise the DFMEA at the lowest hierarchical level (e.g., a secondary requirement or subfunction), then estimate the effect of each failure mode at the next hierarchal level (e.g. a primary-functional requirement), and so on. The FMEA is a bottom-up approach not a top-down one and usually does not reveal all higher-levels potential failures.

Prior to embarking on any FMEA exercise, we advise the Six Sigma black or green belt to book a series of FMEA meetings in advance, circulate all relevant information ahead of the FMEA meetings, clearly define objectives at the start of each meeting, adhere to effective roles, and communicate effectively.

A service DFMEA can be conducted by following the steps in Section 11.2. Here we highlight any peculiar steps for such an exercise. The fundamental steps in the DFMEA to be taken by the DFSS team are:

1. Constructing the project boundary (scope) bounded by a functional mapping of the team's choice. Steps or activities, subprocesses, and processes are at different hierarchal levels. The team will start at the lowest hierarchical level and proceed upward. The relative information from the lower levels is input to the appropriate higher level in the respective steps. Here the IDEF3 example of Chapter 8 is used for illustration in Figures 11.2 and 11.3. Revisit the functional mapping, using a technique or a combination of techniques from Chapter 8 that fully describes interfaces at the all levels of the hierarchy within the scope. Interfaces will include controlled and uncontrolled inputs like environmental factors, deterioration, and customer usage.
2. Identifying the potential failures for all functional requirements at all hierarchical levels in the map. The team needs to identify all potential ways in which the design may fail. For each functional requirement in the map, the team will brainstorm the design failure modes. Failure modes describe how each functional entity in the map may initially fail prior to the end of its intended life or introduce unwanted side effects. The potential design failure mode is the way in which a service design entity in the functional map may fail to deliver its array of functional requirements.
3. Studying the failure causes and effects. The causes are generally categorized as conceptual weaknesses due to axiom violations (El-Haik, 2005). Another type is due to the mean effects of noise factors and their interaction with the design parameters, for example, unit-to-unit production deficiencies, such as when a step is not within specifications. Assembly errors, that is, steps that are delivered to specifications but with assembly process errors, may propagate to the higher levels in the map. In addition, material variation, environment, and operator usage are the premier noise factors for team's consideration.
4. The direct consequence of a failure is the failure mode on the next higher hierarchical level, the customer and regulations. Potential failure causes can be analyzed by tools like fault tree analysis (FTA), cause-and-effect diagrams, and cause-and-effect matrices. Two golden rules should be followed in cause

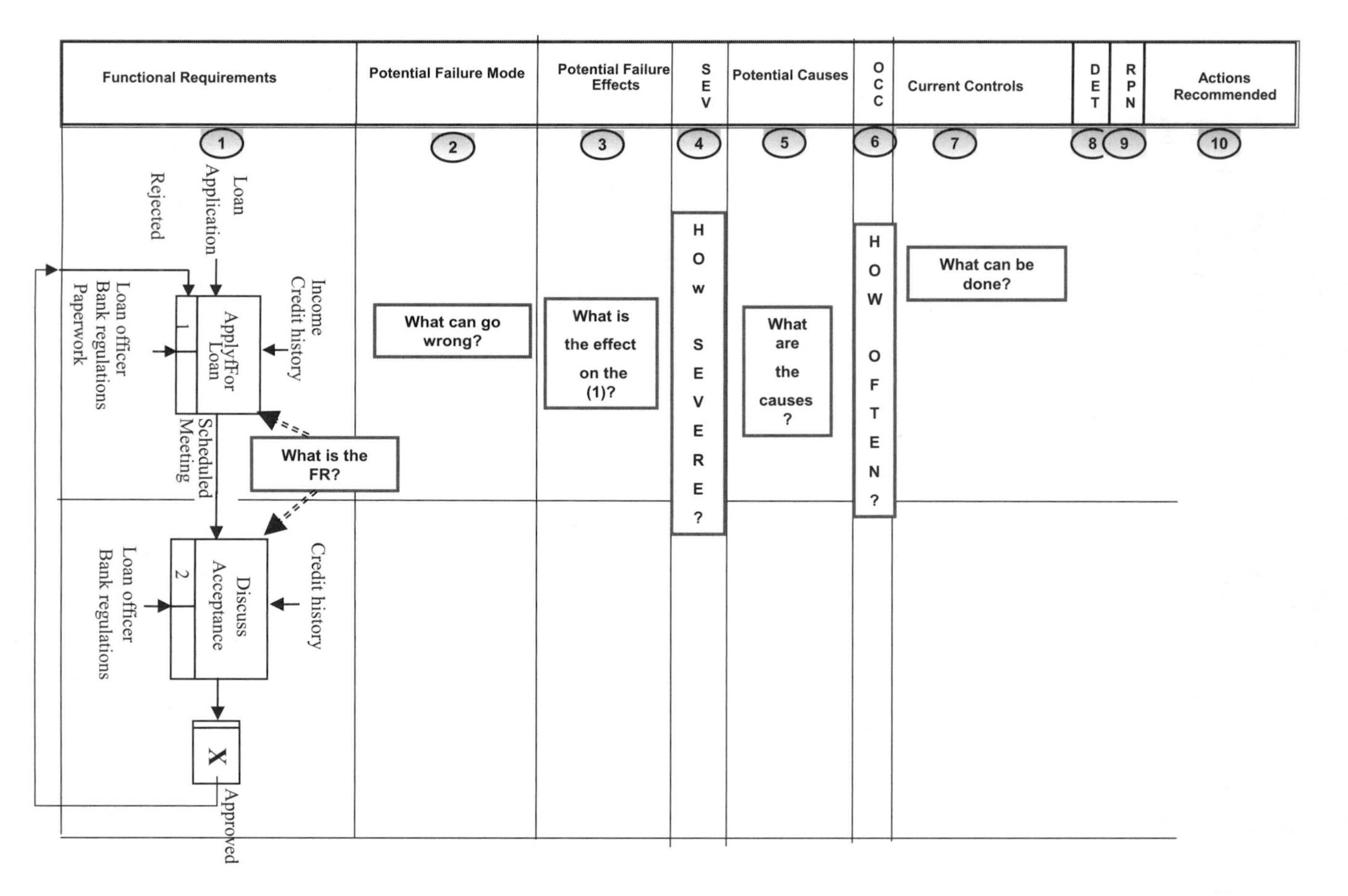

Figure 11.2 Service DFMEA of Chapter 8 “buying house” example, Level 2.

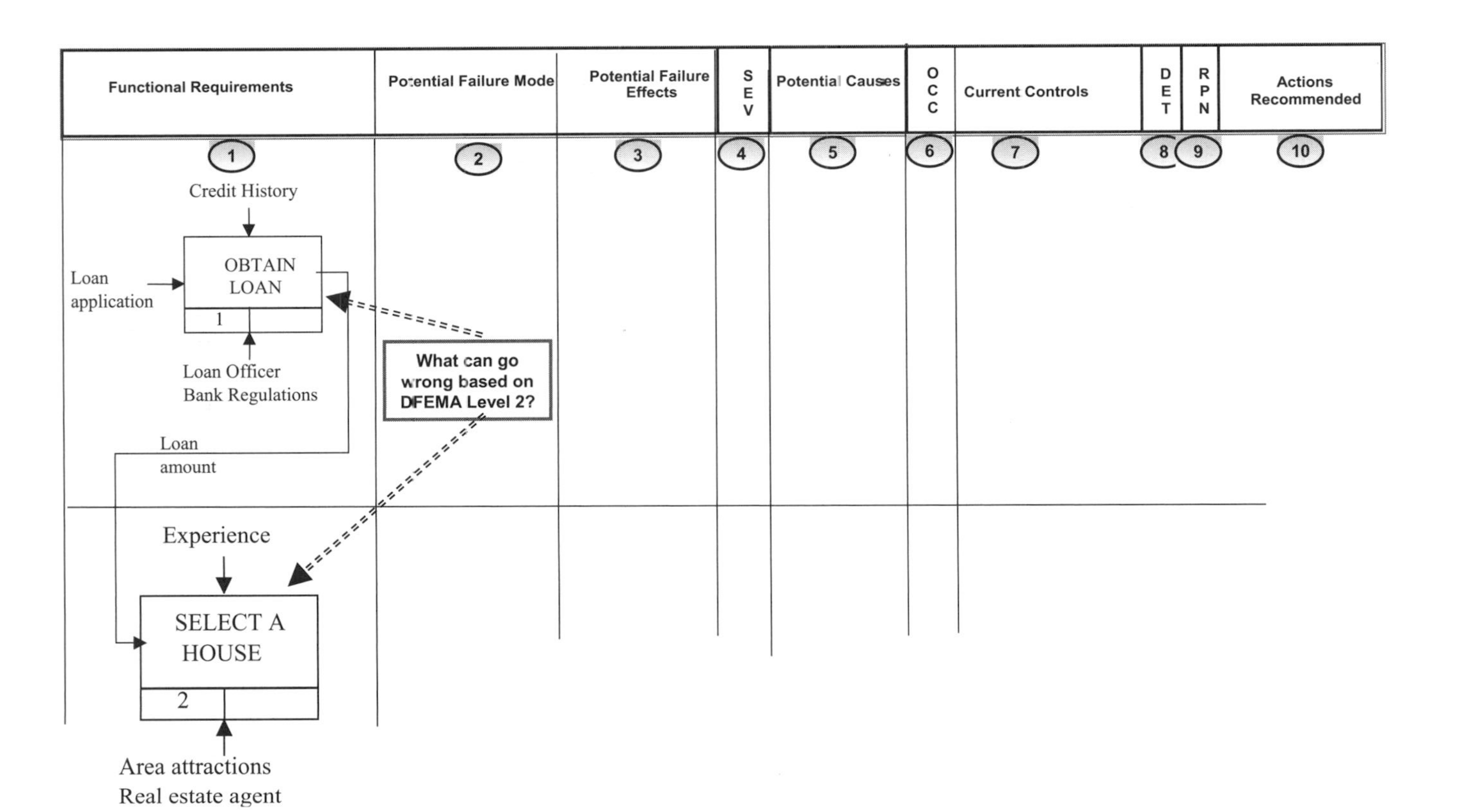

Figure 11.3 Service DFMEA of Chapter 8 "buying house" example, Level 1.

identification: the team should start with modes with the highest severity ratings, and the team should try to go beyond the first-level cause to second- or third-level causes.

5. Ranking of potential failure modes using the RPN numbers, so that actions can be taken to address them. Each failure mode must be considered in terms of severity of the consequences, detection, and occurrence of its causes. The service FMEA ratings are listed in Table 11.7.
6. Classifying any special cause as a "critical" or "significant" characteristic that will require controls. When the failure mode is given a severity rating greater than the specific critical rating, then a potential "critical" characteristics[1] (design parameter) may exist. Critical characteristics usually affect safety or compliance to regulations. When a failure mode–cause combination has a severity rating in the range below the "critical" threshold, then a potential "significant" characteristic exists. "Significant" implies significant relative to some critical-to characteristics in the QFD (see Chapter 7). Both types of classification are input to the PFMEA and are called "special" characteristics. Special characteristics require "special controls", that is, additional effort (administrative, measurement, overdesign, etc.) beyond the normal control. Robust design methodology is usually used to identify the "special" characteristics. In this regard, the team should investigate the effect of noise factors on design failure. Noise factors and their relation to the FMEA and the DFSS roadmap are depicted in Figure 11.4. To mitigate the failure risk, the team should use the DFSS road map to select a sound strategy.
7. Deciding on "design controls" as the methods to detect failure modes or their causes. There are two types of controls. Type I controls are those designed to prevent the cause or failure mechanism or mode and its effect from occurring. Type I controls also address reducing the occurrence. Type II controls address detection of cause and mechanisms of failure modes, either by analytical or physical methods, before the item is released to production.
8. Identifying and managing corrective actions. Based on the RPN numbers, the team moves to decide on the corrective actions. The corrective action strategies include:
   - Transferring the risk of failure to other systems outside the project's scope
   - Preventing failure altogether (e.g., design Poka-yoke)
   - Mitigating risk of failure by:
     a. Reducing "severity" (most difficult)
     b. Reducing "occurrence" (redundancy, mistake-proofing)
     c. Increasing the "detection" capability (e.g., brainstorming sessions, concurrent design, use top-down failure analysis like FTA)

[1]Manufacturing industry uses ∇ (inverted delta) to indicate "critical" characteristics. The "critical" and "significant" terminology started in automotive industry.

Table 11.7 Service FMEA ratings

Severity		Occurrence		Detection	
Rating	Description	Rating	Description	Rating	Description
1	Minor—Unreasonable to expect this will be noticed in the process, or impact any process or productivity. No or negligible effect on product function. The customer will probably not notice it.	1–2	Remote—Probability of failure less than 0.02% of total (< 200 PPM).	1–2	Remote—Likelihood of defect being shipped is remote (< 199 PPM).
2–3	Low—Very limited effect on local process, no downstream process impact. Not noticeable to the system but slightly noticeable in product (subsystem and system).	3–5	Low—Probability of failure from 0.021 to 0.5% of total (201 PPM to 5000 PPM).	3–4	Low—Likelihood of defect being shipped is low (20 PPM to 1000 PPM).
4–6	Moderate—Effects will be throughout the process. May require unscheduled rework. May create minor damage to equipment. Customer will notice immediately. Effect on subsystem or product performance deterioration.	6–7	Moderate—Probability of failure from 0.5 to 2% of total (5001 PPM to 20000 PPM).	5–7	Moderate—Likelihood of defect being shipped is moderate (1001 PPM to 20000 PPM).
7–8	High—May cause serious disruptions to downstream processes. Major rework. Equipment, tool or fixture damage. Effect on major product system but not on safety or government regulated item.	8–9	High—Probability of failure from 2% to 10% of total (20001 PPM to 100000 PPM).	8–9	High—Likelihood of defect being shipped is high (20001 PPM to 100000 PPM).
9–10	Extremely high—Production shut down. Injury or harm to process or assembly personnel. Effect on product safety or involves non-compliance with government regulated item.	10	Very high—Probability of failure greater than 10% of total (>100001 PPM).	10	Very high—Likelihood of defect being shipped is very high (> 100000 PPM).

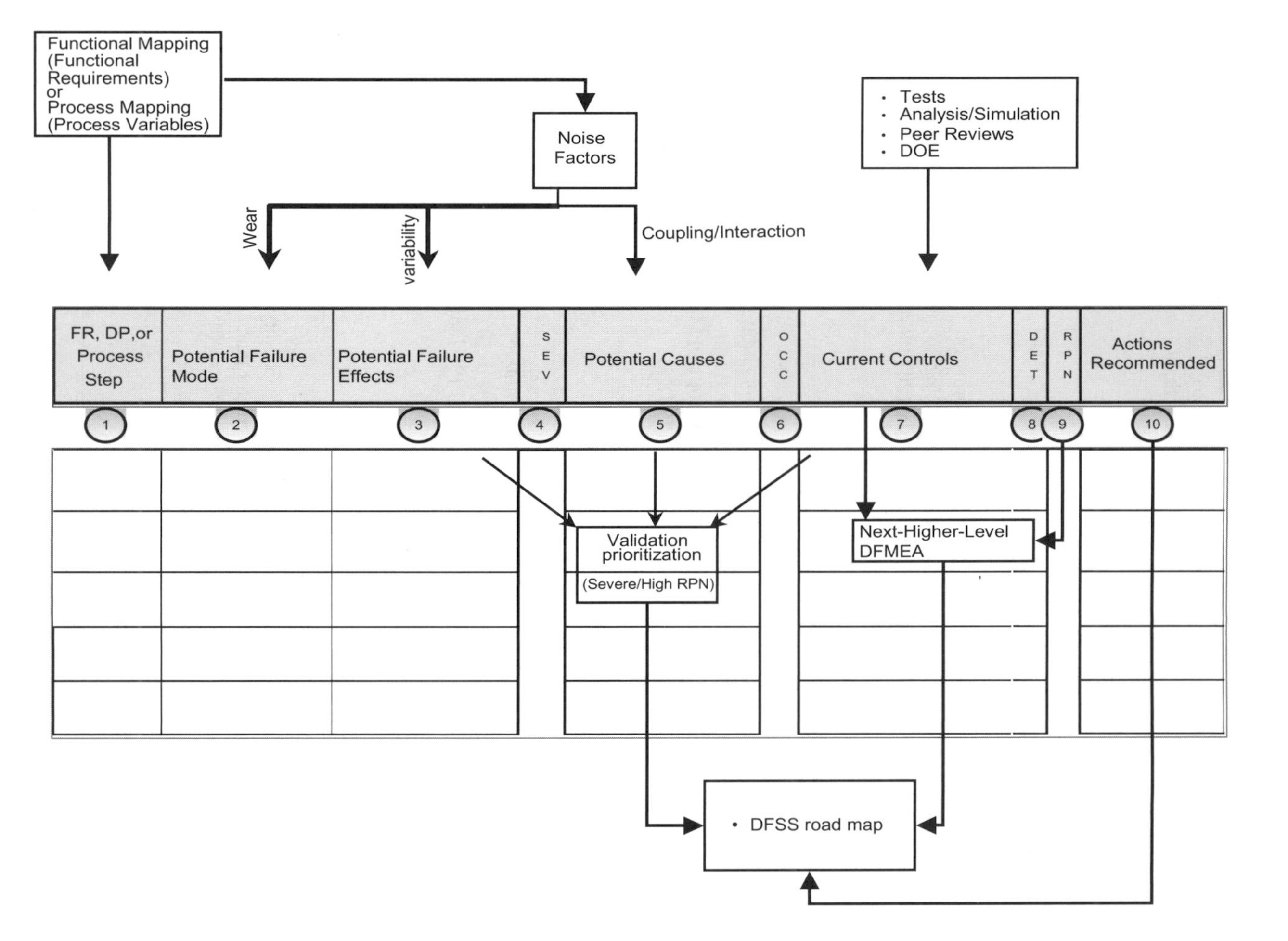

Figure 11.4 FMEA interfaces in the DFSS road map.

9. Review analysis, document and update the DFMEA. The DFMEA is a living document and should be reviewed and managed on an ongoing basis. Steps 1–8 should be documented in the appropriate media. The potential failure modes at any level can be brainstormed by leveraging existing knowledge such as process engineering, mapping, and simulation analysis, historical failure databases of the baseline design, possible designed-in errors, and physics of failures, if applicable. For comprehensiveness and as a good practice, the black belt should instruct the DFSS team members to always maintain and update their specific list of failures modes.

11.4 PROCESS FMEA (PFMEA)

The activities in the PFMEA are similar to the DFMEA; however, the focus here is on process failures. The fundamental steps in the PFMEA are the same as those of a DFMEA. Steps peculiar to PFMEA are presented in this section.

As a first step, the DFSS team should already have the project scope defined with the constructed project process map. The team can maximize the design quality by preventing all processes failures. Similar to DFMEA, the team should start with the lowest hierarchical level, the step level, and utilize the process map to cover all processes. The team then proceeds to the next hierarchal level, the subprocess level, finishes all subprocess PFMEA, and proceeds upward.

The process map should fully describe the design interfaces at all levels of the hierarchy within its scope. Interfaces will include controlled inputs, the subprocess and process variables, uncontrolled inputs, and noise factors like processing, operator, and assembly errors and variations. The macro-level process map is leveraged by the micromappings. Micromappings exhibit detailed operations like transportation, inspection stations, and cycle times. The relative information from the lower levels is input to the next-higher hierarchical level PFMEA, where appropriate, in the respective steps.

The team must revisit the process map at all hierarchical levels where defined. The task here is to make sure that all process variables in the process map end up being hosted by some process or subprocess domain (see Figure 6.1), setting the stage for addressing the potential failures for each hierarchical level in the process map. Having gone through the corresponding hierarchical-level DFMEA, the team must identify all potential ways in which the design may fail due to all processes failures. For each requirement or design parameter in the map, the team will brainstorm the process failure modes. Failure modes describe how each hierarchical entity in the structure may initially fail prior to the end of its intended life. The potential process failure mode is the way in which a processed service entity in the structure may fail to deliver its array of design parameters and functional requirements.

For proper failure treatment, the team should study the failure causes and effects. The causes are generally categorized as weaknesses due to inherent process weakness caused by conceptual vulnerability in the design functional and/or process maps and failures due to the effect of noise factors. In this cause of failure, a service is produced within the design specification. Nevertheless, it still cannot be delivered or function as intended or requested by the customer. The other type of failures may be caused by noise factors such as variation and deficiencies due mainly to incapable processes, material variation, machine deterioration, and environment and operator error.

The effect of a failure is the direct consequence of the failure mode in the current process step in terms of rework or inefficiency, the next upstream processes, and, ultimately, the customer. Potential failure causes can be analyzed at the process level by tools like fault tree analysis (FTA), cause-and-effect diagrams, and cause-and-effect matrices. To address the causes, the team should start with modes with highest severity ratings and try to go beyond the first-level cause to second- or third-level causes. The team should ask the following process questions:

1. What incoming source of variations could cause this process to fail to deliver its mapped functional requirements?
2. What could cause the process to fail, assuming that the incoming inputs are correct and to specifications?
3. If the process fails, what are the consequences for operator health and safety, machinery, the component itself, the next downstream processes, the customer, and regulations?

The classification of any process variable as a "special" characteristic that will require controls like operator safety characteristics, indicates that it is a process parameter that does not affect service but may have an impact on safety or government regulations applicable to process operation. Another category of process "special" characteristics is the "high-impact" characteristics. This type of process characteristic occurs when the condition of being outside of the specification tolerance severely affects the operation of the process itself or subsequent operations and does not affect service being processed. Both types of classification are input to the PFMEA and are called "special" characteristics. The PFMEA handling of controls and corrective actions parallels the service DFMEA.

PFMEA should be conducted according to the process map. It is useful to add the PFMEA and DFMEA processes to the design project-management charts. The PERT or CPM approaches are both advisable. The black belt should schedule short meetings (less than two hours) with clearly defined objectives. Intermittent objectives of a FMEA may include tasks like measurement system evaluation, process capability verification, and conducting exploratory DOEs or simulation studies. These activities consume resources and time, introducing sources of variability to the DFSS project closure cycle time.

11.5 FMEA INTERFACES WITH OTHER TOOLS

Fault Tree Analysis (FTA), a top-down approach, can enhance the understanding of safety-related and catastrophic failures. FTA, like FMEA, helps the DFSS team answer the "what if?" questions. This tool increases the design team's understanding of their creation by identifying where and how failures may occur. In essence, FTA can be viewed as a mathematical model that graphically uses deductive logic gates (e.g., AND, OR, etc.) to combine events that can produce the failure or the fault of interest. The objective is to emphasize the lower-level faults that directly or indirectly contribute to high-level failures in the DFSS project structures. Facilitated by the mapping development, FTA must be performed as early as possible, particularly with regard to safety-related failures as well as Design for Serviceability.

11.5.1 Cause-and-Effect Tools

The cause-and-effect diagram, also know as the "fishbone" or "Ishikawa" diagram, and the cause–effect matrix are two commonly used tools that help the DFSS team in their FMEA exercise. The cause-and-effect diagram classifies the various causes thought to affect the operation of the design, indicatiing the cause-and-effect relations among them with arrows.

The diagram is formed from the causes that can result in the undesirable failure mode. The causes are the independent variables and the failure mode is the dependent variable. An example is depicted in Figure 11.5. In this example, a company that manufactures consumer products suffers from warranty costs, measured as a percentage of sales. The current performance is above the budgeted upper specification limit. Customers tend to be very sensitive to warranty issues and equipment downtime, which results in a dissatisfaction. Figure 11.5 shows several causes for the cost.

The cause-and-effect matrix is another technique that can be used to identify failure causes. In the columns, the DFSS team can list the failure modes. The team then proceeds to rank each failure mode numerically using the RPN numbers. The team uses brainstorming to identify all potential causes that can impact the failure modes and lists these along the left hand side of the matrix. It is useful practice to classify theses causes as design weaknesses or noise factor effects by type (e.g. variability and production environment). The team then rates, numerically, the effect of cause of each failure mode within the body of the matrix. This is based on the experience of the team and any available information. The team then cross-multiplies the rate of the effect by the RPN to total each row (cause). These totals are then used to analyze and determine where to focus the effort when creating the FMEA. By grouping the causes according to their classification (weakness, environmental effect, production unit-to-unit effect, wear effect, etc.), the team will develop several hypotheses about the strength of these types of causes to devise their attack strategy.

11.5.2 Quality Systems and Control Plans

Control plans are the means to sustain any DFSS project findings. However, these plans are not effective if not implemented within a comprehensive quality operating system. A solid quality system can provide the means through which your project

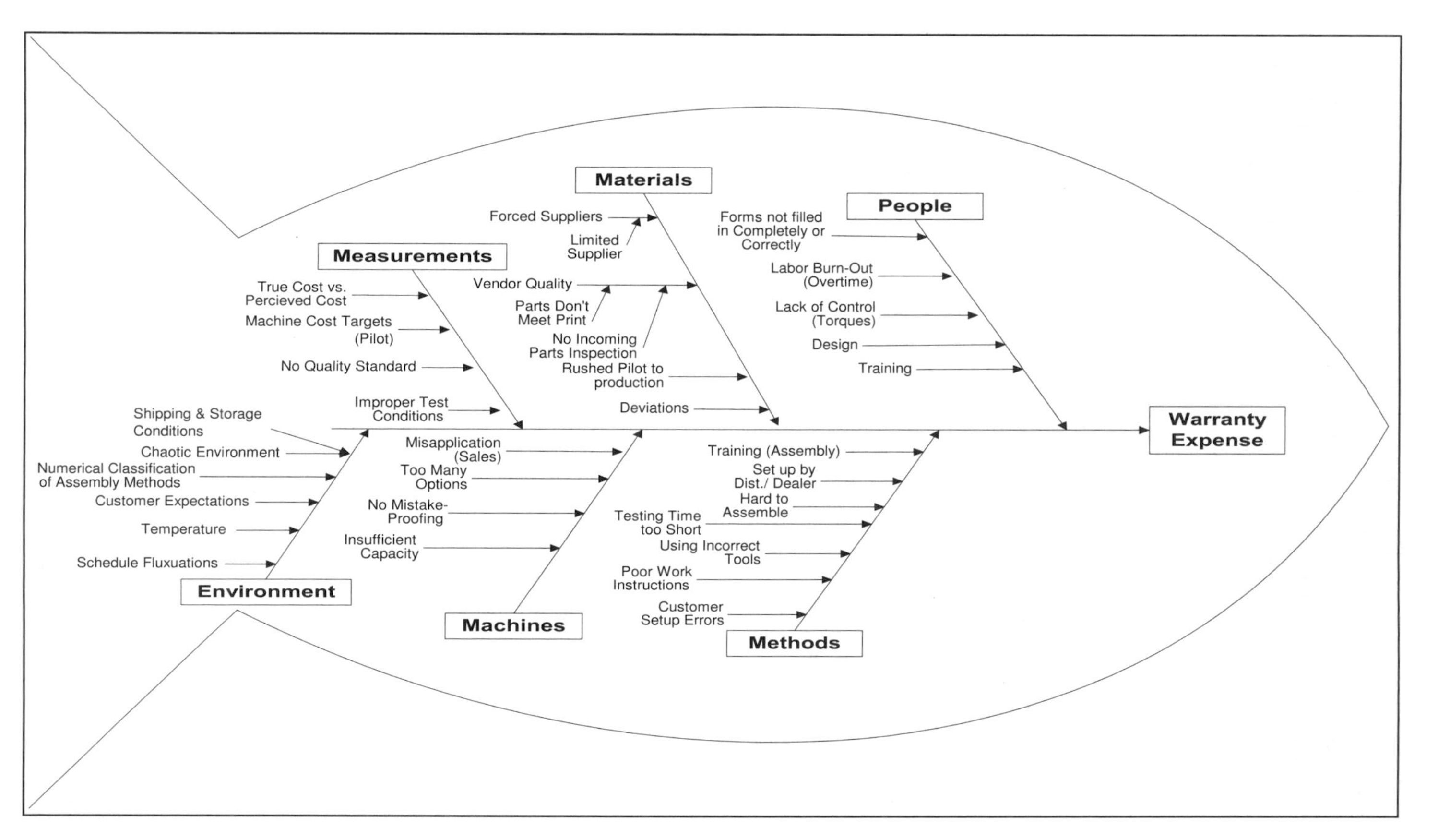

Figure 11.5 Warranty cost fishbone diagram.

will sustain its long-term gains. Quality system certifications are becoming a customer requirement and a trend in many industries. The validate phase of the ICOV DFSS algorithm requires that a solid quality system be employed in the DFSS project area.

The quality system objective is to achieve customer satisfaction by preventing nonconformity at all stages from design through service. A quality system is the Six Sigma deploying company's agreed-upon method of doing business. It is not to be confused with a set of documents that are meant to satisfy an outside auditing organization (i.e., ISO 900x). That is, a quality system represents the actions, not the written words of a company. The elements of an effective quality system include the quality mission statement, management reviews, company structure, planning, design control, data control, purchasing-quality-related functions (e.g., supplier evaluation and incoming inspection), design of product and process structures for traceability, process control, preventive maintenance, process monitoring and operator training, capability studies, MSA, audit functions, inspection and testing, service, statistical analysis, and standards.

Automated or manual control methods are used for both design (service or product) as well as their processes. Control methods include tolerancing, mistake-proofing (Poka-yoke), statistical process control charting [SPC, like X-bar&R or X&MR charts (manual or automatic), p&np charts (manual or automatic), c&u charts (manual or automatic), etc.], with or without warning, trend signals applied to control the process variables or monitor the design parameters, standard operating procedures (SOP) for detection purposes, and short-term inspection actions. In applying these methods, the DFSS team should revisit operator training to ensure proper control functions and to extract historical long-term and short-term information.

Control plans are the living documents of the manufacturing, assembly, or production environment. They are used to document all process control methods, as suggested by the FMEA or other DFSS algorithm steps like optimization and robust design studies. The control plan is a written description of the systems for controlling services, products, and processes. The control plan should be updated to reflect changes of controls based on experience gained over time.

12

FUNDAMENTALS OF EXPERIMENTAL DESIGN

12.1 INTRODUCTION

In the mid 1920s, a British statistician named Ronald Fisher put the finishing touches on a method for making breakthrough discoveries. Fisher's method, now known as design of experiments (DOE), has become a powerful tool for engineers, researchers, and Six Sigma practitioners.

Sir Ronald Fisher first used design of experiments as a research design tool to improve farm yields in early 1930s. Fisher was a geneticist working on improving crop yields in England using supervised field trials, fertilizers, and seed varieties as experimental factors. In his studies, Fisher encountered issues such as uncontrollable variation in the soil from plot to plot and the limited number of plots available for any given trial. Fisher solved these issues by varying the fertilizers or seed varieties used in the field. This action minimized the effects of soil variation in the analysis of the plot yields. Fisher also developed the correct method for analyzing designed experiments called *an*alysis *o*f *va*riance (ANOVA). This analysis method breaks up the total variation in the data into components from different sources. This analysis of variance delivers surprisingly precise results when applied to a small, well-structured matrix. Today, these components are called *signals* and *noise,* and the estimated effects are calculated as the *sum of squares* and the *signal-to-noise ratio*. There is a signal component for each controlled source (factor) in the experiment and a noise component representing variations not attributable to any of the controlled variations. ANOVA using the signal-to-noise ratio provides precise allocations of the effects of the factors and interactions. The ANOVA solutions developed for these problems work just as well in today's Six Sigma applications as they did in 20th century agriculture.

Design of experiments provides a powerful tool within the DFSS road map to accomplish breakthrough improvements in products, services, or process efficiency and effectiveness by optimizing the fulfillment of CTSs, FRs, and DPs. In each case, we create a response variable (y) and vary the factors that can cause a change in the performance of y. We can have CTS = f(FRs), or CTS = f(DPs), or FR = f(DPs), or DP = f(PVs) (see Section 8.5.2). It is important to note that most experimentation occurs in the relationships between DPs and PVs; if these do not yield breakthrough performance, then the design team should revisit the FRs and look at options. The most proper utilization of DOE is in the optimize phase within the road map of Figure 5.1. Optimization in a DOE sense implies finding the proper settings of influential factors that enable the DFSS team to shift the mean and reduce the variation of their design requirements or responses. DOE has been available for decades but its penetration in industry has been limited. Besides ignorance and lack of proper training, implementation has been resisted because of the discipline required as well as the use of statistical techniques. Scientist, managers, and service designers usually fear employment of statistics. Six Sigma Black and Green Belts play a big role in helping their teams and satellite members overcome such emotional barriers.

Traditionally, the approach to experimentation in a process required changing only one factor at a time (OFAT). Soon, it was found that the OFAT approach is incapable of detecting *interactions*[1] among the factors, which is a more probable and possible event than most professionals think. Therefore, design of experiments is also called statistically designed experiments. The purpose of the experiment and data analysis is to establish and detail the transfer functions[2] between outputs (e.g., design parameters) and experimental factors (e.g., process variables and noise factors) in a mapping that usually uses a P-diagram as illustrated in Figure 12.1.

A transfer function is the means of optimization and design detailing and is usually documented in the scorecard. A transfer function is treated as a living entity within the DFSS methodology that passes through its own life cycle stages. A transfer function is first identified using the proper mapping method (see Chapter 8) and then detailed by derivation, modeling, or experimentation. The prime uses of the transfer function are optimization and validation. Design transfer functions belonging to the same hierarchical level in the design structures (system, subsystem, or component) should be recorded in that hierarchical-level scorecard (see Chapter 6). A scorecard is used to record and optimize the transfer functions. The transfer function concept is integrated with other DFSS concepts and tools like mapping, robust design, and P-diagrams within the rigor of design scorecards (Yang & El-Haik, 2003).

In Fischer's experiments, the output, or response variable, y, was usually the yield of a certain farm crop. Controllable factors, $\mathbf{x} = (x_1, x_2, \ldots, x_n)$ were usually the "farm variables," such as the amount of various fertilizers applied, watering patterns, selection of seeds, and so on. Uncontrollable factors, $\mathbf{z} = (z_1, z_2, \ldots, z_p)$ could be soil types, weather patterns, and so on (see Figure 12.1) . In early agricultural experiments, the experimenter wanted to find the cause-and-effect relationship between the yield and controllable factors. That is, the experimenter wanted to know

[1]Interaction could be two level, three level, and so on. In a two-level interaction, a change in the response (output) of factor A occurs when factor B changes.

[2]Recall that DOE is a transfer-function-detailing method as presented in Chapter 6.

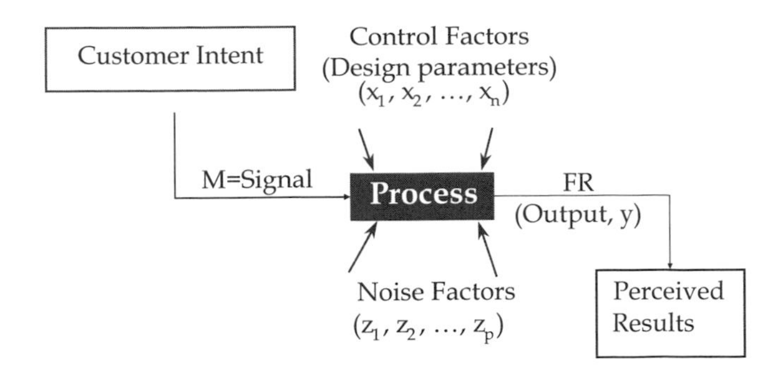

Figure 12.1 The P-diagram.

how different types of fertilizers, their application quantities, the watering pattern, and types of seeds, would influence the yield of the crop.

Today, DOE implementation has expanded to service industries. For example, SW Bell conducted a DOE to design a better telephone bill by varying a dozen factors including color, alignment, orientation, and shading. The new bill generated a 78% favorable rating (compared to 48% for the old bill) and a $2 million savings in postage. In advertising, Crayola designed and implemented a DOE to attract people to their new Internet site via an e-mail to teachers and parents. They found a combination of factors that yielded a 350% script improvement (Almquist, 2001). The Federal Aerospace Agency (FAA) used a fractional factorial DOE setup to minimize communication delays in the New York traffic control system utilizing simulation. Factors such as location of navigation signals and size of jets were employed (Box et al., 1978). On the marketing side, a candy producer used DOE to design a marketing strategy. Factors such as advertising medium, design of the package, candy bar size, and free samples were included (Holland & Cravens, 1973). John Deere Engine Works in Waterloo, Iowa, used DOE to improve the adhesion of its highly identifiable green paint onto aluminum. In the process, the company has discovered how to eliminate an expensive chromate-conversion procedure. Eastman Kodak in Rochester, New York, learned via DOE that it needed only to retool an existing machine instead of making a huge capital purchase for a new one. The solution meant improved light-sealing film-pack clips used by professional photographers, reducing setup time from 8 hours to 20 minutes while reducing scrap by a factor of 10.

The above are some examples of classical DOE case studies. Besides classical DOE, the Taguchi methods are another route for quality optimization strategies; it builds robustness into the design entity during the optimization phase. The Taguchi method is a combination of sound engineering design principles and Taguchi's version of DOE, which is called an orthogonal array experiment. In this chapter, we introduce both schools of DOE for transfer function detailing (also see Chapter 6). Other aspects of DOE methods within DFSS can be found in Yang and El-Haik (2003).

12.2 HYPOTHESIS TESTING

When testing and observing changes in processes, how does one know that the observed response was due to something that we controlled (changed) rather than a random effect from a host of environmental noise factors? At what point are we confident that the measured change is significant? These are some of the questions that hypothesis testing helps provide answers to.

Hypothesis testing is a method of inferential statistics that is aimed at testing the viability of a hypothesis about a certain population parameter in the light of experimental data. This hypothesis is usually called the null hypothesis. In the most common use of hypothesis testing, the null hypothesis is put forward and it is determined whether the available data are strong enough to reject it. If the data are very different from what would be expected under the assumption that the null hypothesis is true, then the null hypothesis is rejected. If the data are not greatly at variance with what would be expected under the assumption that the null hypothesis is true, then the null hypothesis is not rejected. Failure to reject the null hypothesis is not the same thing as accepting the null hypothesis.

In DOE, hypothesis testing is essential for comparing if–then factorial settings (design parameter or process variables) and, therefore, design alternatives. Two or more design alternatives can be compared with the goal of identifying the optimum design alternative relative to some requirements. For example, in the context of simulation modeling (see Chapter 14), the null hypothesis is often the reverse of what the simulation analyst actually believes about the model performance. Thus, the null hypothesis is put forward to allow the collected simulation data to contradict it and, hopefully, discredit it. If the Six Sigma Black or Green Belt has proposed a new design alternative to a certain production process, he or she would be interested in testing experimentally whether the proposed design works better than the existing production process. To this end, the team would design an experiment comparing the two methods of production. The hourly throughput of the two processes could be collected and used as data for testing the viability of the null hypothesis. The null hypothesis would be that there is no difference between the two methods (the estimated population means of the throughput for the two production processes, μ_1 and μ_2, are identical). In such case, the analyst would be hoping to reject the null hypothesis and conclude that the proposed method he or she developed is a better approach.

The symbol H_0 is used to indicate the null hypothesis, where "null" refers to the hypothesis of no difference. For the example just given, the following would designate the null hypothesis: H_0: $\mu_1 = \mu_2$, or H_a: $\mu_1 \neq \mu_2$. The alternative hypothesis (H_a) is simply that the mean throughput of the proposed method (μ_1) is higher than that of the current production process (μ_2), that is, H_a: $\mu_1 > \mu_2$.

It is possible for the null hypothesis to be that the difference (d) between population means is of a particular value (H_0: $\mu_1 - \mu_2 = d$), or the null hypothesis could be that the population mean is of a certain standard value (H_0: $\mu = \mu_0$).

The test of hypothesis implies defining a certain design parameter, process variables, or functional requirement as a random variable with a valid measurement system, derived test statistics, and an associated distribution, as well as a determined rejection area. Rejection area reflects a 100% $(1-\alpha)$ confidence level in the hypothesis decision. The Greek symbol α represents the risk associated with the party affected by rejecting the null hypothesis. Data is collected by random sampling techniques. Hypothesis testing requires normality as an assumption. In mean hypothesis testing, most tests involve comparisons of a mean performance with a certain value or with another process mean of the same random variable. For more than two populations, DOE is used to test such hypotheses of means equality, that is, H_0: $\mu_1 = \mu_2 = \ldots = \mu_a$, where a is the number of different populations (settings) of a random variable, say A, that will be tested. When the variance of the random variable (σ^2) is known, the Z score, Z_0, or the Z value under true H_0, is used as a test statistic for the null hypothesis H_0: $\mu = \mu_0$, assuming that the observed population is normal or the sample size is large enough so that the central limit theorem applies. Z_0 is computed as follows:

$$Z_0 = \frac{\overline{X} - \mu_0}{\sigma/\sqrt{n}} \tag{12.1}$$

The 100% $(1-\alpha)$ confidence interval on the true population mean is

$$\overline{X} - Z_{\alpha/2}\sigma/\sqrt{n} \le \mu \le \overline{X} + Z_{\alpha/2}\sigma/\sqrt{n} \tag{12.2}$$

We use the absolute value of Z_0 because the normal distribution is symmetrical and this is shown as $|Z_0|$. Therefore, the null hypothesis H_0: $\mu = \mu_0$, would be rejected if $|Z_0| > Z_{\alpha/2}$ when H_a: $\mu \neq \mu_0$, $Z_0 < -Z_\alpha$ when H_a: $\mu < \mu_0$, and $Z_0 > Z_\alpha$ when H_a: $\mu > \mu_0$. For the null hypothesis H_0: $\mu_1 = \mu_2$, Z_0 is computed as follows:

$$Z_0 = \frac{\overline{X}_1 - \overline{X}_2}{\sqrt{\dfrac{\sigma_1^2}{n_1} + \dfrac{\sigma_2^2}{n_2}}} \tag{12.3}$$

The null hypothesis H_0: $\mu_1 = \mu_2$ would be rejected if $|Z_0| > Z_{\alpha/2}$ when H_a: $\mu_1 \neq \mu_2$, $Z_0 < -Z_\alpha$ when H_a: $\mu_1 < \mu_2$, and $Z_0 > Z_\alpha$ when H_a: $\mu_1 > \mu_2$.

When the variance parameter (σ^2) is unknown, which is typically the case in real-world applications, the t statistic, t_0, is used as a test statistic for the null hypothesis H_0: $\mu = \mu_0$. The test statistic, t_0, is computed as follows:

$$t_{0,n-1} = \frac{\overline{X} - \mu_0}{s/\sqrt{n}} \tag{12.5}$$

where n is the sample size. The 100% $(1-\alpha)$ interval on the true population mean is

$$\overline{X} - t_{\alpha/2,n-1}s/\sqrt{n} \le \mu \le \overline{X} + t_{\alpha/2,n-1}s/\sqrt{n} \tag{12.6}$$

The null hypothesis H_0: $\mu = \mu_0$ would be rejected if $|t_0| > t_{\alpha/2,n-1}$ when H_a: $\mu \neq \mu_0$, $t_0 < -t_{\alpha,n-1}$ when H_a: $\mu < \mu_0$, and $t_0 > t_{\alpha,n-1}$ when H_a: $\mu > \mu_0$. For the null hypothesis H_0: $\mu_1 = \mu_2$, t_0 is computed as

$$t_{0,n_1-n_2-2} = \frac{\overline{X}_1 - \overline{X}_2}{\sqrt{\dfrac{s_1^2}{n_1} + \dfrac{s_2^2}{n_2}}} \tag{12.7}$$

The degrees of freedom $n_1 - n_2 - 2$ implies n_1 and n_2 samples where drawn for population 1 and population 2, respectively. Similarly, the null hypothesis H_0: $\mu_1 = \mu_2$ would be rejected if $|t_0| > t_{\alpha/2,n_1-n_2-2}$ when H_a: $\mu_1 \neq \mu_2$, $t_0 < -t_{\alpha,n_1-n_2-2}$ when H_a: $\mu_1 < \mu_2$, and $t_0 > t_{\alpha,n_1-n_2-2}$ when H_a: $\mu_1 > \mu_2$.

Although the null hypotheses discussed so far have all involved the testing of hypotheses about one or more population means, null hypotheses can involve any parameter such as an experiment investigating the variance (σ^2) of two populations, or the proportion between two variables, or the correlation between two variables. As an example, the correlation between job satisfaction and performance on the job would test the null hypothesis that the population correlation (ρ) is 0. Symbolically, H_0: $\rho = 0$.

Sometimes it is required to compare more than two alternative systems. In such cases, most practical studies rely on conducting multiple paired comparisons using several paired-*t* confidence intervals, as discussed above. Other statistical methods can still be used to compare more than two design alternatives with respect to a given performance measure, that is, functional requirement, design parameter, or process variable. Design of experiments is an advanced statistical method that is often utilized for comparing a much larger number of alternative systems. This is the subject of this chapter.

In all cases of hypothesis testing, a significance test is performed to determine if an observed value of a statistic differs enough from a hypothesized value of a parameter (null hypothesis) to draw the inference that the hypothesized value of the parameter is not the true value. A significance test consists of calculating the probability of obtaining a statistic that differs from the null hypothesis (given that the null hypothesis is correct) more than the statistic obtained in the sample. If this probability is sufficiently low, then the difference between the parameter and the statistic is considered to be "statistically significant." The significance level is used in hypothesis testing as follows. First, the difference between the results of the experiment and the null hypothesis is determined. Then, assuming the null hypothesis is true, the probability of obtaining a difference that is as large or larger is computed. Finally, this probability is compared to the significance level (α). If the probability is less than or equal to the significance level, then the null hypothesis is rejected and the outcome is said to be statistically significant. The lower the significance level, the more the data must diverge from the null hypothesis to be significant. Therefore, the 0.01 level is more conservative than the 0.05 level.

There are two kinds of errors that can be made in significance testing: Type I error (α), in which a true null (there is no difference) can be incorrectly rejected; and Type II error (β), in which a false null hypothesis (there is a difference) can fail to be rejected. A Type II error is only an error in the sense that an opportunity to reject the null hypothesis correctly was lost. It is not an error in the sense that an incorrect conclusion was drawn, since no conclusion is drawn when the null hypothesis is not rejected. Table 12.1 summarized the two types of test error.

Table 12.1 The two types of test errors

Statistical decision	True state of null hypothesis (H_0)	
	H_0 is true	H_0 is false
Reject H_0	Type I error (α)	Correct
Accept H_0	Correct	Type II error (β)

Type I errors are generally considered more serious than Type II errors since they result in drawing conclusions that the null hypothesis is false when, in fact, it is true. The experimenter often makes a trade-off between Type I and Type II errors. The more an experimenter protects him or herself against Type I errors by choosing a low level, the greater the chance of a Type II error. Requiring very strong evidence to reject the null hypothesis makes it very unlikely that a true null hypothesis will be rejected. However, it increases the chance that a false null hypothesis will not be rejected, thus lowering the test power. The probability of a Type I error (α) is called the significance level and is set by the experimenter. The significance level (α) is commonly set to 0.05 and 0.01, the latter being more conservative since it requires stronger evidence to reject the null hypothesis at the 0.01 level then at the 0.05 level. Hypothesis test power is the probability of correctly rejecting a false null hypothesis. Power is therefore defined as $1 - \beta$, where β is the Type II error probability. It is the probability that the data gathered in an experiment will be sufficient to reject the null hypothesis. If the power of an experiment is low, then there is a good chance that the experiment will be inconclusive. There are several methods for estimating the test power of an experiment before the experiment is conducted. If the power is too low, then the experiment can be redesigned by changing one of the factors that determine the power, such as the sample size, the process standard deviation (σ), or the size of difference between the means of the tested processes.

12.3 CLASSICAL DOE BASICS

In DOE, we deliberately change experimental factors (ones we can control) and observe their effects on the output responses, the design requirements. Experimental runs are randomly conducted to prevent trends and to allow the factor effects to reveal their true, unbiased significance. Randomness is a very important aspect of classical DOE. In a classical DOE study, data collection and result interpretation are all dependant on this assumption. The data obtained in the experiment will be used to fit empirical models relating an output, y, with the experimental factors, the xs. Mathematically, we are trying to find the following transfer function relationship:

$$y = f(x_1, x_2, \ldots, x_n) + \varepsilon \tag{12.8}$$

where ε (epsilon) is experimental error or experimental variation. The existence of ε means that there may not be an exact functional relationship between y and (x_1, x_2,

$\ldots, x_n$). This is because the uncontrollable factors ($z_1, z_2, \ldots, z_p$) will influence the requirement y but are not accounted for in Equation (12.1), and there are experimental and measurement errors on both y and ($x_1, x_2, \ldots, x_n$) in the experiment. A DOE study within a DFSS study road map will follow a multiple step methodology, described in the following subsections.

12.3.1 DOE Study Definition

DOE is used in research as well as in product and service optimization settings, although sometimes for very different objectives. The primary purpose in scientific research is usually to show the statistical significance of an effect that a particular factor exerts on the dependent variable of interest (a design requirement denoted as "y"). In an optimization DOE, the primary objective is usually to extract the maximum amount of *unbiased* information regarding the factors affecting a process or product from as few observations as possible to minimize cost. In the research application, DOE techniques are used to uncover the interactive nature of the application that is manifested in higher-order interactions (those involving three or more factors). In a robust design variability experiment (see Chapter 13) , interaction effects are often regarded as a "nuisance" as they only complicate the process of identifying significant factors.

In either case, this is not trivial. The team needs to decide on the objective of the DOE study or studies within their DFSS project road map. Do they want to reduce defects? Is their purpose to improve current service performance? What is the study scope within the mapping (see Chapter 8)? Do they work on a process or a subprocess? Is one DOE sufficient?

To develop an overall DOE, we suggest the following steps:

1. Define the problem and set the objectives
2. Select the responses
3. Select the factors and levels
4. Identify noise variables
5. Select the DOE design
6. Plan the experiment with regard to
   - Resources
   - Supplies
   - Schedule
   - Sample size
   - Risk assessment

The following process should be followed:

- Understand the current state of the DFSS project by reviewing and developing an understanding of the technical domains that are active in the process or processes being optimized. A review of the physical laws, process and systems behaviors, and underlying assumptions is appropriate at this stage.

- Develop a shared vision for optimizing service and process mappings. The team is advised to conduct a step-by-step review of the DOE methodology to understand and reinforce the importance of each strategic activity associated with experimentation and to facilitate consensus on the criteria for completion of each activity.
- The DFSS team will need an appreciation for what elements of experimental plan development will require the greatest time commitment. The team will also need to discuss the potential impact associated with compromise of the key DOE principles, if any.
- Plan time allocation. The team needs to develop a work plan that includes timing for several individual DOE sequences, if required.
- Plan time for measurement system(s) verification.
- Develop a P-diagram for preliminary experimentation to determine the most important sources of variation, including the noise factors.
- Plan for selection of experimental factors and logistical considerations for building test samples, and so on.
- Allocate a budget to perform the DOE plan. Develop cost estimation and seek resource approval to conduct the DOE test plan. Plan budgeting for multiple optimization experiments.
- Add something about risk consideration. What effect will experimental conditions have on yield, scrap, rework, and safety? Will it affect the end user or will it be contained internally? What will happen to the experimental units after the experiment? Will they be scrapped or be reused?

12.3.2 Selection of the Response

The purpose of describing the process in terms of its inputs and outputs is to structure the development of the DOE strategy. The description provided by process mapping provides an effective summary of what level of process optimization are taking place, what measurement approach the DOE optimization will be based on, and what major sources of variation influence the process. As such, the following actions are suggested:

- Revisit the definition of the responses in the DOE by developing a characterization of boundary conditions and describe the process(es) delivering them in terms of the inputs and outputs using a P-diagram. The description is also a characterization of the basic function of the system, which has been identified in Chapter 8 using the mapping techniques of the team choice. The response can be generalized as a measurement characteristic related to the process mappings of a service process map (e.g., a functional requirement or a design parameter). It may be helpful to list various measurement characteristics, which can be viewed as alternative measurement approaches.

- After study definition, the team needs to select the requirements that will be optimized. In this example, we will explore the case of a single requirement (y). In selecting a response variable, the DFSS team should select the left-hand side of a defined transfer function of their study as depicted in the project process and functional mappings. DOE will be used to detail their design and to provide other useful information about the design under study. It is desirable for the y to be a continuous variable characteristic of the design, making data analysis much easier and meaningful, a variable that can be easily and accurately measured.

12.3.3 Choice of DOE Factors, Levels, and Ranges

In general, every step used in a process allows for the adjustment of various factors affecting the resultant quality of the service produced. Experimentation allows the design team to adjust the settings in a systematic manner and to learn which factors have the greatest impact on the resultant functional requirement or design parameter. Using this information, the settings can be constantly improved until optimum quality is obtained.

Factors can be classified as control or noise. Control factors are design parameters or process variables that are freely specified by the design team using their knowledge about the concept design and the technology that is being developed for the purpose of DOE optimization.

There are two kinds of factors: continuous and discrete. A continuous factor can be expressed over a defined real number interval with one continuous motion of the pencil. For example, weight, speed, and price are continuous factors. A discrete factor is also called a categorical variable or attribute variable. For example, types of marketing strategy, types of delivery methods (face to face or via a third party), and types of operating system are discrete factors. Historic information and brainstorming by the team will facilitate a P-diagram listing of potential factors belonging to the categories (response, control, and noise)[3] and aid in structuring the development of a strategy for DOE.

A key aspect of DFSS philosophy is that during the design stage, inexpensive parameters can be identified and studied, and can be combined in a way that will result in performance that is insensitive to uncontrollable sources of variation. The team's task is to determine the combined best settings (parameter targets) for each of the control parameters that have been judged by the design team to have the potential to improve the output(s) of interest. The selection of factors will be done in a manner that will enable target values to be varied during experimentation with no major impact on service cost. The greater the number of potential control factors that are identified, the greater the opportunity for optimization of the functional output in the presence of noise factors.

[3]In robust design classification, another category, not considered here, is the "signal" factor. A signal is a parameter that is controlled by the customer/user of the service (or by the output from another process element) to express the intended value of the service or product response.

Noise factors cause the functional requirement (response "y") to deviate from the intended performance or target desired by the customer. Noise factors can be classified into external sources (usage and environment), unit-to-unit sources (production and supplier variation), and deterioration sources (wear-out or the general effects of usage over time). The concept to account for noise factors was advanced by Taguchi within his robustness methodology and will be explored in chapter 13. For a classical DOE setting, the subject of this chapter, it is important to have the design team conduct such categorization of input factors (both control and noise) and document such classification in the P-diagram. Selected factors of both categories deemed to be experimentation candidates by the team are considered as factorial variables in the same array of DOE testing. The idea is to look for interaction between the two categories in order to reduce the variation on the design functional requirements (DOE responses).

In a DOE study, each experimental factor will be changed at least once, that is, each factor will have at least two settings. Otherwise, that factor will not be a variable but rather a fixed factor in the experiment. The number of settings of a factor in the experiment is called levels. For a continuous factor, the levels often correspond to different numerical values. For example, two levels of a product price could be given as \$190 and \$195. For continuous factors, the range of a variable is also important. If the range of variable is too small, then we may miss lots of useful information. If the range is too large, then those extreme values might yield infeasible experimental runs or defective or unuseable output. For a discrete variable, the number of levels is often equal to "the number of useful choices." For example, if the color of the product in a marketing DOE is the factor, then the number of levels depends on how many preferred choices there are and which levels the team want to test in this experiment. For example, a retailer may use a designed experiment to quantify the effect of factors such as price, advertising cost, display space (in square footage), and location on profit and revenue. The first three factors are continuous, whereas the location is categorical. The choice of the number of levels in an experiment depends on time and cost considerations. The more levels we have in experimental factors, the more information we will get from the experiment, but there will be more experimental runs, leading to higher cost and longer time to conclusion.

12.3.4 Select DOE Strategy

The choice of DOE objectives (recall Section 12.3.1) has a profound effect on whether to conduct a research, optimization, or screening DOE. In science and research, the primary focus is on experimental designs with up to, perhaps, five factors and more emphasis is focused on the significance of interactions. However, experimentation in industry is broader in implementation, with many factors whose interaction effects cannot be evaluated. The primary focus of the discussion is placed on the derivation of unbiased main effect (and, perhaps, two-way) estimates with a minimum number of observations.

Although DOE is used primarily for optimization in the DFSS road map, there are many situations in which the knowledge is not profound and the team may resort to exploring factor relations via a screening-type DOE. In this type of DOE, the objective is to segregate the vital few influential factors while withholding factors known to have little effect. It is not uncommon that there are very many different factors that may potentially be important. Special designs (e.g., Plackett–Burman designs and Taguchi orthogonal arrays) have been developed to screen such large numbers of factors in an efficient manner, that is, with the least number of observations necessary. For example, you can design and analyze an experiment with n factors and only $n + 1$ runs; you will be able to estimate the main effects for each factor and, thus, you can quickly identify which ones are important and most likely to yield improvements in the process under study.

A confirmation-type DOE is used to prove the significance of what was found in the screening DOE, and usually follows screening DOEs. In most service cases, the confirmation type is used to detail design transfer functions as well as optimization. When a transfer surface rather than a transfer function is desired (see Chapter 6), a special optimization technique called response surface methodology is usually used to find the optimum design in the design space. In these cases, a verification type of DOE should be performed to confirm the predictability of the transfer function or surface for all responses of interest and for which the process remains optimum or robust under use conditions.

The purpose of DOE strategy is to coordinate all the knowledge about the process and/or service under development into a comprehensive experimentation and data-collection plan. The plan should be designed to maximize research and development efficiency through the application of a sound testing array, functional requirements (responses), and other statistical data analysis.

The DFSS team is encouraged to experimentally explore as many factors as feasible to investigate the functional performance potential of the design being adopted from the characterize phase within the project map. Transferability of the improved functional performance to the customer environment will be maximized as a result of the application of a sound optimization strategy during data collection.

Data from the optimization experiment will be used to generate an analytical model, which will aid in improving the design sigma level in the selected DOE response. The validity of this model and the resulting conclusions will be influenced by the experimental and statistical assumptions made by the team. For example, what assumptions can be made regarding the existence of interactions between factor main effects? What assumptions can be made (if any) regarding the underlying distribution of the experimental data? What assumptions can be made regarding the effect of nuisance factors on the variance of the response?

12.3.5 Develop a Measurement Strategy for DOE

The objective of developing a measurement strategy is to identify validated measurement system(s) in order to observe the output of the process that is being developed or improved. The measurement strategy is the foundation of any experimentation effort. In optimization DOE, the team should revisit the objective and express (in quantifiable terms) the level of functional performance that is to be expected from the DOE. An opportunity statement for optimized performance, translated in terms of total cost or time-to-market effects, will help rationalize cost versus quality trade-offs should they arise. Note that binary variables (pass/fail) or any discrete (attribute) data are symptoms of poor functional choices and are not experimentally efficient for optimization.

The team needs to validate that the measurement approach has a high strength of association to changing input conditions for the service under design. The way to determine and quantify this association is to use a correlation analysis. The design requirement (y) is measured over a range of input parameters values. The input variable can be any important design parameter, and the analysis can be repeated for additional parameters in an effort to determine the degree of correlation. In this context, measurement repeatability improves with increasing positive correlation. Interpretation of the quality of a measurement system can be accomplished by applying confidence intervals to a plot of y data calculated using data statistics. When statements about the data are bounded by confidence intervals, a much clearer picture can be derived that influences conclusions about what the data are saying. The ability to make precise statements from the data increases with increasing sample sizes. A black belt can calculate the appropriate sample size for a desired confidence region using statistical methods reviewed in Chapter 2. The following formula can be used to approximate the number of replications[4] (n) that are needed to bring the confidence interval half-width (h_w) down to a certain specified error amount at a certain significance level (α):

$$n = \left[\frac{(Z_{a/2})s}{h_w} \right]^2 \tag{12.9}$$

where s is the sample standard deviation.

In analyzing a measurement system, the team needs to assure its repeatability and reproducibility (R&R). Refer to Chapter 2 for a discussion of measurement systems. The team should analyze and verify that the repeatability error of the measurement system is orders of magnitude smaller than the tolerance of interest for the specimens to be measured. Otherwise, it will become difficult to measure the actual effect of design parameter changes during experimentation because the effect on the system response will be masked by excessive measurement error. Taking repeated samples and using the average for a response could reduce the effect of measurement error.

12.3.6 Experimental Design Selection

There are varying considerations that enter into the different types of designs. In the most general terms, the goal is always to allow the design team to evaluate in an unbiased (or least-biased) way the consequences of changing the levels of a particular

[4]To replicate is to run each combination of factor levels in the design more than once. This allows the estimation of the so-called pure error in the experiment.

factor, that is, regardless of how other factors were set. In more technical terms, the team attempts to generate designs in which main effects are *unconfounded* among themselves and, in some cases, even unconfounded with the interaction of factors.

Experimental methods are finding increasing use in the service industry to optimize processes. Specifically, the goal of these methods is to identify the optimum settings for the different factors that affect a process.

The general representation of the number of experiments is l^k where l is the number of levels of decision variables and k is the number of factors. Using only two levels of each control factor (low and high) often results in 2^k factorial designs. The easiest experimental method is to change one design factor while the other factors are kept fixed. Factorial design, therefore, looks at the combined effect of multiple factors on system performance. Fractional and full factorial DOE are the two types of factorial design. The major classes of designs that are typically used in experimentation are:

- 2^k full factorial
- $2^{(k-p)}$ (two-level, multifactor) designs screen designs for large numbers of factors
- $3^{(k-p)}$ (three-level, multifactor) designs (mixed designs with two- and three-level factors also supported)
- Central composite (or response surface) designs
- Latin square designs
- Taguchi robust design analysis (see Chapter 13)
- Mixture designs and special procedures for constructing experiments in constrained experimental regions

Interestingly, many of these experimental techniques have "made their way" from the product arena into the service arena, and successful implementations have been reported in profit planning in business, cash-flow optimization in banking, and so on.

The type of experimental design to be selected will depend on the number of factors, the number of levels in each factor, and total number of experimental runs that can be afforded. In this chapter, we are primarily considering full factorial designs (2^k type) and fractional factorial designs (2^{k-p} type) on the classical DOE side. If the number of factors and levels are given, then full factorial experiments will need more experimental runs, and thus be more costly, but they also provide more information about the design under study. The fractional factorial experiments will need fewer runs, and thus be less costly, but they will also provide less information about the design. This will be discussed in subsequent sections.

12.3.7 Conduct the Experiment

Classical DOE of random effects (the subject of this chapter) uses runs (observations) that are created either by running trials of factor combinations or by using representative analytical modeling to simulate the desired DOE combinations. The experimental samples should be as representative of production as reasonable (as close to the production pedigree as possible). The sample set utilization described below suggests how many samples are required. The test sample combinations are defined by the DOE methods in order to preserve orthogonality, so the factors are independent of each other, and are prepared accordingly. Experimental conditions should be clearly labeled for identification and imposed noise factors considered. A data collection form will help to organize data collection during experimentation. In addition, it is important to record and identify the specific experimental combination for each sample. This information can be used to recreate the actual conditions and to regenerate the data if necessary. The team should also record the actual run order to assist in the recreation of conditions if this should become necessary later in the analysis. To verify the level and precise combination of factor conditions during each experimental trial, a record is made of the actual conditions for each experimental sample at the time the response data were collected. When running the experiment, we must pay attention to the following (Yang & El-Haik, 2003):

1. Check the performance of measurement devices first.
2. Check that all planned runs are feasible.
3. Watch out for process drifts and shifts during the run.
4. Avoid unplanned changes (e.g., swap operators halfway through).
5. Allow some time (and back-up material) for unexpected events.
6. Obtain buy-in from all parties involved.
7. Preserve all the raw data
8. Record everything that happens.
9. Reset all factors to their original state after the experiment.

12.3.8 Analysis of DOE Raw Data

There are several statistical methods for analyzing designs with random effects. The ANOVA technique provides numerous options for estimating variance components for random effects, and for performing approximate hypothesis tests using F-tests based on synthesized error terms.

Statistical methods will be used. A large portion of this chapter is dedicated to how to analyze the data from a statistically designed experiment. From the analysis of experimental data, we are able to identify the significant or insignificant effects and interactions, the arguments of the transfer function. Not all the factors are the same in terms of their effects on the output. When the DFSS team changes the level of a factor, if its impact on the response is relatively small, in comparison with inherited experimental variation due to uncontrollable noise factors and experimental error, then this factor might be an insignificant factor. Otherwise, if a factor has a large impact on both the response mean and variance then it might be a significant factor. Sometimes, two or more factors may have interactions; in this case, their effects on the output will be complex. However, it is also possible that none of the experimental factors will be found to be significant. In this case, the experiment is inconclusive in finding influential factors yet it leaves us with the knowledge of factors that are not significant. This situation may indicate that we may have missed important factors in the experiment. DOE data analysis can identify significant and insignificant factors by using ANOVA. In addition, the DFSS team will be able to rank the relative importance of factor effects and their interactions using ANOVA with a numerical score.

As a typical output, a DOE data analysis provides an empirical mathematical transfer function relating the output, y, to experimental factors. The form of the transfer function could be linear or polynomial with significant interactions. DOE data analysis can also provide graphical representations of the mathematical relationship between experimental factors and output, in the form of main effect charts and interaction charts.

Additionally, if there were an ideal direction of goodness for the output, for example, if y were the satisfaction of customers, then the direction of goodness for y would be higher the better. By using the mathematical transfer function model, DOE data analysis identifies the best setting of experimental factors, which will achieve the best possible result for the output.

An advantage to using DOE for optimization includes the ability to use statistical software to develop the transfer function to make predictions about any combination of the factors in between and slightly beyond the different levels, and to generate types of informative two- and three-dimensional plots. In addition, the design team needs to be aware of the fact that DOEs do not directly compare results against a control or standard. They evaluate all effects and interactions and determine if there are statistically significant differences among them. Statistical confidence intervals can be calculated for each response optimized. A large effect might result, but if the statistical error for that measurement is high, then that effect is not believable. In other words, we get a large change in output but we cannot be certain that it was due to anything that we did.

12.3.9 Conclusions and Recommendations

Once the data analysis is completed, the DFSS team can draw practical conclusions about their study. If the data analysis provides enough information, we might be able to make recommendations for some changes to the design to improve its robustness and performance. Sometimes, the data analysis cannot provide enough information and additional experiments may be needed to be run. When the analysis of the experiment is complete, one must verify whether the conclusions are good or not. These are called validation or confirmation runs. The interpretation and conclusions from an experiment may include a "best" setting to use to meet the goals of the experiment. Even if this "best" setting were included in the experiment, it should be run again as part of the confirmation runs to make sure nothing has changed and that the response values are close to their predicted values. Typically, it is very desirable to have a stable process. Therefore, one should run more than one test at the "best" settings. A minimum of three runs should be conducted. If the time between actually running the experiments and conducting the confirmation runs is more than the average time span between the experimental runs, the DFSS team must be careful to ensure that nothing else has changed since the original data collection. If the confirmation runs do not produce the results expected, then they team needs to verify that they have the correct settings for the confirmation runs, revisit the transfer function model to verify the "best" settings from the analysis, and verify that they had the correct predicted value for the confirmation runs. Otherwise, the transfer function model may not predict very well in the defined design space. Nevertheless, the team will still learn from the experiment and they should use the information gained from this experiment to design another follow-up or substitute experiment.

12.4 FACTORIAL EXPERIMENT

In the DFSS project road map (see Chapter 5), experimental design is usually conducted for optimization, that is, to produce a credible transfer function or surface plot. This usually means determining how an output of a process responds to a change in some factors. In addition, a DOE is conducted following the optimization phase to validate or confirm the obtained transfer function. In either case, careful planning (see Section 12.2) at the outset of the DOE will save a great deal of resources.

If the response (y) under consideration is dependant on one factor, the DOE strategy is simple—conduct an experiment by varying the factor and measure the corresponding response at these values. A transfer function is obtained by fitting a line or a curve to the experimental observations. In a classical DOE, the best factor levels are used to obtain a certain level of information, the predictive equation, or the transfer function. The proper choice of experimental conditions increases the information obtained from the experiment. The recognition and measurement of interactions is of real value in the study of processes in which interaction is common. A DOE can also be used to reduce the resources needed in experimenting by eliminating redundant observations and ending up with approximately the same amount of information from fewer experimental runs.

Most experiments involve two or more experimental factors. In this case, factorial designs are the most frequently used designs. By a factorial design, we mean that all combinations of factor levels will be tested in the experiment. For example, if we have two factors in the experiment, say factor A and B, each having multiple levels, and if A has a levels and B has b levels, then in a factorial experiment we are going to test all ab combinations. In each combination, we may duplicate the experiment several times, say, n times. Then there are n replicates in the experiment. If $n = 1$, then the experiment is called single replicate. Therefore, for two factors, the total number of experimental observations (runs) is equal to abn. For example, in an experiment of two factors, factor A and factor B, with each factor at two levels, the number of runs equals $2 \times 2 \times n = 4n$. In general, a two-factor factorial experiment (factors A and B) has the arrangement of Table 12.2. Each cell of Table 12.2 corre-

Table 12.2 General arrangement for a two-factor factorial design

	Factor B			
Factor A	1	2	...	b
1	Y_{111}	Y_{121}	...	Y_{1b1}
	Y_{112}	Y_{122}	...	Y_{1b2}
2	Y_{211}	Y_{221}	...	Y_{2b1}
	Y_{212}	Y_{222}	...	Y_{2b2}
⋮	⋮	⋮	⋮	⋮
a	Y_{a11}	Y_{a21}	...	Y_{ab1}
	Y_{a12}	Y_{a22}	...	Y_{ab2}

sponds to a distinct factor level combination, known as a *treatment* in DOE terminology.

12.5 MATHEMATICAL TRANSFER FUNCTION MODELS

If we denote A as x_1 and B as x_2, then one possible mathematical transfer function model is

$$y = f_1(x_1) + f_2(x_2) + f_{12}(x_1, x_2) + \varepsilon \tag{12.10}$$

Here $f_1(x_1)$ is the main effect of A, $f_2(x_2)$ is the main effect of B, and $f_{12}(x_1, x_2)$ is the interaction of A and B.

12.6 WHAT IS INTERACTION?

An interaction between factors occurs when the change in response from one level to another level of one factor is not the same as the change in response at the same two levels of a second factor, that is, the effect of one factor is dependent upon a second factor. Interaction plots are used to compare the relative strength of the effects across factors.

If there is no interaction, then a transfer function with two parameters can be written as

$$y = f_1(x_1) + f_2(x_2) + \varepsilon \tag{12.11}$$

where $f_1(x_1)$ is a function of x_1 alone, and $f_1(x_2)$ is a function of x_2 alone. We call above model an additive model (see Chapter 6). However, if the interaction effect is not equal to zero, then we do not have an additive transfer function model. Let us look at the following example.

Example 12.1: Marketing Case Study (McClave et al., 1998)

Most short-term supermarket strategies such as price reductions, media advertising, and in-store displays are designed to increase unit sales of particular products temporarily. Factorial designs such as the following have been described in the *Journal of Marketing Research* that evaluate the effectiveness of such strategies. Factor A is the "Price" at the following levels: regular, reduced, and cost to supermarket, that is, $a = 3$. Factor B is the "Display Level" at the following levels: normal display space, normal space plus end of aisle display, and twice the normal display space, that is, $b = 3$. A complete factorial design based on these two factors involves nine treatments. Each treatment is applied three times ($n = 3$) to a particular product at a particular supermarket. Each application lasts a full week and the dependant variable (y) of interest is "unit sales for the week." To minimize treatment carry over effects, each treatment is preceded and followed by a week in which the product is priced at its regular price and is displayed in its normal manner. Table 12.3 reports the data collected.

The values in the table are the number of units sold in the treatment application week. For example, if the supermarket applied the Price = "Regular" and Display = "Normal" values the units sold are 989, 1025, and 1030 for each of the three weeks. By simply plotting the response versus different factor level combinations, we produce the main effect and interaction plots of mean data shown in Figure 12.2 and Figure 12.3, respectively.

For example, in Figure 12.3, the data mean at the Price = "Cost to supermarket" and Display = "Normal Plus" treatment is calculated as (2492 + 2527 + 2511)/3 = 2510 units sold. Clearly, the effect of factor A is linear, but this is not the case for factor B, as reflected by the curvature at the "Normal Plus" level of factor B in Figure 12.3.

Lack of interaction can be depicted in an interaction plot by parallel lines. The greater the departure of the lines from the parallel state, the higher the degree of interaction. Interaction can be synergistic or nonsynergistic. In the former, the effect of taking both factors together is more than the added effects of taking them separately. This is reversed for the nonsynergistic case. In both cases, the corresponding interaction plots are not parallel.

Figure 12.4 depicts different interaction plot scenarios. In Case 1, the response is constant at the three levels of factor A but differs for the two levels of factor B. Thus, there is no main effect of factor A but a factor B main effect is present. In

Table 12.3 Supermarket marketing strategy example

Display \ Price	Regular	Reduced	Cost to Supermarket
Normal	989, 1025, 1030	1211, 1215, 1182	1577, 1559, 1598
Normal Plus	1191, 1233, 1221	1860, 1910, 1926	2492, 2527, 2511
Twice Normal	1226, 1202, 1180	1516, 1501, 1498	1801, 1833, 1852

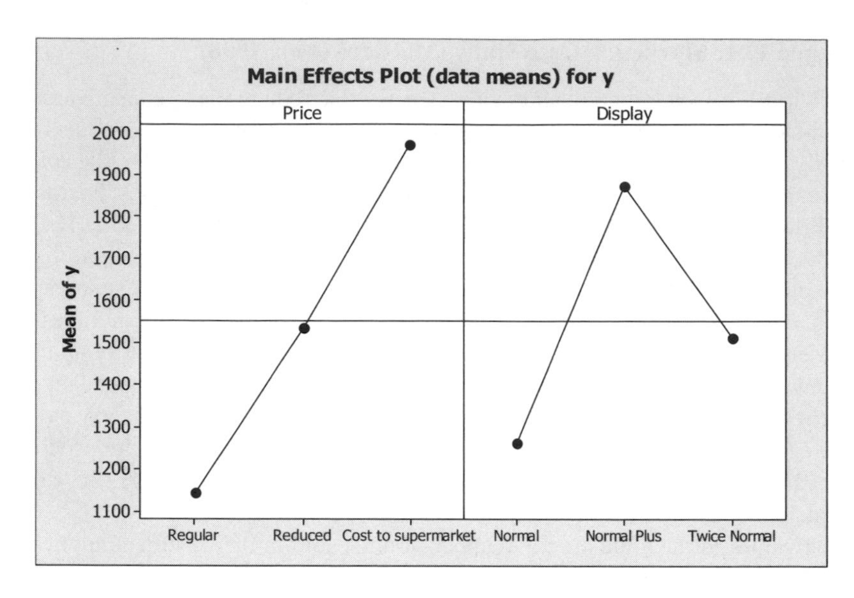

Figure 12.2 Main effect plot of factor A = "Price" and factor B = "Display."

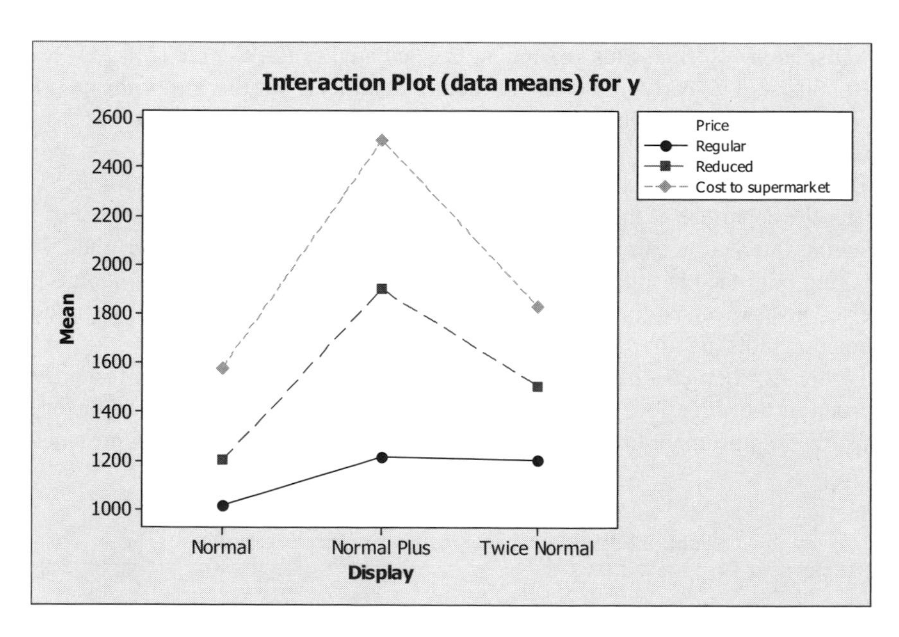

Figure 12.3 Interaction plot for factor A = "Price" and factor B = "Display."

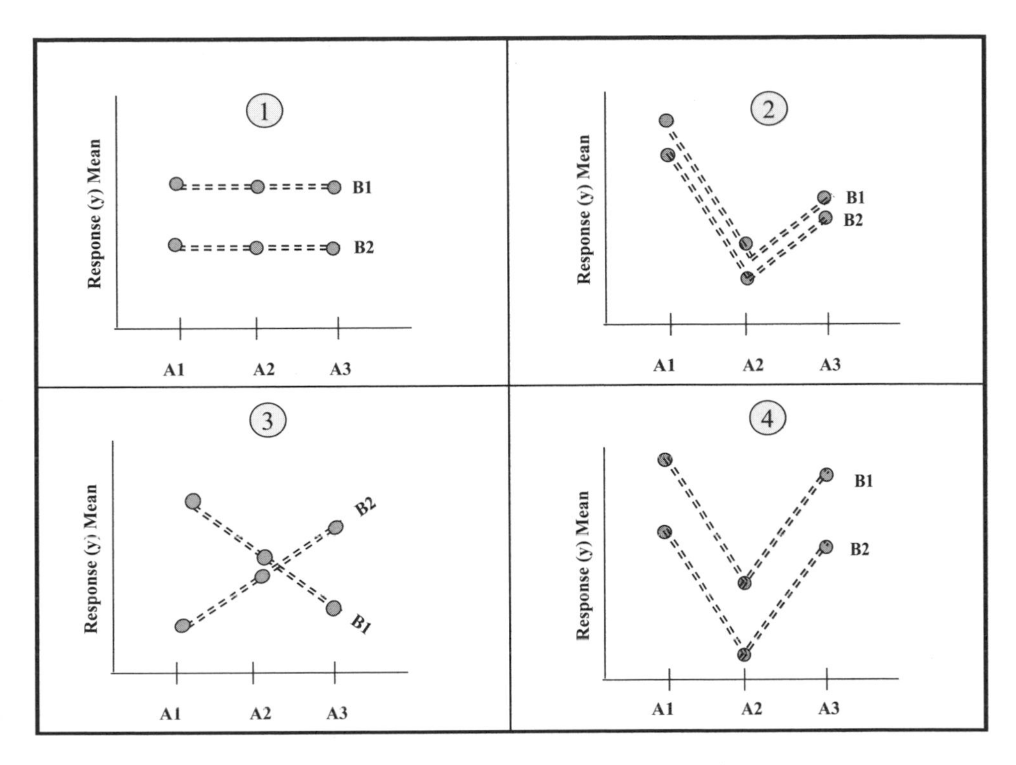

Figure 12.4 Different interaction plot scenarios.

Case 2, the mean responses differ at the levels of factor A, but the means are almost equal at the levels of factor B. In Case 3 and Case 4 both factors affect the response (y). In Case 3, the mean response between factor B levels varies with factor A levels. The effect on response depends on factor B and, therefore, the two factors interact. In Case 4, the effect of factor A on response (y) is independent of factor B, that is, the two factors do not interact.

12.7 ANALYSIS OF VARIANCE (ANOVA)

Analysis of variance (ANOVA)[5] is used to investigate and model the relationship between a response variable (y) and one or more independent factors. In effect, analysis of variance extends the two-sample t-test for testing the equality of two population means to a more general null hypothesis comparing the equality of more than two means versus them not all being equal. ANOVA includes procedures for fitting ANOVA models to data collected from a number of different designs and

[5]ANOVA differs from regression in two ways: the independent variables are qualitative (categorical), and no assumption is made about the nature of the relationship (that is, the model does not include coefficients for variables).

graphical analysis for testing equal variance assumptions, confidence interval plots, and graphs of main effects and interactions.

For a set of experimental data, the data most likely varies due to changes in experimental factors, but some of the variation might be caused by unknown or unaccounted for factors, experimental measurement errors, or variation within the controlled factors themselves.

There are several assumptions that need to be satisfied for ANOVA to be credible:

1. The probability distributions of the response (y) for each factor level combination (treatment) is normal.
2. The response (y) variance is constant for all treatments.
3. The samples of experimental units selected for the treatments is random and independent.

The ANOVA method produces the following:

1. A decomposition of the total variation of the experimental data to its possible sources (the main effect, interaction, or experimental error)
2. A quantification of the variation due to each source
3. Calculation of significance, that is, which main effects and interactions have significant effects on response (y) data variation.
4. A transfer function when the factors are continuous variables (noncategorical in nature)

12.7.1 ANOVA Steps for a Two Factors, Completely Randomized Experiment (Yang & El-Haik, 2003)

Step 1. Decompose the total variation in the DOE response (y) data to its sources (treatment sources: factor A, factor B, factor A × factor B interaction, and error). The first step of ANOVA is the "sum of squares" calculation that produces the variation decomposition. The following mathematical equations are needed:

$$\bar{y}_{i..} = \frac{\sum_{j=1}^{b} \sum_{k=1}^{n} y_{ijk}}{bn} \quad \text{(Row average)} \tag{12.12}$$

$$\bar{y}_{.j.} = \frac{\sum_{i=1}^{a} \sum_{k=1}^{n} y_{ijk}}{an} \quad \text{(Column average)} \tag{12.13}$$

$$\bar{y}_{ij.} = \frac{\sum_{k=1} y_{ijk}}{n} \quad \text{(Treatment or cell average)} \tag{12.14}$$

$$\bar{y}_{...} = \frac{\sum_{i=1}^{a}\sum_{j=1}^{b}\sum_{k=1}^{n} y_{ijk}}{abn} \quad \text{(Overall average)} \tag{12.15}$$

It can be shown that

$$\begin{aligned}\underbrace{\sum_{i=1}^{a}\sum_{j=1}^{b}\sum_{k=1}^{n}(y_{ijk}-\bar{y}_{...})^2}_{SS_T} &= \underbrace{bn\sum_{i=1}^{a}(\bar{y}_i-\bar{y}_{...})^2}_{SS_A} + \underbrace{an\sum_{j=1}^{b}(\bar{y}_j-\bar{y}_{...})^2}_{SS_B} \\ &+ \underbrace{n\sum_{i=1}^{a}\sum_{j=1}^{b}(\bar{y}_{ij}-\bar{y}_i-\bar{y}_j+\bar{y}_{...})^2}_{SS_{AB}} + \underbrace{\sum_{i=1}^{a}\sum_{j=1}^{b}\sum_{k=1}^{n}(y_{ijk}-\bar{y}y_{ij})^2}_{SS_E}\end{aligned} \tag{12.16}$$

Or simply

$$SS_T = SS_A + SS_B + SS_{AB} + SS_E \tag{12.17}$$

As depicted in Figure 12.5, SS_T denotes the "total sum of squares," which is a measure for the "total variation" in the whole data set; SS_A is the "sum of squares" due to

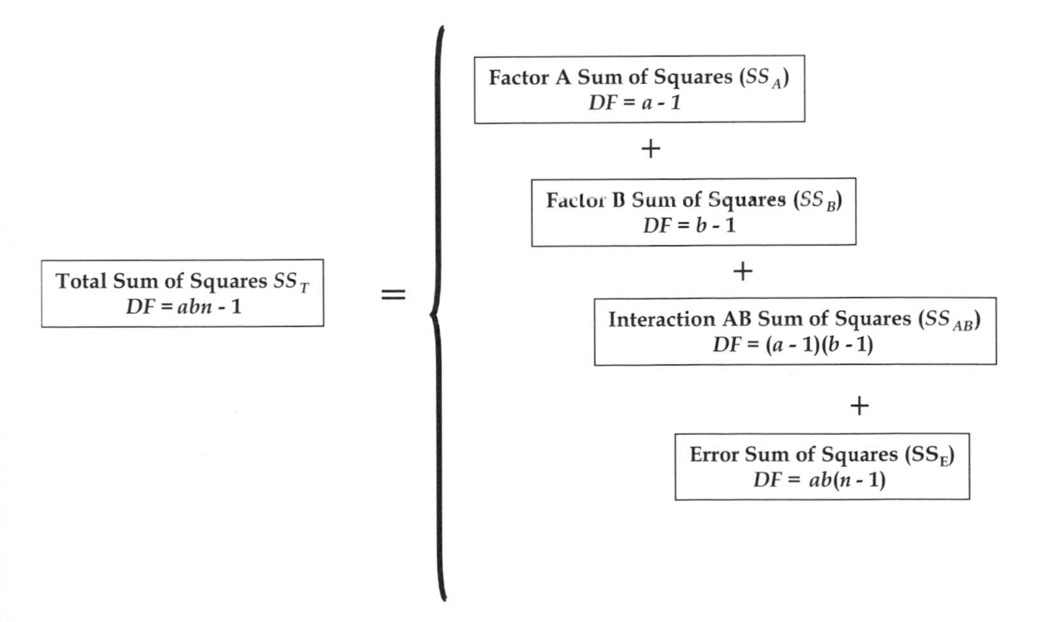

Figure 12.5 ANOVA variation decomposition.

factor A, which is a measure of total variation caused by the main effect of A; SS_B is the sum of squares due to factor B, which is a measure of total variation caused by the main effect of B; SS_{AB} is the sum of squares due to the interaction of factors A and B (denoted as AB), which is the measure of variation caused by interaction; and SSE is the sum of squares due to error, which is the measure of total variation due to error.

Step 2. Test the null hypothesis with regard to the significance of the factor A mean effect and the factor B mean effect as well as their interaction. The test vehicle is the mean square calculation. The mean square of a source of variation is calculated by dividing the source of variation sum of squares by its degrees of freedom.

The actual amount of variability in the response data depends on the data size. A convenient way of expressing this dependence is to say that the sum of squares has degrees of freedom (DF) equal to its corresponding variability source data size reduced by one. Based on statistics, the number of degrees of freedom associated with each sum of squares is shown in Table 12.4.

Test for Main Effect of Factor A
H_0: No difference among the a mean levels of factor A ($\alpha_{A1} = \alpha_{A2} = \ldots = \alpha_{Aa}$)
H_a: At least two factor A mean levels differ

Test for Main Effect of Factor B
H_0: No difference among the mean levels of factor B ($\alpha_{B1} = \alpha_{B2} = \ldots = \alpha_{Ba}$)
H_a: At least two factor B mean levels differ

Test for Main Effect of Factor A × Factor B Interaction
H_0: Factor A and factor B do not interact in the response mean
H_a: Factor A and factor B interact in the response mean

Step 3. Compare the F-test of the mean square of the experimental treatment sources to the error to test the null hypothesis that the treatment means are equal. If the test results in nonrejection of the null hypothesis, refine the experiment by increasing the number of replicates n or adding other factors. Otherwise, the response is unrelated to the two factors.

In the F-test, the F_0 will be compared, with F-critical defining the null hypothesis rejection region values with appropriate degree of freedom. If F_0 is larger than the critical value, then the corresponding effect is statistically significant. Several

Table 12.4 Degrees of freedom for two-factor factorial design

Effect	Degree of freedom (DF)
A	$a - 1$
B	$b - 1$
AB interaction	$(a - 1)(b - 1)$
Error	$ab(n - 1)$
Total	$abn - 1$

statistical software packages, such as Minitab™, can be used to analyze DOE data conveniently, or spreadsheet packages like Excel™ can also be used.

In ANOVA, a sum of squares is divided by its corresponding degree of freedom to produce a statistic called the "mean square" that is used in the F-test to see if the corresponding effect is statistically significant. An ANOVA is often summarized in a table similar to Table 12.5.

Test for Main Effect of Factor A

Test statistic: $F_{0,a-1,ab(n-1)} = MS_A/MS_E$, with a numerator DF equal to $(a-1)$ and denominator DF equal to $ab(n-1)$.

H_0 hypothesis rejection region: $F_{0,a-1,ab(n-1)} \geq F_{\alpha,a-1,ab(n-1)}$, with numerator DF equal to $(a-1)$ and denominator DF equal to $ab(n-1)$.

Test for Main Effect of Factor B

Test statistic: $F_{0,b-1,ab(n-1)} = MS_B/MS_E$, with numerator DF equal to $(b-1)$ and denominator DF equal to $ab(n-1)$.

H_0 hypothesis rejection region: $F_{0,b-1,ab(n-1)} \geq F_{\alpha,b-1,ab(n-1)}$, with numerator DF equal to $(b-1)$ and denominator DF equal to $ab(n-1)$.

Test for Main Effect of Factor A x Factor B Interaction

Test statistic: $F_{0,(a-1)(b-1),ab(n-1)} = MS_{AB}/MS_E$, with a numerator DF equal to $(a-1)(b-1)$ and denominator DF equal to $ab(n-1)$.

H_0 hypothesis rejection region: $F_{0,(a-1)(b-1),ab(n-1)} \geq F_{\alpha,(a-1)(b-1),ab(n-1)}$, with numerator DF equal to $(a-1)(b-1)$ and denominator DF equal to $ab(n-1)$.

The interaction null hypothesis is tested first by computing the F-test of the mean square for interaction versus the mean square for error. If the test results in nonrejection of the null hypothesis, then proceed to test the main effects of the factors. If the test results in rejection of the null hypothesis, we conclude that the two factors interact in the mean response (y). If the test of interaction is significant, a multiple

Table 12.5 ANOVA table

Source of variation	Sum of squares	Degree of freedom	Mean squares	F_0
A	SS_A	$a-1$	$MS_A = \dfrac{SS_A}{a-1}$	$F_0 = \dfrac{MS_A}{MS_E}$
B	SS_B	$b-1$	$MS_B = \dfrac{SS_B}{b-1}$	$F_0 = \dfrac{MS_B}{MS_E}$
AB	SS_{AB}	$(a-1)(b-1)$	$MS_{AB} = \dfrac{SS_{AB}}{(a-1)(b-1)}$	$F_0 = \dfrac{MS_{AB}}{MS_E}$
Error	SS_E	$ab(n-1)$		
Total	SS_T	$abn-1$		

comparison method such as Tukey's grouping procedure can be used to compare any or all pairs of the treatment means.

Next, test the two null hypotheses that the mean response is the same at each level of factor A and factor B by computing the F-test of the mean square for each factor main effect versus the mean square for error. If one or both tests result in rejection of the null hypothesis, we conclude that the factor affects the mean response (y). If both tests result in nonrejection, an apparent contradiction exists. Although the treatment means apparently differ, the interaction and main effect tests have not supported that result. Further experimentation is advised. If the test for one or both main effects is significant, use a multiple comparison such as Tukey's grouping procedure to compare the pairs of the means corresponding to the levels of the significant factor(s).

The results and data analysis methods discussed above can be extended to the general case in which there are a levels of factor A, b levels of factor B, c levels of factor C, and so on, arranged in a factorial experiment. There will be $abc \ldots n$ total number of trials if there are n replicas. Clearly, the number of trials needed to run the experiment will increase very fast with the increase of number of factors and the number of levels. In practical application, we rarely use general full factorial experiments for more than two factors. Two-level factorial experiments are the most popular experimental methods.

Example 12.2: Marketing Case Study (from Example 12.1) ANOVA

The ANOVA table is given in Table 12.6. In this example, there are three effects: factor A or "Price", factor B or "Display," and factor × factor B ("Price" × "Display") interaction. The larger the sum of squares of a treatment effect, the more variation is caused by that effect, and the more important that effect is. In this example, the sum of squares for "Price" is 3,089,054 unit^2, by far the largest. However, if different effects have different degrees of freedom, then the results might be skewed. The F-test is a better measure of relative importance. In this example, the F-test for "Price" is 3121.8, for "Display" it is 1709.37, and for the "Price" × "Display" interaction it is 258.07. So, clearly, "Price" (factor A) is the most important factor. In DOE, we also use the P-value to determine if an effect is statistically significant. The most commonly used criterion is to compare P-values with a 0.05 significance level. If the P-

Table 12.6 ANOVA table of Example 12.2

Source of variation	Sum of squares	Degree of freedom	Mean squares	F_0	P-value
Factor A: Price	3,089,054	2	1,544,527	3121.89	0.0
Factor B: Display	1,691,393	2	845,696	1709.37	0.0
Price × Display	510,705	4	127,676	258.07	0.0
Error	8905	18	495		
Total	5,300,057	26			

value is less than 0.05, then that treatment effect is significant. In this example, the P-values for "Price" and "Display" are 0.0, both smaller than 0.05, so the main effects of both "Price" and "Display" are statistically significant. It is also the case for "Price" × "Display" interaction, with a 0.0 as a P-value indicating a significant effect.

This example was analyzed using Minitab™, which generated the interaction and main effects of this example in Figure 12.2 and Figure 12.3. From the interaction plot (Figure 12.3), it is clear that the three line effects are not parallel, so there is an interaction effect. In main effect chart, we can clearly see that "Price" at the "Cost to supermarket" level gives much higher units sold. For "Display," the "Normal Plus" level gives highest units sold. Overall, for achieving the highest product sold units, the supermarket needs to sell it at the cost and display it in a normal-plus space. The ε (error) is 8905, which is 0.1% of the total, so we may conclude that the three factors are statistically significant and provide practical leverage in the response.

12.8 2^K FULL FACTORIAL DESIGNS

In many cases, it is sufficient to consider the factors affecting a process at two levels. For example, the shoppers for a product may be male or female, marketing strategy may either be set the price a little higher or a little lower than a target, and so on. The black belt would like to determine whether any of these changes affect the results of a process. The most intuitive approach to study these factors would be to vary the factors of interest in a full factorial design, that is, to try all possible combinations of levels. Such an experiment is called a 2^k, an experiment with k factors each with two levels, that is, the number of treatment combinations in a two-level full factorial of k factors is $2 \times 2 \ldots 2 = 2^k$. If there are n replicas of each treatment combination, then the total number of experimental trials is $2^k n$. Because there are only two levels for each factor, we call them the "low" and "high" levels. For example, if a factor is "Price," with two levels, "Regular" and "Regular Plus," then "Regular" is the low level and "Regular Plus" is the high level.

The two-level factorial design is the most popular designs because it is a full factorial design with the least number of runs, an ideal situation for screening experiments. The two-level factorial design is the basis of the fractional factorial design that will be discussed next.

The number of necessary runs in a full 2^k experiment will increase geometrically. For example, if the black belt wants to study seven factors, the necessary number of runs in the experiment would be $2^7 = 128$. To study 10 factors he or she would need $2^{10} = 1{,}024$ runs in the experiment. Because each run may require time-consuming and costly setting and resetting of the process, it is often not feasible to require that many different runs for the experiment. Under these conditions, fractional factorials are used that sacrifice interaction effects so that main effects may still be computed correctly.

The standard layout for a two-level design uses a binary notation with +1 and –1 denoting the "high level" and the "low level," respectively, for each factor. For example, Table 12.7 describes an experiment in which four trials (or runs) were con-

Table 12.7 2^4 experimental design

Run number	$A(X1)$	$A(X2)$
1	−1	−1
2	+1	−1
3	−1	+1
4	+1	+1

ducted with each factor set to high or low during a run, according to whether the matrix had a +1 or −1 setting for the factor during that trial. The substitution of +1 and −1 for the factor levels is called *coding*. This aids in the interpretation of the coefficients and mathematically helps fit a transfer function experimental model. Also note in Table 12.7 the use of the functional argument of X1 and X2 that denotes the coding of a hypothetical factor A and factor B, respectively.

12.8.1 Full 2^k Factorial Design Layout

If the experiment had more than two factors, there would be additional columns in the layout matrix corresponding to the extra factors—a column per factor. In a full 2^k factorial design, the number of distinct experimental runs $N = 2^k$. For example, if $k = 4$, then, $N = 2^4 = 16$; and if $k = 5$, then $N = 2^5 = 32$; and so on.

Table 12.8 gives a standard layout for a 24 factorial experiment. The run number is sequenced by standard order, which is featured by the sequence −1 +1 −1 +1 . . .

Table 12.8 Experimental layout for a 2^4 design

	Factors				Replicas				
Run No.	A	B	C	D	1	2	. . .	n	Response Total*
1	−1	−1	−1	−1					(1)
2	1	−1	−1	−1					*a*
3	−1	1	−1	−1					*b*
4	1	1	−1	−1					*ab*
5	−1	−1	1	−1					*c*
6	1	−1	1	−1					*ac*
7	−1	1	1	−1					*bc*
8	1	1	1	−1					*abc*
9	−1	−1	−1	1					*d*
10	1	−1	−1	1					*ad*
11	−1	1	−1	1					*bd*
12	1	1	−1	1					*abd*
13	−1	−1	1	1					*cd*
14	1	−1	1	1					*acd*
15	−1	1	1	1					*bcd*
16	1	1	1	1					*abcd*

*Computed by adding the replica row in a given run.

for A, −1 −1 +1 +1 for factor B, −1 −1 −1 −1 and +1 +1 +1 +1 for factor C, and so on. In general, for a 2^k experiment, the first column starts with −1 and alternates in sign for all 2^k runs; the second column starts with −1 repeated twice, and then alternates with two in a row of the opposite sign until all 2^k places are filled. The third column starts with −1 repeated four times, then four repeats of +1s and so on. The *i*th column starts with 2^{i-1} repeats of −1 followed by 2^{i-1} repeats of +1 and so on. This is known as the Yates Standard Order.

Every run (simultaneously experimenting with combinations of factorial levels) can also be represented by the symbols in the last column of the table, where the symbol depends on the corresponding levels of each factor. For example, for run number 2, A is at high level (+1), and B, C, and D are at low level (−1), so the symbol is *a*, which means that only A is at high level. For run 15, B, C, and D are at the high level, so we use *bcd*. For the first run, all factors are at a low level, so we use (1) here. In data analysis, we need to compute the "total" for each run, which is the sum of all replicas for that run. We often use like symbols to represent the totals. There could be n replicas, and when $n - 1$ it is called single replica.

12.8.2 2^k Data Analysis

For a 2^k full factorial experiment, the numerical calculations for ANOVA, main effect chart, interaction chart, and mathematical transfer function model become easier, in comparison with the general full factorial experiment. Below we give a step-by-step procedure for the entire data analysis in Example 12.3.

Example 12.3 (Magner et al., 1995)

Studies have indicated that the intensity of employees' negative reactions to unfavorable managerial decisions can be influenced by their perception of the fairness of the decision process. For example, employees who receive unfavorable budgets should have milder negative reactions if they participated in the budgetary process than had they not participated. Twenty-four managers served as subjects in a 2^2 factorial experiment with factors "Budgetary Participation" (yes [−1] or no [+1]) and "Budget Favorability" (favorable [−1] or unfavorable [+1]). Six managers were included in each of four treatments of the design. Each manager was asked to evaluate the trustworthiness of the superior who was responsible for setting the budget. A well-established interpersonal trust scale was used. The scale ranges from 4 (low trust) to 28 (high trust). Table 12.9 shows the resulting data.

Step 1: Preparation. First construct the *analysis matrix* for the problem. The analysis matrix is a matrix that has not only all columns for factors, but also the columns for all interactions. The interaction columns are obtained by multiplying the corresponding columns of factors involved. For example, in a 2^2 experiment, the analysis matrix is as shown in Table 12.10, where the AB column is generated by multiplying the A and B columns.

Next, attach experimental data to the analysis matrix.

Table 12.9 Example 12.3 2^2 setup

Factor B: Favorability \ Factor A: Participation	Yes			No		
Favorable	19	22	25	17	16	20
	18	25	23	24	23	21
Unfavorable	23	28	20	16	12	15
	26	18	23	18	15	14

Step 2: Compute Contrasts. The contrast of each factor is calculated by multiplying the factor column coefficient by the corresponding total and summing them. In Table 12.11, the column coefficients for A (first column) are (–1, +1, –1, +1), the total is [(1), a, b, ab] = (132, 121, 138, 90). Therefore,

$$\text{Contrast}_{\text{A}} = -(1) + a - b + ab = -132 + 121 - 138 + 90 = -59$$

Similarly,

$$\text{Contrast}_{\text{B}} = -(1) - a + b + ab = -132 - 121 + 138 + 90 = -25$$

$$\text{Contrast}_{\text{AB}} = (1) - a - b + ab = 132 - 121 - 138 + 90 = -37$$

Contrasts are the basis for many subsequent calculations.

Step 3: Compute Effects. "Effects" include both main effects and interaction effects. All effects are computed by the following formula:

$$\text{Effect} = \frac{\text{Contrast}}{2^{k-1}n} \tag{12.18}$$

where N is the total number of runs. The definition for any main effect, for example, the main effect of A, is

$$A = \bar{y}_{A^+} - \bar{y}_{A^-} \tag{12.19}$$

Table 12.10 Analysis Matrix

Run number	A	B	AB
1	–1	–1	(–1) × (–1) = +1
2	+1	–1	(+1) × (–1) = –1
3	–1	+1	(–1) × (+1) = –1
4	+1	+1	(+1) × (+1) = +1

Table 12.11 Analysis matrix and data for Example 12.3

				Response						
Run No.	A	B	AB	1	2	3	4	5	6	Total
1	–1	–1	1	19	22	25	18	25	23	(1) = 132
2	1	–1	–1	17	16	20	24	23	21	a = 121
3	–1	1	–1	23	28	20	26	18	23	b = 138
4	1	1	1	16	12	15	18	15	14	ab = 90

that is, the average response for A at high level minus the average of response for A at low level. By Equation (12.18),

$$A = \frac{\text{Contrast}_A}{2^{k-1}n} = \frac{-59}{2^{2-1}\,6} = -4.917$$

Similarly,

$$B = \frac{\text{Contrast}_B}{2^{k-1}n} = \frac{-25}{2^{2-1}\,6} = -2.083$$

$$AB = \frac{\text{Contrast}_{AB}}{2^{k-1}n} = \frac{-37}{2^{2-1}\,6} = -3.083$$

Step 4: Compute SS (Sum of Squares). The sum of squares is the basis for the analysis of variance computation. The formula for the sum of squares is

$$SS = \frac{\text{Contrast}^2}{2^k n} = \frac{\text{Contrast}^2}{Nn} \tag{12.20}$$

Therefore,

$$SS_A = \frac{\text{Contrast}_A^2}{2^2 n} = \frac{(-59)^2}{4 \times 6} = 145.042$$

$$SS_B = \frac{\text{Contrast}_B^2}{2^2 n} = \frac{(-25)^2}{4 \times 6} = 26.042$$

$$SS_{AB} = \frac{\text{Contrast}_{AB}^2}{2^2 n} = \frac{(-37)^2}{4 \times 6} = 57.042$$

For completing ANOVA, we also need SS_T and SS_E, in a 2^2 factorial design:

$$SS_T = \sum_{i=1}^{2} \sum_{j=1}^{k} \sum_{k=1}^{n} y_{ijk}^2 - \frac{y_{...}^2}{Nn} \tag{12.21}$$

where y_{ijk} is actually each individual response, and $y_{...}$ is the sum of all individual responses. In the above example,

$$SST = 19^2 + 22^2 + \ldots + 14^2 - (19 + 22 + \ldots + 14)^2/24 = 10051 - (481)^2/24 = 410.958$$

SS_E can be calculated by:

$$SS_E = SS_T - SS_A - SS_B - SSA_B$$

In the above example,

$$SS_E = SS_T - SS_A - SS_B - SSA_B = 410.958 - 145.042 - 26.042 - 57.042 = 182.832$$

Step 5: Complete ANOVA Table. The ANOVA table computation is the same as that of general factorial design. Minitab™ or other statistical software can calculate the ANOVA table conveniently. In above example, the ANOVA table computed by Minitab™ is as shown in Table 12.12. Clearly, with a critical F-test value, $F_{0.05,1,20} = 4.35$, both the main effect A as well as the interaction AB are statistically significant. Factor B (favorability) is not significant. This is reflected in Figure 12.6.

Step 6: Plot Main Effect and Interaction Charts for all Significant Effects. For any main effect, for example, main effect A, the main effect plot is actually the plot of $\bar{y}_{A^-}$ and $\bar{y}_{A^+}$ versus the levels of A. The interaction chart is plotted by charting all combinations of $\bar{y}_{A^-B^-}$, $\bar{y}_{A^-B^+}$, $\bar{y}_{A^+B^-}$, and $\bar{y}_{A^+B^+}$. For example, $\bar{y}_{A+} = (ab + a/12) = 17.583$ and $\bar{y}_{A^-} = [b + (1)/12] = 22.5$, and son on. The construction of the interaction plot is left as an exercise.

Step 7: Validate Assumptions. ANOVA assumes that the data collected are normally and independently distributed with the same variance in each treatment combination or factor level. These assumptions can be evaluated by the residuals defined as $e_{ijl} = y_{ijl} - \bar{y}_{ij\cdot}$. That is, the residual is just the difference between the observations and the corresponding treatment combination (cell) averages (Hines and Montgomery, 1990).

Table 12.12 ANOVA table of Example 12.3

Source of variation	Sum of squares	Degree of freedom	Mean squares	$F_{0,1,20}$	P-value
Factor A: Participation	145.042	1	145.042	15.866	0.001
Factor B: Favorability	26.042	1	26.042	2.849	0.107
Factor A × Factor B	57.042	1	57.042	6.240	0.021
Error	182.832	20	9.1416		
Total	410.958	23			

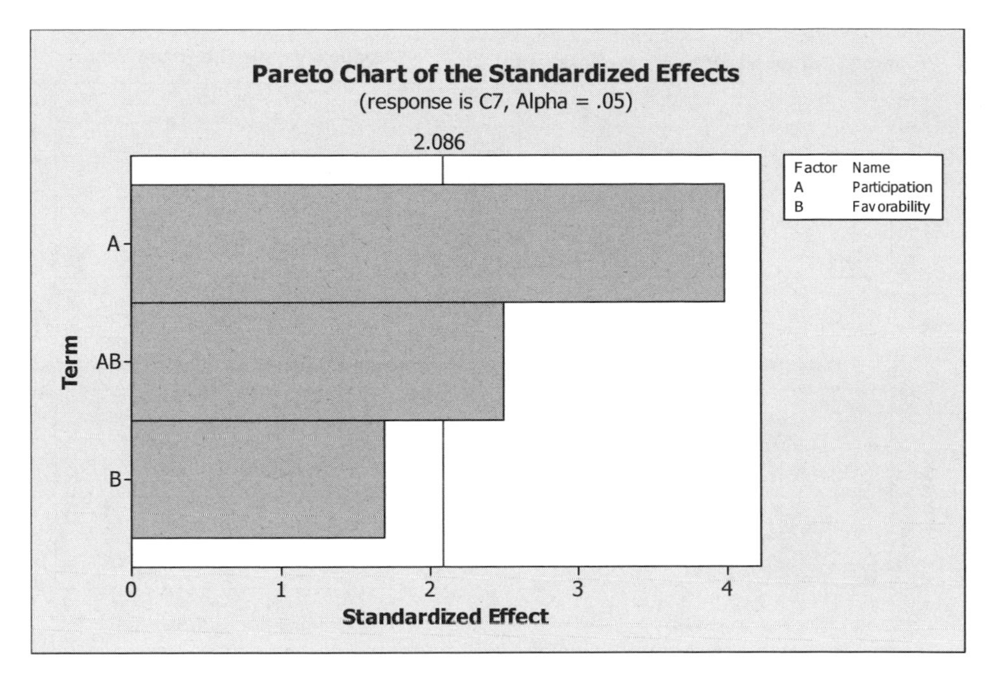

Figure 12.6 Example 12.3 Pareto chart of effects.

Plotting the residuals on normal probability paper and looking for a straight-line fit can check the normality assumption. To check the assumption of equal variance at each factor level, we plot the residual against the factor levels and compare the spread in the residuals. It is also useful to plot the residuals against fitted values or treatment combination averages. The variability in the residuals should not in any way depend on the value of cell averages. When a pattern appears in the plots, indicating nonnormality, it suggests the need for transformation, that is, analyzing the data in a different metric or dimension. In some problems, the dependency of residuals scatter in fitted values is very important. We would like to select the level that improves y in the direction of goodness; however, this level may also cause more variation in y from observation to observation.

Plotting the residuals against the time or run order in which the DOE was performed can check the independence assumption. A pattern in this plot, such as a sequence of positive and negative residuals, may indicate that the observations are not independent. This indicates that the time or run order is important or factors that change over time are important and have not been included in the DOE.

Figure 12.7 presents the residual analysis of Example 12.3. The normal probability plot of these residuals does not appear to deviate from normality. The graph of residuals versus fitted values does not reveal any unusual pattern.

Figure 12.8 depicts the Box plot of the response data. It reflects a variability difference between the levels of factor A, the "Participation." This warrants hy-

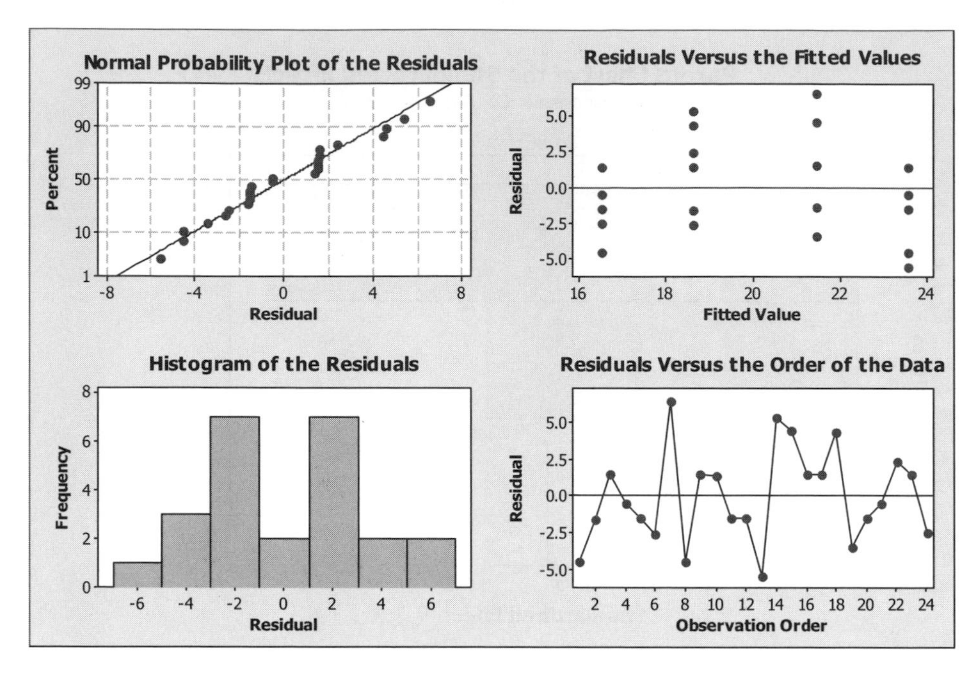

Figure 12.7 ANOVA assumption diagnostics.

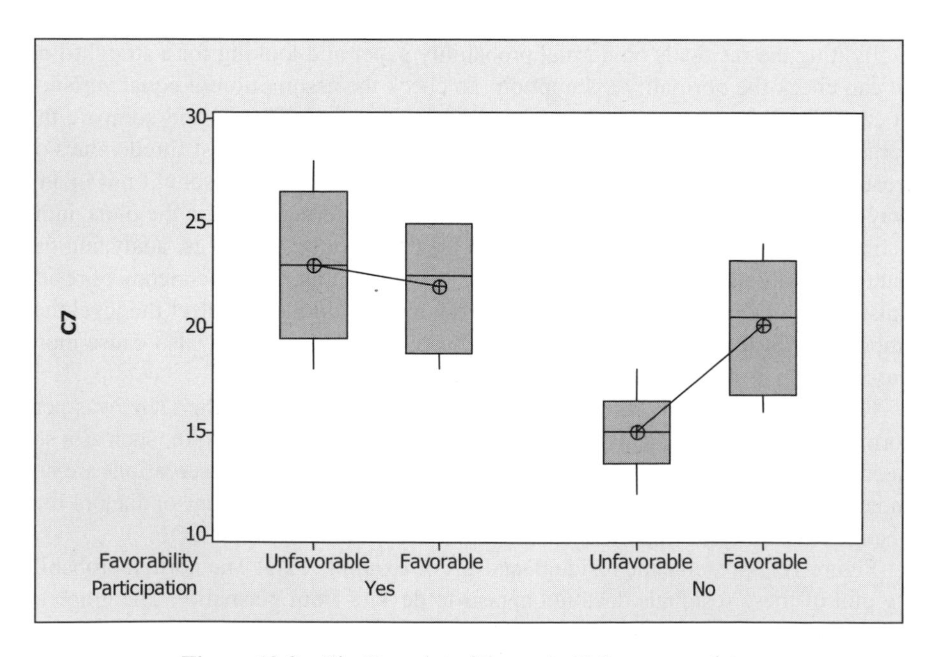

Figure 12.8 The Box plot of Example 12.3 response data.

pothesis testing on the variance. See Appendix for comments on DOE graphical analysis.

Step 8: Establish Transfer Function Model. Similar to regression, ANOVA involves the statistical analysis of the relationship between a response and one or more factors. It is not surprising to formulate the ANOVA as a regression analysis. We can establish a regression transfer function model for the DOE data. Here are the rules:

1. Only significant effects are included in the model. For the above example, since A and AB are significant, all factors are included (without B, we cannot have AB).
2. We use the mathematical variable x_1 to express A, x_2 for B, x_3 for C, and so on; and the mathematical variables x_1x_2 for AB interaction, x_1x_3 for AC interaction, $x_1x_2x_3$ for ABC interaction, and so on.

The model for Example 12.3 is given by $y = a_0 + a_1x_1 + a_2x_2 + a_{12}x_1x_2 + \varepsilon$, where a_0 = average of all responses ($\bar{y}_{...}$), a_i = (ith factor effect)/2, and ε is the error term (see Section 6.4). For example, a_1 = A/2 = –4.917/2 = –2.459 in Example 12.3. The transfer function fitted model is given by $\hat{y} = 20.042 - 2.459x_1 - 1.0415\,x_2 - 1.5415\,x_1x_2$, where x_1 and x_2 are coded values. This is an approximate (fitted) theoretical transfer function, hence the $\hat{y}$ notation (see Chapter 6).

Step 9: Determine Optimal Settings. Depending on the objective of the problem, we can determine the optimum setting of the factor levels by examining the main effects plot and the interaction plot. If there is no interaction, the optimal setting can be determined by looking at one factor at a time. If there are interactions, then we have to look at the interaction plot first. For Example 12.3, since the AB interaction is significant, we have to find the optimal settings by looking into the AB interaction. From the interaction plot, assuming the trustworthiness response with a "larger the better" direction of goodness, A should be set at low level, that is, "Yes" for Budgetary Participation. Factor B (Budget Favorability) was not significant, however, due to the negative nonsynergistic interaction, it is recommended to get the managers involved in the budget setting regardless of the responsible supervisor. This will work better with "Unfavorable" budget scenarios.

12.8.3 The 2^3 Design

A two-level, full factorial design of three factors is classified as a 2^3 design. This design has eight runs. Graphically, we can represent the 2^3 design by the cube shown in Figure 12.9. The arrows show the direction of increase of the factors. The numbers 1 through 8 at the corners of the design box reference the standard order of the runs. In general, cube plots can be used to show the relationships among two to eight factors for two-level factorial or Plackett–Burman designs.

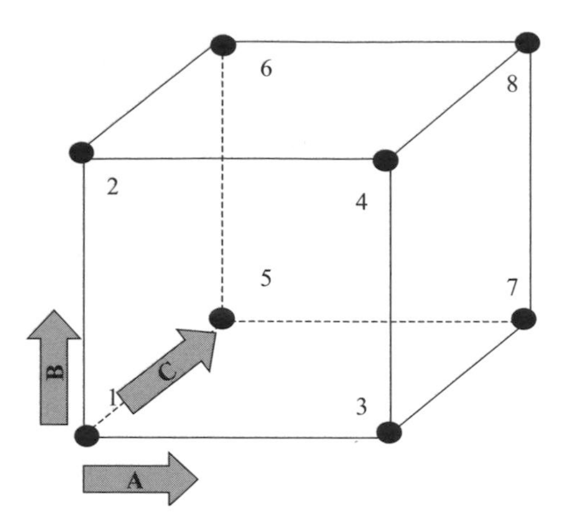

Figure 12.9 A 2^3 full factorial design with factors A, B, and C.

Example 12.3 2^k DOE in Simulation

In simulation applications (see Chapter 14), several model control factors can each be set to two levels (low and high), a 2^k factorial design arrangement. For example, assume that we have three control factors in a certain production system: buffer size, conveyor speed, and number of forklifts. An experimental design is required in order to study the effect of the three control factors on the throughput of the production system. Two levels of each control factor can be defined: buffer size (5,10) in units, conveyor speed (12.5, 15.0) in feet per minute, and number of forklifts (3, 4). In quantitative control factors, it is common to use a minus sign to indicate the low factor level and plus sign to indicate the high factor level. Since we have three factors, each with two levels, a 2^3 factorial design is required in this experiment. The model response (y), which is throughput in this case, is evaluated at the designed treatments. Table 12.13 summarizes the design matrix of the eight factorial designs.

Multiple treatment runs are set at each of the 2^3 factor-level combinations. The average throughput is estimated from the treatment runs to represent a model response (y) of interest to the team at each factor-level combination. For example, y_5 represents the response resulting from running the treatment with low buffer size (5), low conveyor speed (12.5), and high number of forklifts (4).

As the number of control factors increase, full factorial design may lead to running huge numbers of trials, even when using only two levels of each factor. This is particularly critical when running multiple replications at each factor-level combination. For example, experimenting with 10 control factors each with two levels, requires $2^{10} = 1{,}024$ experiments. Running only five replications at this factorial design results in 5,120 runs, which requires significant time and computation. Hence, strategies of factor screening and developing fractional factorial design are often used in order to cope with large numbers of control factors.

Table 12.13 Design matrix for a 2^3 factorial design

Design point combination	Model control Factors: Buffer size (factor A)	Conveyor speed (factor B)	Forklifts number (factor C)	Response (y) throughput
1	5 (–)	12.5 (–)	3 (–)	y_1
2	10 (+)	12.5 (–)	3 (–)	y_2
3	5 (–)	15.0 (+)	3 (–)	y_3
4	10 (+)	15.0 (+)	3 (–)	y_4
5	5 (–)	12.5 (–)	4 (+)	y_5
6	10 (+)	12.5 (–)	4 (+)	y_6
7	5 (–)	15.0 (+)	4 (+)	y_7
8	10 (+)	15.0 (+)	4 (+)	y_8

12.8.4 The 2^3 Design with Center Points

Center points are usually added, as in Figure 12.10, to validate the assumption in a 2^k factorial design concerning the linearity of factorial effects. In a 2^k factorial design, this assumption can be satisfied approximately rather than exactly. The addition of center points of n_C replicas to a factorial design is the team's insurance against unknown curvature (nonlinearity) while providing an error estimate without impacting the factorial-effects estimation.

In a 2^k design, center points are represented by (0, 0, 0, . . . , 0) indicating the design center as in Figure 12.10. For k quantitative factors, let $\overline{y}_F$ be the average of the

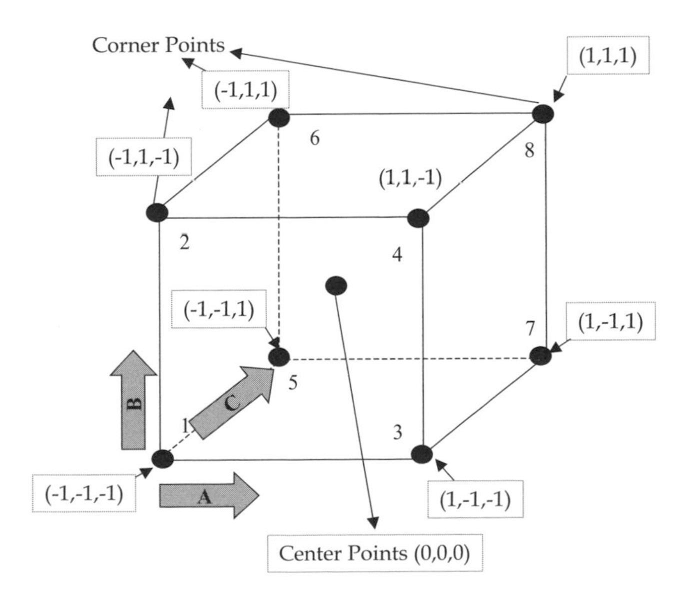

Figure 12.10 A 2^3 full factorial design with factors A, B, and C and center points.

2^k factorial points and let $\bar{y}_C$ be the average of the n_C center points. When the difference $\bar{y}_F - \bar{y}_C$ is large, the center points lie far from the planes passing through the factorial points, and curvature exists. A single degree of freedom is usually given for the sum of squares of nonlinearity or curvature that can be calculated using

$$SS_{\text{Curvature}} = \frac{n_F n_C (\bar{y}_F - \bar{y}_C)}{n_F n_C} \tag{12.22}$$

with $n_C - 1$ degrees of freedom, and n_F is the number of factorial design points. For a 2^3 design, $n_F = 8$. When center points are added to the 2^3 design, a potential transfer function is produced with quadratic terms such as

$$\begin{aligned} y = f(x_1, x_2, x_3) = & \underbrace{a_0 + a_1x_1 + a_2x_2 + a_3x_3}_{\text{Linear terms}} + \underbrace{a_{12}x_1x_2 + a_{13}x_1x_3 + a_{23}x_2x_3}_{\text{Interaction terms}} \\ & + \underbrace{a_{11}x_1^2 + a_{22}x_2^2 + a_{33}x_3^2 + \varepsilon}_{\text{Quadratic terms}} \end{aligned} \tag{12.23}$$

where ε is the error term.

Each of the coefficients a_{11}, a_{22}, and a_{33}, can be tested for curvature significance using the hypothesis $H_0 = a_{11} + a_{22} + a_{33} = 0$, versus H_a: at least one quadratic term coefficient $\neq 0$.

12.9 FRACTIONAL FACTORIAL DESIGNS[6]

As the number of factors k increases, the number of runs specified for a full factorial design can quickly become very large. For example, when $k = 6$, $2^6 = 64$. However, in this six-factor experiment, there are six main effects, say, A, B, C, D, E, and F; 15 two-factor interactions, AB, AC, AD, AE, AF, BC, BD, BE, BF, CD, CE, CF, DE, DF, and EF; 20 three-factor interactions, ABC, ABD, . . . ; 15 four-factor interactions, ABCD, . . . ; six five-factor interactions; and one six-factor interaction.

Fractional factorial design is a good solution for experimental design with limited resources. The fractional factorial design procedure usually starts by selecting a subset of combinations to test. The subset is selected by focusing on the main effects of control factors and some interactions of interest. The subset size represents 2^{k-p} of the 2^k possible factorial designs. Determining the p subset, which can be 1, 2, 3, and so on, depends on the level of confounding in 2^{k-p} fractional factorial designs, where one effect is confounded with another if the two effects are determined using the same formula. Factor screening provides a smaller treatment subset by screening out factors with little or no impact on process performance, using methods such as Plackett–Burman designs, supersaturated design, group-screening designs, and frequency-domain methods.

[6]See Chapter 12 of Yang and El-Haik (2003).

In general, practitioners have found that in an optimization experiment, higher-order interaction effects that involve three factors or more are rarely found to be significant. Usually, some main effects and two-factor interactions are significant. However, in a 2^6 experiment, out of 63 main effects and interactions, 42 of them are higher-order interactions and only 21 of them are main effects and two-factor interactions. As k gets larger, the overwhelming proportion of effects in the full factorials will be higher-order interactions. Since those effects are most likely to be insignificant, a lot of experimental resources in full-factorial designs are wasted. That is, as the number of factors k increases, most of data obtained in the full factorial is used to estimate higher-order interactions, which are most likely to be insignificant.

Fractional factorial experiments are designed to greatly reduce the number of runs and to use the information from the experimental data wisely. Fractional experiments only require a fraction of the runs of a full factorial. For two-level experiments, they only use ½, ¼, ⅛, . . . of the runs from a full factorial. Fractional factorial experiments are designed to estimate only the main effects and two-level interactions, and not to estimate three-factor and other higher-order interactions.

12.9.1 The 2^{3-1} Design

Detailed accounts of how to design 2^{k-p} experiments can be found in Box, Hunter, and Hunter (1978), Montgomery (1991), or Gunst and Hess (1989), to name only a few of the many textbooks on this subject. In general, it will successively use the highest-order interactions to generate new factors.

Consider a two-level, full-factorial design for three factors, namely, the 2^3 design. Suppose that the experimenters cannot afford to run all eight treatment combinations but they can afford four runs. If a subset of four runs ($p = 1$), is selected from the full-factorial design, then it is a 2^{3-1} design. Now let us look at Table 12.14, where the original analysis matrix of a 2^3 design is divided into two portions. In this table, we simply rearrange the rows such that the highest interaction ABC's contrast coefficients are all +1s in first four rows and all –1s in second four rows.

Table 12.14 2^{3-1} design

Treatment combination	Factorial effects							
	I	A	B	C	AB	AC	BC	ABC
A	+1	+1	–1	–1	–1	–1	+1	+1
B	+1	–1	+1	–1	–1	+1	–1	+1
C	+1	–1	–1	+1	+1	–1	–1	+1
abc	+1	+1	+1	+1	+1	+1	+1	+1
Ab	+1	+1	+1	–1	+1	–1	–1	–1
Ac	+1	+1	–1	+1	–1	+1	–1	–1
Bc	+1	–1	+1	+1	–1	–1	+1	–1
(1)	+1	–1	–1	–1	+1	+1	+1	–1

The second column in this table is called the identity or I column because it is a column with all +1s. If we select the first four runs as our experimental design, it is called a fractional factorial design with the defining relation I = ABC, where ABC is called the generator.

In Table 12.14, we can find that since all the contrast coefficients for ABC are +1s, we will not be able to estimate the effect of ABC at all. For other main effects and interactions, the first four runs have an equal number of +1s and –1s so we can calculate their effects. However, we find that the contrast coefficients of factor A are identical as those of BC interaction, the contrast coefficients of factor B are exactly the same as those of AC, as are those of C and AB. Since the effects are computed based on contrast coefficients, there is no way we can distinguish the effect of A and BC, B and AC, and C and AB. For example, when we estimate the effect of A, we are really estimating the combined effect of A and BC. This "mix up" of main effects and interactions is called *aliasing* or *confounding*. All alias relationships can be found from the defining relation I = ABC. If we simply multiply A at both sides of the equation, we get AI = AABC. Since multiplying identical columns will give an I column, the above equation becomes A = BC. Similarly, we can get B = AC and C = AB. The first half fraction based on I = ABC is called the principal fraction. If we use the second half of Table 12.12, the defining relationship will be I = –ABC. Because all ABC coefficients are equal to –1s, we can easily determine that A = –BC, B = –AC, and C = –AB. Therefore, A is aliased with –BC, B is aliased with –AC, and C is aliased with –AB. We will completely lose the information about the highest-order interaction effect, and we will also partially lose some information about lower-order interactions.

12.9.2 Half Factional 2^k Design

The half fractional 2^k design is also called 2^{k-1} design because it has $N = 2^{k-1}$ runs. We can use the definition relationship to lay out the experiment. Here we give the procedure to lay out the 2^{k-1} design and illustrate that with an example.

Step 1: Compute $N = 2^{k-1}$ and Determine the Number of Runs. For example, for $k = 4$, $N = 2^{k-1} = 2^3 = 8$.

Step 2: Create a Table with N Runs and Lay Out the First k – 1 Factors in Standard Order. For example, for $k = 4$, the factors are A, B, C, and D; and the first $k - 1 = 3$ factors are A, B, and C. In Table 12.15, we lay out the first three columns with A, B, and C in standard order.

Step 3: Use Defining Relation to Create the Last Column. In above example, if we use I = ABCD as the defining relation, then D = ABC. We then can get the D column by multiplying the coefficients of A, B, and C columns in each row. In the above example, I = ABCD, we can derive the following alias relationships:

A = BCD, B = ACD, C = ABD, D = ABC
AB = CD, AC = BD, AD = BC

Table 12.15 2^{3-1} standard order table

	Factors				Replicas				
Run No.	A	B	C	D	1	2	. . .	n	Response total
1	–1	–1	–1	–1					(1)
2	1	–1	–1	–1					*a*
3	–1	1	–1	–1					*b*
4	1	1	–1	–1					*ab*
5	–1	–1	1	–1					*c*
6	1	–1	1	–1					*ac*
7	–1	1	1	–1					*bc*
8	1	1	1	–1					*abc*

Unlike 2^{3-1} design, the main effects are not aliased with two-factor interactions, but two-factor interactions are aliased with each other. If we assume that three-factor interactions are not significant, then main effects can be estimated free of aliases. Though both 2^{3-1} and 2^{4-1} are half fractional factorial designs, 2^{4-1} has less confounding than 2^{3-1}. This is because their *resolutions* are different.

12.9.3 Design Resolution

A 2^{k-p} design means that we study overall k factors; however, p of those factors were generated from the interactions of a full 2^{k-p} factorial design. As a result, the design does not give full resolution; that is, there are certain interaction effects that are confounded with other effects. In general, a design of resolution R is one in which no main effects are confounded with any other interaction of order less than R – 1. In a resolution III (R = III) design, no main effects are confounded with any other interaction of order less than R – I = III – I = II. Thus, main effects in this design are confounded with two-way interactions and, consequently, all higher-order interactions are equally confounded.

The resolution is defined as the length of the shortest "word" in a defining relation. For example, the defining relation of a 2^{3-1} design is I = ABC. There are three letters in the defining relation ("word"), so it is a resolution III design. The defining relation of a 2^{4-1} design is I = ABCD. There are four letters in the defining relation, so it is a resolution IV design. Resolution describes the degree to which estimated main effects are aliased (or confounded) with estimated two-level interactions, three-level interactions, and so on. Higher-resolution designs have less severe confounding, but require more runs.

A resolution IV design is "better" than a resolution III design because we have a less-severe confounding pattern in the resolution IV design than in the resolution III situation, so higher-order interactions are less likely to be significant than lower-order interactions. A higher-resolution design for the same number of factors will, however, require more runs. In two-level fractional factorial experiments, the following three resolutions are most frequently used:

- Resolution III Designs. Main effects are confounded (aliased) with two-factor interactions.
- Resolution IV Designs. No main effects are aliased with two-factor interactions, but two-factor interactions are aliased with each other.
- Resolution V Designs. No main effect or two-factor interaction is aliased with any other main effect or two-factor interaction, but two-factor interactions are aliased with three-factor interactions.

12.9.4 Quarter Fraction of 2^k Design

When the number of factors, k, gets larger, 2^{k-1} designs will also require many runs. Then, a smaller fraction factorial design is needed. A quarter fraction of factorial design is also called a 2^{k-2} design. For a 2^{k-1} design, there is one defining relationship. Each defining relationship is able to reduce the number of runs by half. For a 2^{k-2} design, two defining relationships are needed. If P and Q represent the generators chosen, then I = P and I = Q are called generating relations for the design. Also, because I = P and I = Q, I = PQ. I = P = Q = PQ is called the complete defining relation.

Consider a 2^{6-2} design. In this design, there are six factors, say, A, B, C, D, E, and F. For a 2^{6-1} design, the generator would be I = ABCDEF, and we would have a resolution VI design. For a 2^{6-2} design, if we choose P and Q to have five letters, for example, P = ABCDE, Q = ACDEF, then PQ = BF. From I = P = Q = PQ, the complete defining relation is I = ABCDE = ACDEF = BF. We will only have resolution II! In this case, even the main effects are confounded so, clearly, it is not a good design. If we choose P and Q to be four letters, for example P = ABCE, Q = BCDF, then PQ = ADEF and I = ABCE = BCDF = ADEF. This is a resolution IV design. Clearly, it is also the highest resolution that a 2^{6-2} design can achieve. Now we can develop a procedure to lay out the 2^{k-2} design, and illustrate that with an example.

Step 1: Compute* N = 2^{k-2} *and Determine the Number of Runs. For example, for $k = 6$, $N = 2^{k-2} = 2^4 = 16$.

Step 2: Create a Table with* N *Runs, and Lay Out the First* k – 2 *Factors in Standard Order. For example, for $k = 6$, the factors are A, B, C, D, E, and F. The first $k - 2 = 4$ factors are A, B, C, and D. We will lay out the first four columns with A, B, C, and D in standard order (Table 12.16).

Step 3: Use Defining Relation to Create the Last Two Columns. In Table 12.16, if we use I = ABCE as the defining relation, then E = ABC and I = BCDF, then F = BCD.

Example 12.4 (See Yang and El-Haik, 2003)

The manager of a manufacturing company is concerned about the large number of errors in invoices. An investigation is conducted to determine the major sources of

Table 12.16 2^{6-2} design standard order

Run No.	Factors					
	A	B	C	D	E = ABC	F = BCD
1	–1	–1	–1	–1	–1	–1
2	1	–1	–1	–1	1	–1
3	–1	1	–1	–1	1	1
4	1	1	–1	–1	–1	1
5	–1	–1	1	–1	1	1
6	1	–1	1	–1	–1	1
7	–1	1	1	–1	–1	–1
8	1	1	1	–1	1	–1
9	–1	–1	–1	1	–1	1
10	1	–1	–1	1	1	1
11	–1	1	–1	1	1	–1
12	1	1	–1	1	–1	–1
13	–1	–1	1	1	1	–1
14	1	–1	1	1	–1	–1
15	–1	1	1	1	–1	1
16	1	1	1	1	1	

the errors. Historical data are retrieved from a company database that contains information on customers, types of products, size of shipments, and so on. The investigation group identified four factors relating to the shipment of products and defined two levels for each factor (see Table 12.17). The group then set up a 2^{4-1} factorial design to analyze the data, and the data from the last two quarters were studied and percentage errors in invoices recorded. For data analysis of two-level fractional factorial experiments, we can use the same step-by-step procedure for the two-level full factorial experiments, except that N should be the actual number of runs, not 2^k. By using Minitab™, we get the following results shown in Tables 12.18, 12.19, and 12.20.

By referring to Section 6.4, the transfer function is given by

$$\hat{y} = 0.2864 + \underbrace{C - 3.750L + 4.750T}_{\text{Linear terms}} + \underbrace{1.250CT - 0.750CL + 2.500CS}_{\text{Interaction terms}}$$

Table 12.17 Factors and levels for Example 12.4

Factor	Level	
Customer (C)	Minor (–1)	Major (+1)
Customer location (L)	Foreign (–1)	Domestic (+)
Type of product (T)	Commodity (–1)	Specialty (+)
Size of shipment (S)	Small (+)	Large (+)

Table 12.18 Experiment layout and data for Example 12.4

Factors				Percentage of error	
C	L	T	S = CLT	Quarter 1	Quarter 2
–1	–1	–1	–1	14	16
1	–1	–1	1	19	17
–1	1	–1	1	6	6
1	1	–1	–1	1.5	2.5
–1	–1	1	1	18	20
1	–1	1	–1	24	22
–1	1	1	–1	15	17
1	1	1	1	21	21

The alias structure is as follows:

$I + C \times L \times T \times S$
$C + L \times T \times S$
$L + C \times T \times S$
$T + C \times L \times S$
$S + C \times L \times T$
$C \times L + T \times S$
$C \times T + L \times S$
$C \times S + L \times T$

Based on the Pareto effect plot (Figure 12.11), factor T (type of product) and L (location) as well as interaction CS (customer and size of shipment) are the top three effects (Figure 12.12). However, since CS and TL are aliased, this could also be due to the effect of TL. By using some common sense, the team thinks that TL (types of products and location of customer) interaction is more likely to have a significant effect (Figure 12.13).

Table 12.19 Estimated effects and coefficients for % error (coded units) for Example 12.4

Term	Effect	Coefficient	SE coefficient	$T_{0,1,8}$	P-value
Constant	15.000	0.2864	52.37	0.00	0.000
C	2.000	1.000	0;2864	3.49	0.008
L	–7.500	–3.750	0.2864	–13.09	0.000
T	9.500	4.750	0.2864	16.58	0.000
S	16.00	1	16.00	12.19	0.008
$C-L$	–1.500	–0.750	0.2864	–2.62	0.031
$C-T$	2.500	1.250	0.2864	4.36	0.002
$C-S$	5.000	2.500	0.2864	8.73	0.000

Table 12.20 Analysis of variance for % error using adjusted SS for tests for Example 12.4

Source of variation	Sum of squares	Degree of freedom	Mean squares	$F_{0,1,8}$	P-value
C	16.00	1	16.00	12.19	0.008
L	225.00	1	225.00	171.43	0.000
T	361.00	1	361.00	275.05	0.000
S	16.00	1	16.00	12.19	0.008
$C \times L$	9.00	1	9.00	6.86	0.031
$C \times T$	25.00	1	25.00	19.05	0.002
$C \times S$	100.00	1	100.00	76.19	0.000
Error	10.50	8	10.50	1.31	
Total	762.50	15			

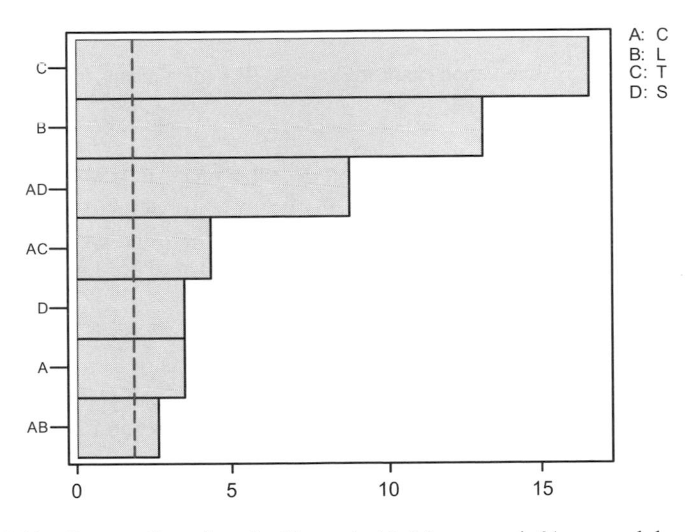

Figure 12.11 Pareto effect chart for Example 12.4 (response is % error, alpha = 0.10).

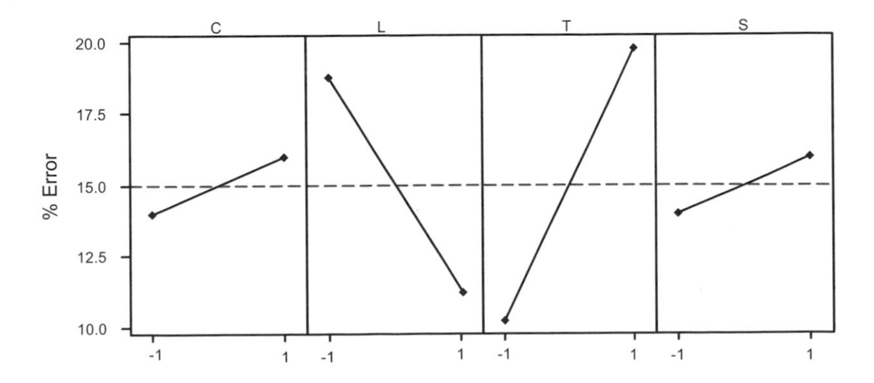

Figure 12.12 Main effect chart of Example 12.4 (data means for % error).

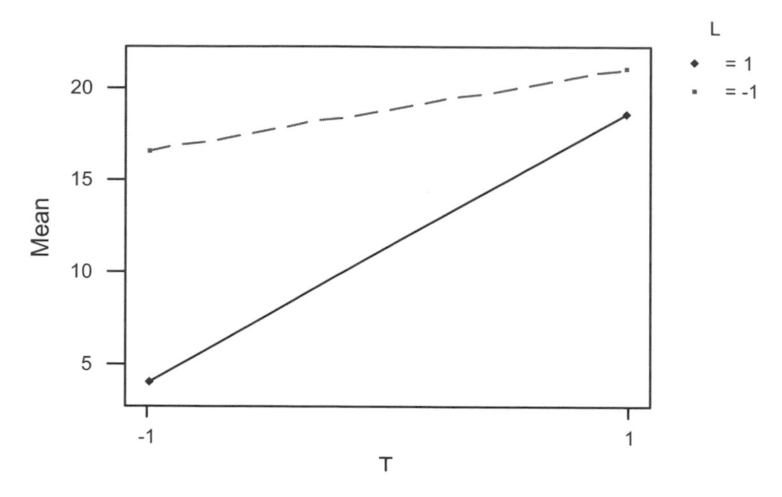

Figure 12.13 Interaction chart of Example 12.4 (data means for % error).

Type of product is found to be most significant effect. There are more invoice errors for specialty products and there are much fewer invoice errors for commodity product, especially for domestics customers—only 5% of invoices have errors. Location of customer is second most significant effect. There are lot more invoice errors for foreign customers, even for commodity products.

A 2^k fractional factorial design having 2^{k-p} runs is called a $1/2^p$ fraction of a 2^k design, or a 2^{k-p} fractional factorial design. These designs need p independent generators. The selection of these p generators should make the resultant design have highest possible resolution. Montgomery (1997) lists many good 2^{k-p} fractional factorial designs.

2^{k-p} designs are the "engines" of design of experiments because they can analyze many factors simultaneously with relative efficiency in a few experimental runs. The experiment design is also straightforward because each factor has only two settings. The simplicity of these designs is a major flaw. That is, the underlying use of two-level factors is the belief that the mathematical relationships between responses and factors are basically linear in nature. This is often not the case, as many variables are related to the responses in nonlinear ways. Another problem of fractional designs is the implicit assumption that higher-order interactions do not matter; but sometimes they do. In these cases, it is nearly impossible for fractional factorial experiments to detect higher-order interaction effects.

The richness of information versus cost of two-level designs is depicted in Figure 12.14, in which RSM is the response surface methodology (See Yang and El-Haik, 2003 for RSM within the product DFSS context). The curves in the figure are chosen to indicate conceptual nonlinearity of both cost and information richness (in terms of the design transfer function) and not to convey exact mathematical equivalence. Clearly, there is trade-off between what you want to achieve and what you are able to pay—the value proposition.

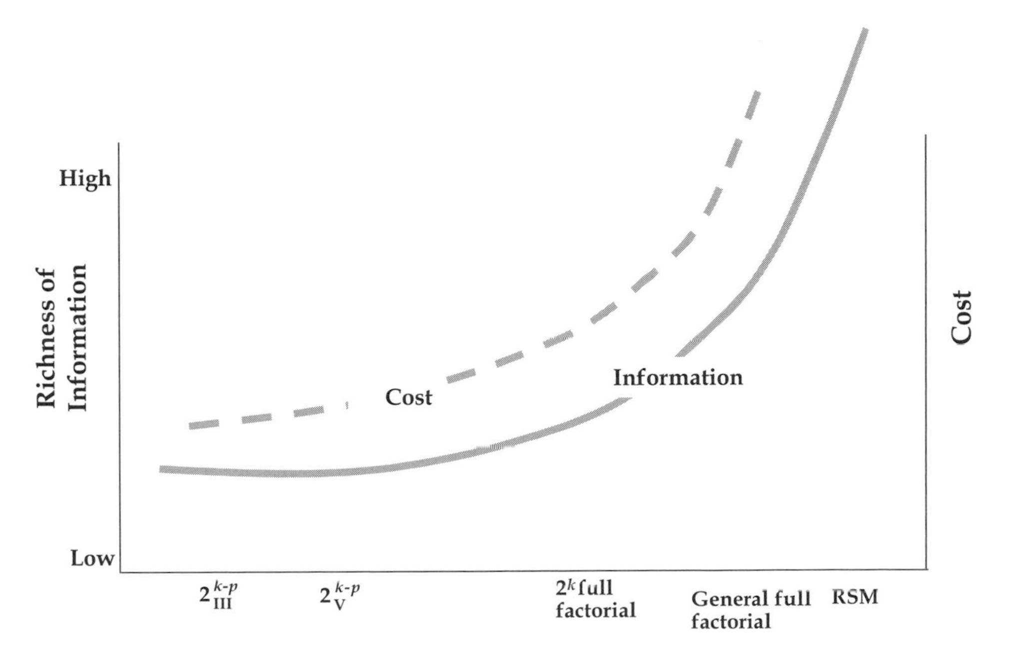

Figure 12.14 Richness of information versus cost of two-level designs.

12.10 OTHER FACTORIAL DESIGNS

12.10.1 Three-Level Factorial Design

These designs are used when factors need to have more than two levels. For example, if the black belt suspects that the effect of the factors on the dependent variable of interest is not simply linear, then the design needs at least three levels in order to test for the linear and quadratic effects and interactions for those factors. The notation 3^k is used. It means that k factors are considered, each at three levels. This formulation is more suitable with factors that are categorical in nature, with more than two categories, such as the scenario of Example 12.1.

The general method of generating fractional factorial designs at three levels (3^{k-p} designs) is very similar to that described in the context of 2^{k-p} designs. Specifically, one starts with a full factorial design, and then uses the interactions of the full design to construct new factors by making their factor levels identical to those for the respective interaction terms (i.e., by making the new factors aliases of the respective interactions).

The three levels are (usually) referred to as low, intermediate, and high levels. These levels are numerically expressed as 0, 1, and 2. One could use the digits –1, 0, and +1, but this may be confused with respect to the two-level designs in which 0 is reserved for center points. Therefore, the 0, 1, 2 scheme is usually recommended. Three-level designs are used to model possible curvature in the transfer function and to handle the case of nominal factors at three levels. A third level for a continuous factor facilitates investigation of a quadratic relationship between the response and each of the factors. Unfortunately, the three-level design is prohibitive in terms of the number of runs and, thus, in terms of cost and effort.

12.10.2 Box–Behnken Designs

For 2^{k-p} designs, Plackett and Burman (1946) developed highly fractionalized designs to screen the maximum number of (main) effects in the least number of experimental runs. Their equivalent for 3^{k-p} designs are the so-called Box–Behnken designs (Box and Draper, 1987). These designs do not have simple design generators (they are constructed by combining two-level factorial designs with incomplete block designs), and suffer from complex confounding of interaction. However, the designs are economical and, therefore, particularly useful when it is expensive to perform the necessary experimental runs. The analysis of these types of designs proceeds basically in the same way as was described in the context of 2^{k-p} designs. However, for each effect, one can now test for the linear effect and the quadratic (nonlinear) effect. Nonlinearity often occurs when a process performs near its optimum.

12.11 SUMMARY

A design of experiment (DOE) is a structured method for determining the transfer function relationship between factors affecting a process and the output of that process (y). In essence, it is a methodology for conducting and analyzing controlled tests to evaluate the effects of factors that control the value of a variable or a group of responses. DOE refers to experimental methods used to quantify indeterminate measurements of factors and interactions between factors statistically, through observance of forced changes made methodically as directed by systematic tables.

There are two main bodies of knowledge in DOE: experimental design and experimental data analysis. Two types of experimental design strategies are discussed in this chapter: full factorial and fractional factorial. Full factorial design is used to obtain more information when affordable, since the size of the experiment will grow exponentially with the number of experimental factors and levels. Fractional factorial design obtains less information from an experiment but the experiment size will grow much slower than that of a full factorial design. In addition, we can adjust the resolution of a fractional factorial design so it can obtain the needed information while keeping the experiment to a manageable size. Therefore, fractional factorial design has become the "work horse" of DOE. DOE is critical in creating orthogonal designs, those in which the factors are independent of each other.

The main DOE data analysis tools include analysis of variance (ANOVA), empirical transfer function model building, and main effects and interaction charts. ANOVA is able to identify the set of significant factors and interactions, and rank the relative importance of each effect and interaction in terms of their effect on design output. Empirical transfer function models, main effect plots, and interaction plots show the empirical relationship between design output and design factors. They can also be used to identify optimal factor level settings and corresponding optimal design performance levels.

APPENDIX

Diagnostic Plots of Residuals

Before accepting a particular ANOVA result or a transfer function that includes a particular number of effects, the black belt should always examine the distribution of the residual values. This is computed as the difference between the predicted values (as predicted by the current model) and the observed values. He or she can compute the histogram for these residual values, as well as probability plots, or generate them from a statistical package such as Minitab™.

The parameter estimates and ANOVA tables are based on the assumption that the residuals are normally distributed. The histogram provides one way to visually check whether this assumption holds.

The normal probability plot is another common tool to assess how closely a set of observed values (residuals in this case) follows a theoretical distribution. In this plot, the actual residual values are plotted along the horizontal X-axis; the vertical Y-axis shows the expected normal values for the respective values, after they have been rank ordered. If all values fall onto a straight line, then one can be satisfied that the residuals follow the normal distribution.

Pareto Chart of Effects

The Pareto chart of effects is often an effective tool for communicating the results of an experiment, in particular to laymen. In this graph, the ANOVA effect estimates are sorted from the largest absolute value to the smallest absolute value. A column represents the magnitude of each effect and, often, a line going across the columns indicates how large an effect has to be to be statistically significant.

Square and Cube Plots

These plots are often used to summarize predicted values for the response variable, given the respective high and low settings of the factors. The square plot will show the predicted values for two factors at a time. The cube plot will show the predicted values for three factors at a time.

Interaction Plots

A general graph for showing the means is the standard interaction plot, in which points connected by lines indicate the means. This plot is particularly useful when there are significant interaction effects in the analysis.

13

SERVICE DESIGN ROBUSTNESS

13.1 INTRODUCTION

In the context of this book, the terms "quality" and "robustness" can be used interchangeably. Robustness is defined as the method of reducing the variation of the functional requirements of a design and keeping them on target as defined by the customer (Taguchi, 1986; Taguchi and Wu, 1986; Phadke, 1989; Taguchi et al., 1989; Taguchi et al., 1999).

Variability reduction has been investigated in the context of robust design (Taguchi, 1986) through methods such as parameter design and tolerance design. The principal idea of robust design is that statistical testing of a product or process should be carried out at the design stage, also referred to as the "off-line stage" very much in line with DFSS. In order to make the service robust against the effects of variation sources in the production and use environments, the design problem is viewed from the point of view of quality and cost (Taguchi, 1986; Taguchi and Wu, 1989; Taguchi et al., 1989; Taguchi et al., 1999; Nair, 1992).

Quality is measured by quantifying statistical variability through measures such as standard deviation or mean square error. The main performance criterion is to achieve the design parameter (DP) target on average, while simultaneously minimizing variability around this target.

Robustness means that a system performs its intended functions under all operating conditions (different causes of variations) throughout its intended life. The undesirable and uncontrollable factors that cause the DP under consideration to deviate from the target value are called "noise factors." Noise factors adversely affect quality, and ignoring them will result in a system not optimized for conditions of use and, possibly, failure. Eliminating noise factors may be expensive. Instead, we seek to reduce the effect of the noise factors on the DP's performance by choosing design parameters and their settings that are insensitive to noise.

Robust design is a disciplined process that seeks to find the best expression of a system or process design. "Best" is carefully defined to mean that the design is the lowest-cost solution to the specification, which itself is based on identified customer needs. Taguchi included design quality as one more dimension of cost. High-quality systems minimize these costs by performing, consistently, on targets, as specified by the customer. Taguchi's philosophy of robust design is aimed at reducing loss due to deviation of performance from the target value, and is based on a portfolio of concepts and measures such as quality loss function (QLF), signal-to-noise (SN) ratio, optimization, and experimental design. quality loss is the loss experienced by customers and society and is a function of how far performance deviates from target. The QLF relates quality to cost and is considered a better evaluation system than the traditional binary treatment of quality, that is, within/outside specifications. The QLF of a design parameter (denoted as y or DP) has two components: mean (μ_y) deviation from targeted performance value (T_y), and variance (σ_y^2). It can be approximated by a quadratic polynomial of the design parameter.

13.2 ROBUSTNESS OVERVIEW

In Taguchi's philosophy, robust design consists of three phases (Figure 13.1). It begins with the *concept design phase* and is followed by *parameter design* and *tolerance design* phases. It is unfortunate to note that the concept design phase has not received the attention it deserves in the quality engineering community, hence the focus on it in this book.

The goal of parameter design is to minimize the expected quality loss by selecting design parameters settings. The tools used are quality loss function, design of experiments, statistics, and optimization. Parameter design optimization is carried out in two sequential steps: minimization of variability (σ_y^2) and mean (μ_y), and adjustment to target T_y. The first step is conducted using the process variables (xs) that affect variability, whereas the second step is accomplished via the design parameters that affect the mean but do not adversely influence variability. The objective is to carry out both steps at low cost by exploring the opportunities in the design space.

Parameter design is the most used phase in the robust design method. The objective is to design a solution entity by making the design parameters (ys) insensitive to variation. This is accomplished by selecting the optimal levels of the design parameters based on testing and using an optimization criterion. Parameter design optimization criteria include the quality loss function the and signal-to-noise ratio (SN).

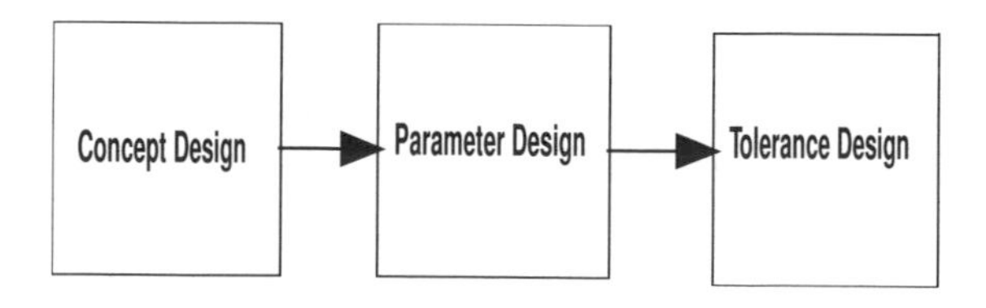

Figure 13.1 Taguchi's robust design.

The optimum levels of the xs are the levels that maximize the SN and are determined in an experimental setup from a pool of economical alternatives. These alternatives assume the testing levels in searching for the optimum.

An output response of a process can be classified as static or dynamic from a robustness perspective. A static entity has a fixed target value. The parameter design phase in the case of the static solution entity brings the response (y) and mean, μ_y, to the target, T_y. On the other hand, the dynamic response expresses a variable target depending on customer intent. In this case, the optimization phase is carried out over a range of useful customer applications called the signal factor. The signal factor can be used to set the y to an intended value. For example, in the loan process of a mortgage, the signal factor is the amount requested.

Parameter design optimization requires the classification of the design parameters (output responses) as *smaller the better* (e.g., minimize customer complaints, reduce operating cost), *larger the better* (e.g., increase profit; maximize satisfaction), *nominal the best* (where keeping the process to a single performance objective is the main concern, e.g., meet delivery commitment), and *dynamic* (where energy related functional performance over a prescribed dynamic rage of usage is perspective, e.g., loan granted versus loan requested).

When robustness cannot be assured by parameter design, we resort to the tolerance design phase. Tolerance design is the last phase of robust design. This involves upgrading or tightening tolerances of some design parameters so that quality loss can be reduced. However, tightening tolerances will usually add cost to the process that controls the tolerance. El-Haik (2005) formulated the problem of finding the optimum tolerances of the design parameters that minimize both quality loss and tolerance control costs.

The important contribution of robust design is the systematic inclusion in the experimental design of noise variables, that is, the variables over which the designer has no or little control. A distinction is also made between internal noise, such as material variability, and environmental noise, which the design team cannot control (e.g., culture, politics). Robust design's objective is to suppress, as far as possible, the effect of noise by exploring the levels of the factors to determine their potential for making the system insensitive to these sources of variation.

Several robust design concepts that apply to service design are presented below.

13.3 ROBUSTNESS CONCEPT #1: QUALITY LOSS FUNCTION

Traditional inspection schemes represent the heart of on-line quality control. Inspection schemes depend on the binary characterization of design parameters, that is, being within the specification limits or outside the specification limits. A process is called conforming if all of its inspected design parameters are within their respective specification limits, otherwise it is nonconforming. This binary representation of the acceptance criteria per a design parameter is not realistic since it characterizes, equally, entities that are marginally off the specification limits and entities that are marginally within these limits. In addition, this characterization does not distinguish the marginally off entities from those that are significantly off. The point here is that it is not realistic to assume that as we move away from the nominal specification in a process, the quality loss is zero as long as we stay within the set tolerance limits. Rather, if the process is not exactly "on target," then loss will result, for example, in terms of customer satisfaction. Moreover, this loss is probably not a linear function of the deviation from nominal specifications but, rather, a quadratic function. Taguchi proposed a continuous and better representation than this dichotomous characterization—the quality loss function (Taguchi and Wu, 1989). The loss function provides a better estimate of how the monetary loss incurred by the production process and customers as an output response (a design parameter) leads to deviation from the targeted performance value, T_y. The determination of the target T_y implies the nominal-the-best and dynamic classifications.

A quality loss function can be interpreted as a method to translate variation and target adjustment to a monetary value. It allows the design teams to perform a detailed optimization of cost by relating technical terminology to economic measures. In its quadratic form, a quality loss is determined by first finding the functional limits,[1] $T_y \pm \Delta y$, of the concerned response. The functional limits are the points at which the process would fail, that is, the design produces unacceptable performance in approximately half of the customer applications. In a sense, these represent performance levels that are equivalent to average customer tolerance. Kapur (1988) continued with this line of thinking and illustrated the derivation of specification limits using Taguchi's quality loss function. Quality loss is incurred due to the deviation, caused by noise factors, in the response (y or FR) as from their intended targeted performance, T_y. Let L denote the quality loss function (QLF) by taking the numerical value of the FR and the targeted value as arguments. By Taylor series expansion[2] at $FR = T$, and with some assumption about the significant of the expansion terms, we have

$$L(FR, T) \cong K(FR - T_{FR})^2 \tag{13.1}$$

Let $FR \in \lfloor T_y - \Delta y, T_y + \Delta y \rfloor$, where T_y is the target value and Δy is the functional deviation from the target (see Figure 13.2). Let A_Δ be the quality loss incurred due to the symmetrical deviation, Δy. Then, by substitution into Equation 13.1 and solving for K,

$$K = \frac{A_\Delta}{(\Delta y)^2} \tag{13.2}$$

In the Taguchi tolerance design method, the quality loss coefficient K can be determined on the basis of losses in monetary terms if the requirement fall outside the

[1]Functional limits or customer tolerance in robust design terminology is synonymous with design range, DR, in the axiomatic design approach terminology. See Section 2.4.

[2]The assumption here that L is a higher-order continuous function such that derivatives exist and are symmetrical around $y = T$.

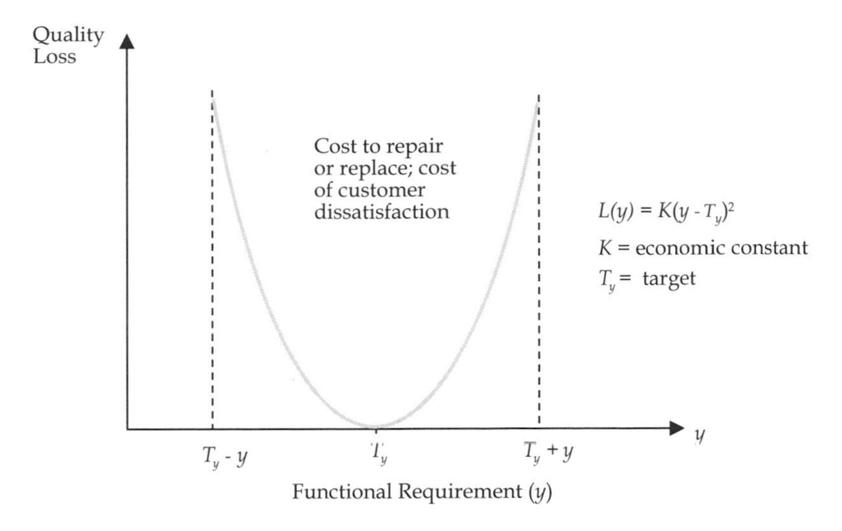

Figure 13.2 Quality loss function.

customer tolerance limits (*design range*) instead of the specification limits usually used in process capability studies for example, the *producer limits.* The specification limits are most often associated with the design parameters. Customer tolerance limits are used to estimate the loss from the customer's perspective or the quality loss to society as proposed by Taguchi. Usually, customer tolerance is wider than manufacturer tolerance. In this chapter, we use with the design range limits terminology. Deviation from this practice will be noted where needed.

Let $f(y)$ be the probability density function (*pdf*) of y. Then, via the expectation operator, E, we have

$$E[L(y, T)] = K[\sigma_y^2 + (\mu_y - T_y)^2] \tag{13.3}$$

Equation (13.3) is very fundamental. Quality loss has two ingredients: loss incurred due to variability σ_y^2, and loss incurred due to mean deviation from target $(\mu_y - T_y)^2$. Usually, the second term is minimized by adjustment of the mean of the critical few design parameters, the critical mean affecting xs. The derivation in Equation (13.3) suits the nominal-the-best classification. Other quality loss function mathematical forms may be found in Chen and Kapur (1989). The following discussion was borrowed from their esteemed paper.

13.3.1 Larger-the-Better Loss Function

For functions like "increase sales" (y = sales), we would like a very large target, ideally $T_y \to \infty$. The requirement (output y) is bounded by the lower functional specifications limit y_l. The loss function is then given by

$$L(y, T_y) = \frac{K}{y^2} \tag{13.4}$$

where $y \geq y_l$. Let μ_y be the average y numerical value of the system range, that is, the average around which performance delivery is expected. Then, by Taylor series expansion around $y = \mu_y$, we have

$$E[L(y, T_y)] = K\left[\frac{1}{\mu_y^2} + \frac{3}{\mu_y^4}\sigma_y^2\right] \tag{13.5}$$

13.3.2 Smaller-the-Better Loss Function

For functions like "reduce complaints," we would like to have zero as the target value. The loss function in this category and its expected values are given in Equations (13.6) and (13.7), respectively.

$$L(y, T) = Ky^2 \tag{13.6}$$

$$E[L(y, T)] = K(\sigma_y^2 + \mu_y^2) \tag{13.7}$$

In the above development as well as in the next sections, the average loss can be estimated from a parameter design or even a tolerance design experiment by substituting the experimental variance S^2 and average $\bar{y}$ as estimates for σ_y^2 and μ_y into the equations above.

Recall the two-settings example in Section 6.5 and Figure 6.9. It was obvious that setting 1 is more robust, that is, it produces less variation in the functional requirement (y) than setting 2 by capitalizing on nonlinearity[3] as well as lower quality, loss similar to scenario of the right-hand side of Figure 13.3. Setting 1 robustness is even more evident in the flatter quadratic quality loss function.

Since quality loss is a quadratic function of the deviation from a nominal value, the goal of the DFSS project should be to minimize the squared deviation or variance of the service around nominal (ideal) specifications, rather than the number of units within specification limits (as is done in traditional SPC procedures).

Several books have recently been published on these methods, for example, Phadke (1989), Ross (1988), and within the context of product DFSS, Yang and El-Haik (2003), to name a few, and it is recommended that the reader refer to these books for further specialized discussions. Introductory overviews of Taguchi's ideas about quality and quality improvement can also be found in Kackar (1986).

[3]In addition to nonlinearity, leveraging interactions between the noise factors and the design parameters is another popular empirical parameter design approach.

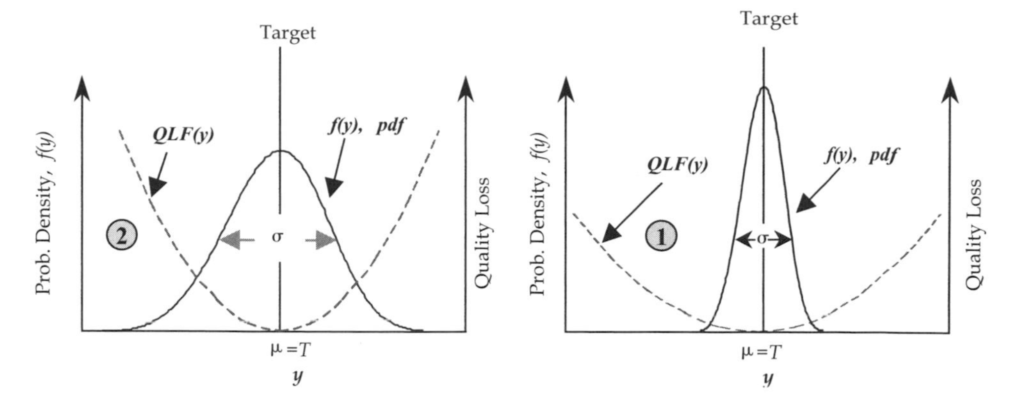

Figure 13.3 The quality loss function scenarios of Figure 6.9.

13.4 ROBUSTNESS CONCEPT #2: SIGNAL, NOISE, AND CONTROL FACTORS

The service of DFSS quality should always respond in exactly the same manner to the signals provided by the customer. When you press the ON button of the TV remote control you expect the TV to switch on. In a DFSS designed TV, the starting process would always proceed in exactly the same manner; for example, three seconds after the remote is pressed, the TV comes to life. If, in response to the same signal (pressing the ON button), there is random variability in this process, then you have less than ideal quality. For example, due to such uncontrollable factors as speaker conditions, weather conditions, battery voltage level, and TV wear, the TV may sometimes start only after 20 seconds and finally not start at all. We want to minimize the variability in the output response to noise factors while maximizing the response to signal factors.

Noise factors are those factors that are not under the control of the design team of a service. In the TV example, these factors include speaker conditions, weather conditions, battery voltage level, and TV wear. Signal factors are those factors that are set or controlled by the customer (end user) of the service to make use of its intended functions.

The goal of a service DFSS project is to find the best experimental settings (in the absence of an off-the-shelf transfer function) of factors under the team's control that are involved in the design, in order to minimize quality loss and, thus, the factors in the experiment represent control factors. Signal, noise, and control factors (design parameters) are usually summarized in a P-diagram similar to that of Figure 12.1.

13.5 ROBUSTNESS CONCEPT #3: SIGNAL-TO-NOISE RATIOS

A conclusion that may be drawn from the previous sections is that quality can be quantified in terms of the respective service response to noise factors and signal factors. The ideal service will only respond to customer signals and be unaffected by random noise factors. Therefore, the goal of the DFSS project can be stated as attempting to maximize the signal-to-noise (SN) ratio for the product. The SN ratios described in the following paragraphs were proposed by Taguchi (1986).

Smaller the better. In cases in which the DFSS team wants to minimize the occurrences of some undesirable service responses, they would compute the following SN ratio:

$$SN = -10 \log_{10}\left(\frac{1}{N} \sum_{n=1}^{N} y_i^2 \right) \tag{13.8}$$

The constant, N, represents the number of observations (that have y_i as their design parameter) measured in an experiment or in a sample. Experiments are conducted and the y measurements are collected. For example, the number of complaints with respect to a home mortgage process could be measured as the y variable and analyzed via this SN ratio. The effect of the signal factors is zero, since zero complaints is the only intended or desired state of the loan process. Note how this SN ratio is an expression of the assumed quadratic nature of the loss function. The factor 10 ensures that this ratio measures the inverse of bad quality; the more complaints in the process, the greater is the sum of the squared number of complaints, and the smaller (i.e., more negative) the SN ratio. Thus, maximizing this ratio will increase quality.

Nominal the best. Here, the DFSS team has a fixed signal value (nominal value), and the variance around this value can be considered the result of noise factors:

$$SN = -10 \log_{10}\left(\frac{\mu^2}{\sigma^2} \right) \tag{13.9}$$

This signal-to-noise ratio could be used whenever ideal quality is equated with a particular nominal value. For example, the time delay with respect to a target commitment in a home mortgage process could be measured as the y variable and analyzed via this SN ratio. The effect of the signal factors is zero, since the target date is the only intended or desired state of the loan process.

Larger the better. Examples of this type of service problem are sales, revenue, profit, and so on. The following SN ratio should be used:

$$SN = -10 \log_{10}\left(\frac{1}{N} \sum_{n=1}^{N} \frac{1}{y_i^2} \right) \tag{13.10}$$

Fraction defective (p). This SN ratio is useful for minimizing scrap (services outside the specification limits), for example, the percent of nonclosed mortgage loans. Note that this SN ratio is identical to the familiar logit transformation:

$$SN = 10 \log_{10}\left(\frac{p}{1-p}\right) \quad (13.11)$$

where p is the proportion defective.

13.6 ROBUSTNESS CONCEPT #4: ACCUMULATION ANALYSIS

Accumulation is the most appealing robust design technique to service due to its treatment of categorical DOE data. In some cases, measurements on a service output response can only be obtained in terms of categorical judgments. For example, consumers may rate a service as excellent, good, average, or below average. In that case, the DFSS team would attempt to maximize the number of excellent or good ratings. Typically, the results of an accumulation analysis are summarized graphically in a stacked bar plot or a line plot.

When analyzing ordered categorical data, ANOVA is not appropriate. Rather, a cumulative plot is produced of the number of observations in a particular category. For each level of each factor, the design team plots the cumulative proportion of the number of categories. Thus, this graph provides valuable information concerning the distribution of the categorical counts across the different factor settings.

13.6.1 Accumulation Analysis Example

Sixteen workers were randomly selected to participate in an experiment to determine the effects of work scheduling and method of payment on attitude toward the job. Two schedules were employed (Factor A), the standard 8:00 am–5:00 pm workday and a modification whereby the worker could decide each day whether to start at 7:00 am or 8:00 am (Level A1); in addition, the worker could choose between a ½ hour and 1 hour lunch period every day (Level A2, current). The two methods of payments (Factor B) were standard hourly rate (Level B1; current) and a reduced hourly rate with an added piece rate based on the worker's production (Level B2). Four workers were randomly assigned to each of the four scheduling-payment treatments, and each competed an attitude test after one month on the job. The design team decided to record the attitude data in the following subjective categories, listed progressively opposite to the direction of goodness:

Category I: Excellent
Category II: Good
Category III: Indifferent
Category IV: Bad

The test scores are shown in Table 13.1.

Step 1: Cumulative Accumulation. Among the four data points (worker attitude) in Experiment 1 three belong to category I, 1 belongs to category II, and none be-

Table 13.1 Worker attitude example

Schedule (A) \ Payment (B)	B1: Hourly rate	B2: Hourly rate and piece rate
A1: 8:00 am–5:00 pm	I, II, I, I (Experiment 1)	IV, III, II, III (Experiment)
A2: Modified	III, II, I, II (Experiment 3)	III, IV, IV, IV (Experiment 4)

long to categories III and IV. Taguchi's accumulation experiment is an effective method for determining optimum control factor levels for categorical data by using accumulation. The number of data points in the cumulative categories for the four experiments are listed in Table 13.2.

For example, the number of data points in the four cumulative categories for Experiment 1 are 3, 4, 4, and 4.

Step 2: Factorial Effect Calculations. In this step, the factor effects are based on the probability distribution by attitude category. To determine the effect of Factor A at the 8:00 am–5:00 pm level (Level A1), the experiment or experiments conducted at that level must be identified and then the data points summed in each cumulative category, as in Table 13.3.

The number of data points in the four cumulative categories for every factor level is tabulated in Table 13.3 by repeating the Step 2 procedure. Note that in Table 13.3, the entry for cumulative category IV is equal to the number of data points for the particular factor level, and that entry is 8 in this case.

Step 3: Probabilities Calculations. The probabilities of the cumulative categories are calculated by dividing the number of data points in each cumulative category by the entry in the last cumulative category for that factor level, which is 8 in this example. The probabilities are calculated and listed in Table 13.4. For example, for Factor A at Level A1 (8:00 am–5:00 pm), the probability of cumulative category I is $\frac{3}{8} = 0.375$, for category II it is $\frac{5}{8} = 0.625$, and so on.

Table 13.2 Accumulation table

	Number of data points by category				Number of data points by cumulative category			
Experiment no.	I	II	III	IV	I	II	III	IV
1	3	1	0	0	3	4	4	4
2	0	1	2	1	0	1	3	4
3	1	2	1	0	1	3	4	4
4	0	0	1	3	0	0	1	4
Sum	4	4	4	4	4	8	11	16

Table 13.3 Factor A (schedule) at Level 8:00 am–5:00 pm (A1) effect calculation

	Categories			
Experiment no.	I	II	III	IV
1	3	4	4	4
2	0	1	3	4
Sum	3	5	7	8

Step 4: Probability Plotting. We want to determine the effect of each factor on the worker attitude probability by category. It is easier to do that with graphical techniques such as bar or line graphs. The latter technique was adopted in this example and is depicted in Figures 13.4 and 13.5.

It is apparent from the Figures 13.4 and 13.5 that Factor B (payment) has a larger effect than Factor A (schedule). In the line plot analysis, for each control factor we need to determine the level whose probability curve is uniformly higher than the curves for the other levels of that factor. A uniformly higher curve means that the particular factor level produces more data points with higher positive attitude; therefore, it is the optimum level. In Figure 13.4 we are looking for a larger probability of category I and smaller probability of category IV. It is apparent that the optimum levels of this experiment are A1 (8:00 am–5:00 pm) and B1 (hourly rate).

Step 5: Prediction. An additive transfer function is used to predict the performance under current baseline and optimum conditions. This function is used to approximate the relationship between the response (y) and factorial levels. Interaction effects are considered errors in this treatment. A verification experiment is conducted and the results are compared with the prediction using the additive transfer function. If the results match the prediction, then the optimum design is considered confirmed. That is, if the predicted and observed signal-to-noise ratios are close to each other, then the additive transfer function can explain the variability in terms of factorial mean effects. Otherwise, additional experimentation might be needed.

Table 13.4 Probabilities calculation

		Number of data points by cumulative categories				Probabilities for the cumulative categories			
Factor	Level	I	II	III	IV	I	II	III	IV
Factor A (schedule)	A1: 8:00 am–5:00 pm	3	5	7	8	0.375	0.625	0.875	1.000
	A2: Modified	1	3	5	8	0.125	0.375	0.625	1.000
Factor B (payment)	B1: Hourly rate	4	7	8	8	0.500	0.875	1.000	1.000
	B2: Hourly rate & piece rate	0	1	4	8	0.000	0.125	0.500	1.000

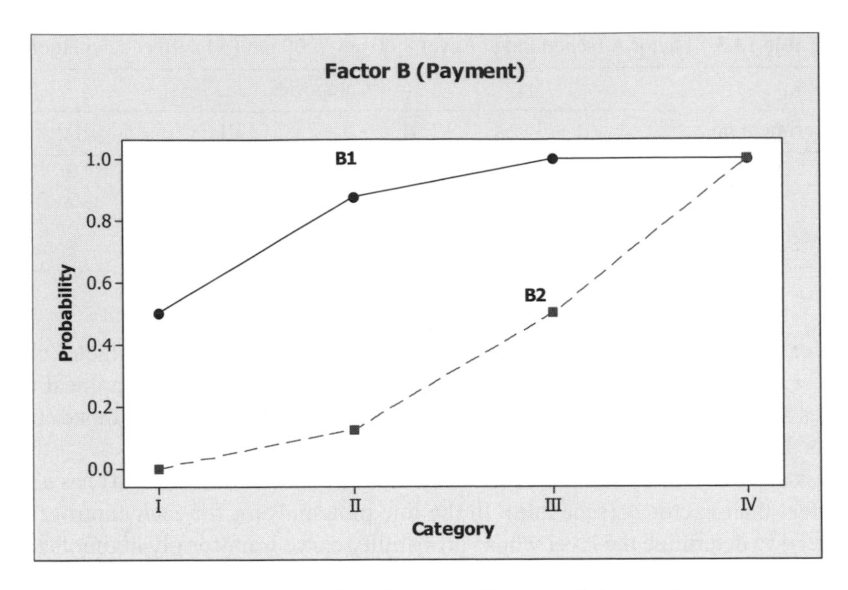

Figure 13.4 Line plots for Factor B accumulation analysis.

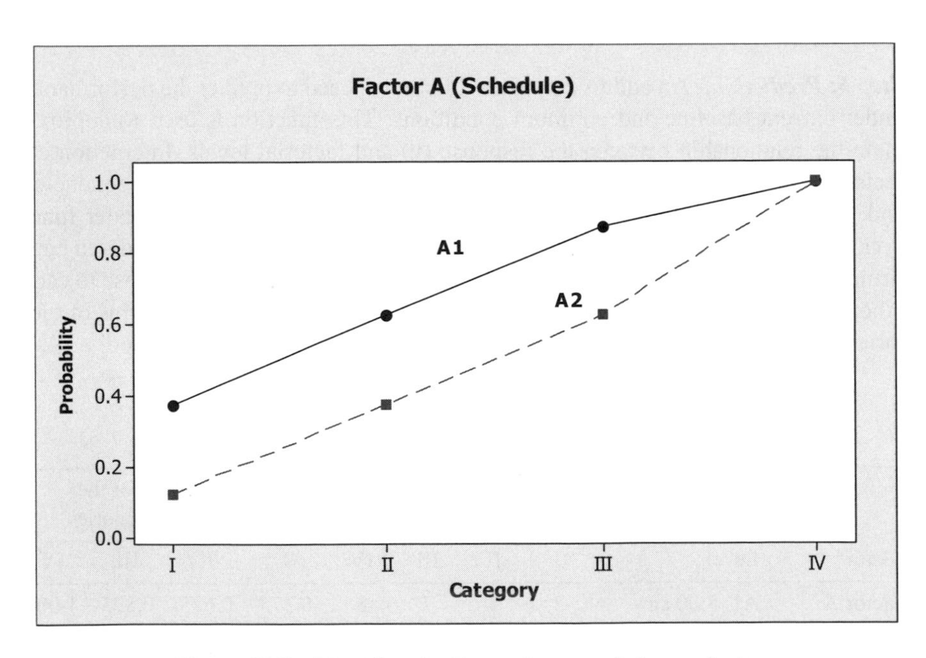

Figure 13.5 Line plots for Factor A accumulation analysis.

In the additive transfer function, only significant factors are included in order not to inflate the predicted improvement as compared to what is actually realized in the confirmation experiment. By neglecting the factors of small sum of squares, this inflation can be eliminated.

In this step, we predict the distribution of attitude under current and optimum conditions. This can be accomplished by the omega transform or the logit transform of the probabilities for cumulative probability p, is given as $\omega(p) = 10 \log_{10}[p/(1 - p)]$, which is similar to the signal-to-noise ratio of fraction defective.

Consider the optimum settings A1 and B1 for the prediction calculations. The average for category I taken over the four experiments is $\mu_{\mathrm{I}} = 4/16 = 0.25$ (see Table 13.2). Referring to Table 13.4, the predicted omega value for category I is:

$$\begin{aligned}\omega_{\mathrm{I}} &= \omega_{\mu_I} + \lfloor \omega_{A1_I} - \omega_{\mu_I} \rfloor + \lfloor \omega_{B1_I} - \omega_{\mu_I} \rfloor \\ &= \omega(0.25) + [\omega(0.375) - \omega(0.25] + [\omega(0.5) - \omega(0.25)] \\ &= -4.77 + [-2.22 + 4.77] + [0 + 4.77] \\ &= 2.55 \text{ dB}\end{aligned}$$

where $\omega(0.25) = 10 \log_{10}[0.25/(1 - 0.25)] = -4.77$ dB. Other omega transform calculations in the above equation are very similar.

By inverse omega transformation, the predicted probability for category I is 0.643. Other probabilities for categories II, III, and IV are obtained analogously. The predicted probabilities for the cumulative categories for optimum settings are listed in Table 13.5. It is clear that the recommended optimum settings give higher probabilities for attitude categories when compared to the starting conditions. The probability of category I is predicted to increase from 0.299 to 0.643 by changing from the current to the optimum condition. Other categories are equal indicating indifference.

13.7 ROBUSTNESS CONCEPT #5: ORTHOGONAL ARRAYS

This aspect of Taguchi robust design methods is the one most similar to traditional techniques. Taguchi developed a system of tabulated designs (arrays) that allows for the maximum number of main effects to be estimated in an unbiased (orthogo-

Table 13.5

	ω for the cumulative categories				Probabilities for the cumulative categories			
Experiment no.	I	II	III	IV	I	II	III	IV
Current (A2, B1)	–3.68	11.00	Infinity	Infinity	0.299	0.926	1.0	1.0
Optimum (A1, B1)	2.55	11.00	Infinity	Infinity	0.643	0.926	1.0	1.0

nal) manner, with a minimum number of runs in the experiment. Latin square designs, 2^{k-p} designs (Plackett–Burman designs, in particular), and Box–Behnken designs also aim at accomplishing this goal. In fact, many of the standard orthogonal arrays tabulated by Taguchi are identical to fractional two-level factorials, Plackett–Burman designs, Box–Behnken designs, Latin square designs, Greco-Latin square designs, and so on.

Orthogonal arrays provide an approach to efficiently designing experiments that will improve the understanding of the relationship between service and process parameters (control factors) and the desired output performance (functional requirements). This efficient design of experiments is based on a fractional factorial experiment, which allows an experiment to be conducted with only a fraction of all the possible experimental combinations of parameter values. Orthogonal arrays are used to aid in the design of an experiment. The orthogonal array will specify the test cases to conduct the experiment. Frequently, two orthogonal arrays are used: a control factor array and a noise factor array, the latter used to conduct the experiment in the presence of difficult to control variation so as to develop a DFSS robust design. This approach to designing and conducting an experiment to determine the effect of design factors (parameters) and noise factors on a performance characteristic is represented in Figure 13.6.

The design parameters or factors of concern are identified in an inner array or control factor array, which specifies the factor level or design parameters. The outer ar-

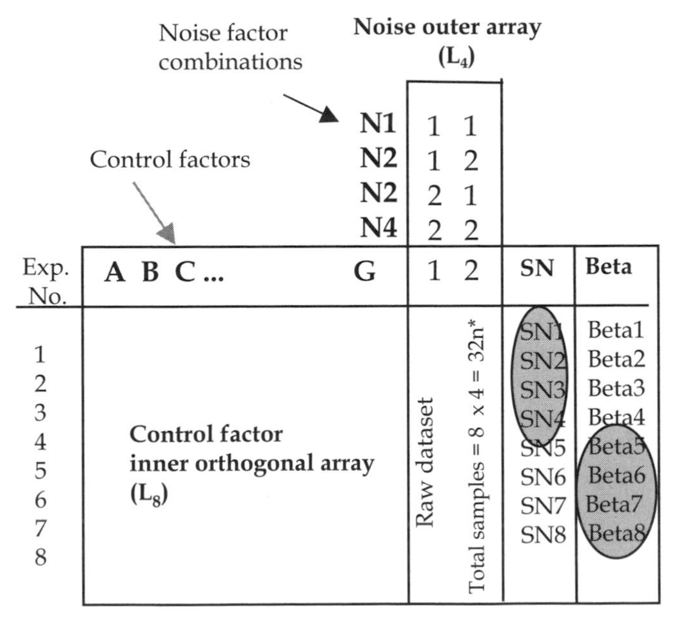

Figure 13.6 Parameter design orthogonal array experiment.

ray or noise factor array specifies the noise factor or the range of variation the service will be exposed to in the production process, and the environment or the conditions it is exposed to. This experimental setup allows for the identification of the design (control) parameter values or factor levels that will produce the best performing, most reliable, or most satisfactory service over the expected range of noise factors.

13.8 ROBUSTNESS CONCEPT #6: PARAMETER DESIGN ANALYSIS

After the experiments are conducted and the signal-to-noise ratio determined for each run, a mean signal-to-noise ratio value is calculated for each design factor level or value. This data is statistically analyzed using analysis of variance (ANOVA) techniques as described in Chapter 12. Very simply, a design (control) factor with a large difference in the signal-to-noise ratio from one factor setting to another indicates that the factor or design parameter is a significant contributor to the achievement of the performance response. When there is little difference in the signal-to-noise ratio from one factor setting to another, this indicates that the factor is insignificant with respect to the response. With the resulting understanding from the experiments and subsequent analysis, the design team can:

- Identify control factors levels that maximize output response in the direction of goodness and minimize the effect of noise, thereby achieving a more robust design.
- Perform the two-step robustness optimization[4]:

  Step 1. Choose factor levels to reduce variability by improving the SN ratio. This is robustness optimization step 1. The level for each control factor with the highest SN ratio is selected as the parameter's best target value. All of these best levels will be selected to produce the "robust design levels" or the "optimum levels" of design combination.

  A response graph summarizing SN gain, similar to Figure 13.7, is usually used. Control factor level effects are calculated by averaging SN ratios, which correspond to the individual control factor levels as depicted by the orthogonal array diagram. In this example, the robust design levels are as follows: Factor A at level 2, Factor C at level 1, and Factor D at level 2, or simply A2C1D2.

  Identify control factors levels that have no significant effect on the functional response mean or variation. In these cases, tolerances can be relaxed and cost reduced. This is the case of Factor B of Figure 13.7.

  Step 2. Select factor levels to adjust mean performance. This is the robustness optimization step 2 and is more suited for dynamic characteristic

[4]Notice that robustness two-step optimization can be viewed as a two-response optimization of the functional requirement (y) as follows. Step 1 targets optimizing the variation (σ_y), and step 2 targets shifting the mean (μ_y) to target T_y. For more than two functional requirements, the optimization problem is called multiresponse optimization.

Control Factors	A	B	C	D
Level 1	0.62	1.82	3.15	0.10
Level 2	3.14	1.50	0.12	2.17
Gain (meas. in dB)	2.52	0.32	3.03	2.07

Figure 13.7 Signal-to-noise ratio response table example.

robustness formulation with sensitivity defined as beta (β). In a robust design, the individual values for β are calculated using the same data from each experimental run, as in Figure 13.6. The purpose of determining the beta values is to characterize the ability of control factors to change the average value of the functional requirement (y) across a specified dynamic signal range, as in Figure 13.8. The resulting beta performance of a functional requirement (y) is illustrated by the slope of a best-fit line in the form of $y = \beta_0 + \beta_1 M$, where β_1 is the slope and β_0 is the intercept of the functional requirement data that is compared to the slope of an ideal function line. A best-fit line is obtained by minimizing the squared sum of error (ε) terms.

In dynamic systems, a control factor's importance for decreasing sensitivity is determined by comparing the gain in SN ratio from level to level for each factor, comparing relative performance gains between each control factor, and then selecting which ones produce the largest gains. That is, the same analysis and selection process is used to determine control factors that can be best used to adjust the mean functional requirement. These factors may be the same ones that have been chosen on the basis of SN improvement, or they may be factors that do not affect the optimization of the SN ratio.

- Most analyses of robust design experiments amount to a standard ANOVA of the respective SN ratios, ignoring two-way or higher-order interactions. How-

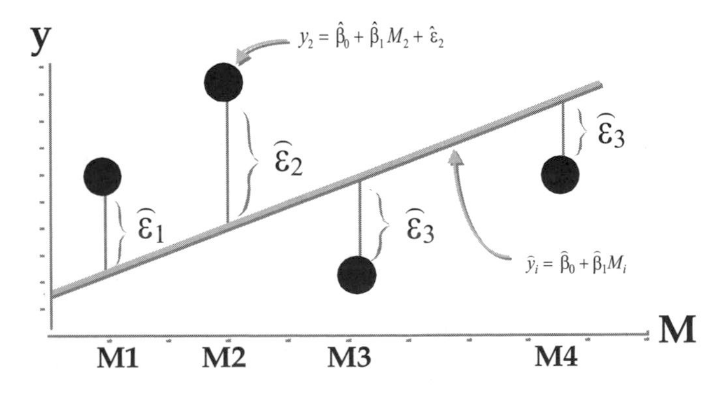

Figure 13.8 Best-fit line of a dynamic robust design DOE.

ever, when estimating error variances, one customarily pools together main effects of negligible size. It should be noted at this point that, of course, all of the designs discussed in Chapter 12 (e.g., 2^k, 2^{k-p}, 3^{k-p}, etc.) can be used to analyze SN ratios. In fact, the many additional diagnostic plots and other options available for those designs (e.g., estimation of quadratic components, etc.) may prove very useful when analyzing the variability (SN ratios) in the design. As a visual summary, a SN ratio plot is usually displayed using the experimental average SN ratio by factor levels. In this plot, the optimum setting (largest SN ratio) for each factor can easily be identified.

For prediction purposes, the DFSS team can compute the expected SN ratio given optimum settings of factors (ignoring factors that were pooled into the error term). These predicted SN ratios can then be used in a verification experiment in which the design team actually sets the process accordingly and compares the resultant observed SN ratio with the predicted SN ratio from the experiment. If major deviations occur, one must conclude that the simple main effect model is not appropriate. In those cases, Taguchi (1986) recommends transforming the dependent variable to accomplish additivity of factors, that is, to make the main effects model fit. Phadke (1989, Chapter 6) also discusses in detail methods for achieving additivity of factors.

13.9 SUMMARY

To briefly summarize, when using robustness methods, the DFSS team first needs to determine the design or control factors that can be controlled. These are the factors in the experiment for which the team will try different levels. Next, they decide to select an appropriate orthogonal array for the experiment. Next, they need to decide on how to measure the quality characteristic of interest. Most SN ratios require that multiple measurements be taken in each run of the experiment, so the variability around the nominal value cannot otherwise be assessed. Finally, the team conducts the experiment and identifies the factors that most strongly affect the chosen signal-to-noise ratio, and they reset the process parameters accordingly.

14

DISCRETE EVENT SIMULATION

14.1 INTRODUCTION

Production and business systems are key building blocks in the structure of modern industrial societies. Companies and industrial firms, through which production and business operations are usually performed, represent the major sector of today's global economy. In the last decade, companies have made continuous improvement of their production and business systems a driver in their strategic planning for the new millennium. To remain competitive, companies have to maintain a high level of performance by maintaining high quality, low cost, short lead times, and high customer satisfaction. Production and business operations have the potential to make or break a company's competitive ability. Therefore, efficient and robust production and business operations become a necessity for survival in the marketplace. These survival instincts are enforced by fierce competition and decreasing business safety margins.

Many industrial engineering (IE) subjects such as operations research and systems modeling offer robust and efficient design and problem-solving tools with the ultimate aim of performance enhancement. Examples of this performance include the yield of a process or a factory, the quality of a product, or the profit of an organization.

Simulation modeling, as a tool for process and system design and improvement, has undergone tremendous development in the last decade, in particular with the launch of Six Sigma methodology. In the DFSS project road map, simulation modeling combined with several statistical tools is most beneficial in the optimize phase (see Figure 5.1). This development can be seein in the growing capabilities of simulation software tools and the application of simulation solutions to a variety of real-world problems in different business arenas. With the aid of simulation, companies are able to design efficient production and business systems, validate and trade off proposed design solution alternatives, solve potential problems, improve systems performance metrics, and, consequently, cut cost, meet targets, and boost sales and profits. In the last decade, simulation modeling has been playing a major role in designing, analyzing, and optimizing processes and systems in a wide range of industrial and business applications.

This chapter provides an introduction to the basic concepts of simulation modeling, with an emphasis on the essentiality of the simulation discipline in serving the increasing needs of companies who seek continuous improvement and optimality of products and processes within their Six Sigma deployment initiatives. The chapter provides an introduction to the concept, terminology, and types of models, along with a description of simulation taxonomy and a justification for utilizing simulation tools in a variety of real-world applications. The objective is to give the Six Sigma teams the background essential to establish a basic understanding of what simulation is all about and to understand the key elements of a typical simulation study.

14.2 SYSTEM MODELING

System modeling includes two important commonly used concepts: system and modeling. It is imperative to clarify these concepts before attempting to focus on their relevance to simulation. This section will introduce the system and modeling concepts and provide a generic classification of the different types of system models.

14.2.1 System Concept

The word "system" is used in a variety of fields. A system is often referred to as a set of elements or operations that are organized and logically related to the attainment of a certain goal or toward the achievement of a certain objective. To reach the desired goal and meet its objective, it is necessary for the system to receive a set of inputs, process them correctly, and produce the required outcomes. Given this definition, we can analyze any system S (see Figure 14.1) around us based on the architecture shown in Figure 14.1.

Each system can be mainly defined in terms of set of inputs (I) and a process (P) that transforms or changes the characteristics of the inputs to produce a set of outputs (O). The process is performed through a set of system elements (E) and relationships (R). An overall goal (G) is often defined to represent the purpose and objective of the system. To sustain the flawless flow and functionality of $I \rightarrow P \rightarrow O$, feedback controls (C) are built into the system inputs, process, and outputs. Thus, building a system follows the method of process mapping and requires:

1. Specifying the set of inputs and their required specifications in order to produce specified outcomes.
2. Listing system elements ($E_1, E_2, E_3, \ldots, E_n$) and defining the characteristics and individual role of each element.

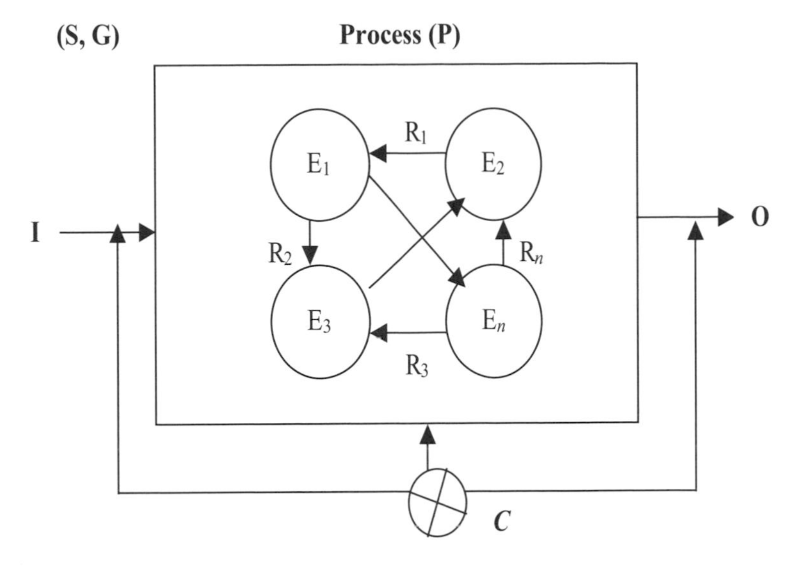

Figure 14.1 System concept.

3. Analyzing and understanding the logical relationships (R_1, R_2, R_3, . . . , R_n) among the defined set of system elements and
4. specifying the set of outcomes that should be produced, and their specifications, in order to reach the specified goal and objective.
5. Specifying the system controls and their rules in monitoring and adjusting the $I \rightarrow P \rightarrow O$ flow to meet the input and output specifications.
6. Realizing the goal or the overall system objective and relating system structure (inputs, elements, relationships, controls, and outputs) to the goal attainment.

This kind of system concept understanding is our gateway to the largest and widest subject of system analysis and design, which represents a key skill for simulation analysts and Six Sigma teams. For any set or arrangement of elements to be called a system, it should contain a defined set of elements with some logical relationships among them, and there should be some kind of goal, objective, or a useful outcome from the interaction of system elements. Typically, transforming system inputs into desired outputs requires processing, which is performed through the resources of system elements and often supported by controls to assure quality and maintain performance.

Applying this definition to an example of a classroom system leads to the following analysis:

1. The set of system elements is defined as follows: S = (tables, chairs, students, instructor, books, whiteboard).

2. The defined elements in S are logically related. For example, chairs are located around the tables, tables face the instructor, students sit on chairs, books are on tables in front of students, the instructor uses the whiteboard to explain the concepts, and so on.
3. The interactions explained in the previous point lead to the goal (G) of learning or educating students on a certain subject.
4. In order to achieve G, students should finish the class with certain qualifications and skills as an outcome (O) of the system.
5. To educate students, learning resources and materials are required as an input set (I) to the system.
6. Finally, the instructor applies regulations and rules of conduct in the classroom, attendance and grading policies, and so on as a sort of control on the education process.

Finally, it is worth mentioning that term “system” includes both products and processes. A product system can be an automobile, a cellular phone, a computer, a calculator, and so on. Any of those products involves the defined components of the system in terms of inputs, outputs, elements, relationships, controls, and goals. Try to analyze all the abovementioned examples from this perspective. On the other hand, a process system can be a manufacturing process, an assembly line, a power plant, a business process such as banking operations, a logistic system, an educational or academic system, and so on. Similarly, any of these processes involves the defined components of the system in terms of inputs, outputs, elements, relationships, controls, and goals. Try to analyze all the abovementioned examples from this perspective too.

14.2.2 Modeling Concept

The word modeling refers to the process of representing a system (a product or process) with a selected replica model that is an easier to understand and less expensive to build compared to the actual model. The system representation in the model implies taking into account the components of the system discussed in the previous section. This includes representing system elements, relationships, goal, inputs, controls, and outputs. Modeling a system, therefore, has two prerequisites:

1. Understanding the structure of the actual (real-world) system and the functionality and characteristics of each system component and relationship. It is imperative to be familiar with the system of interest before attempting to model it and to understand its purpose and functionality before establishing a useful representation of its behavior.
2. Being familiar with different modeling and system representation techniques and methods. This skill is essential when choosing the most appropriate model for representing the real-world system. Once the most appropriate model has been selected, it should also assessed for practicality and feasibility. As

we will discuss in the next section, several model types can be used to create a system model and the selection of the most feasible modeling method is a decision based on economy, attainability, and usefulness.

The key question that the Six Sigma belt needs to ask in this regard is, How does one model a system of interest? The answer to this question is a combination of both art and science in abstracting a real-world system into an operative, representative model. This model should be clear, comprehensive, and accurate so that we can rely on its representation in understanding system functionality, analyzing its different postures, and even predicting its future behavior. As shown in Figure 14.2, system modeling is the process of rendering the actual system into a model that can be used to replace the actual one for analysis and improvement. The elements (objects) of the real-world system are replaced by objects of representation and symbols of indication, including the set of the system's elements (E), element relationships (R), inputs (I), controls (C), and outputs (O) (ERICO). Actual system ERICO elements are mimicked thoroughly in the model, leading to a representation that captures the characteristics of the real-world processes. Of course, such objects are approximated in the model, but the approximation should not overlook key system characteristics.

The science of system modeling is based on learning modeling methodologies and having fluency in analytical skills and logical thinking. The art aspect of modeling involves applying representation, graphical, and abstraction skills to objects, relationships, and structures. It is simply being able to depict system functionality, operation, and interaction of elements, and possessing the capability to map relationships representing behavior. Consider an assembly line. You need to understand its functionality and structure, and try to represent it in a process map diagram. Does your diagram explain everything you need to analyze the assembly line? What else should be quantified and measured? Can your graphical model answer typical questions asked by a production manager?

14.2.3 Types of Models

As mentioned in the previous section, several modeling methods can be used to develop a system model. The Six Sigma black belt will decide on the modeling ap-

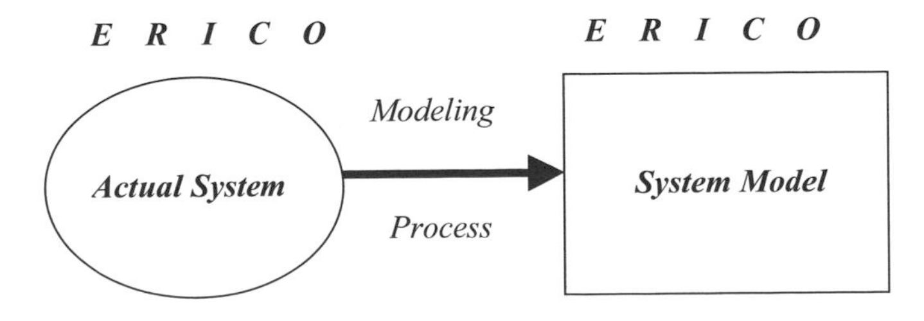

Figure 14.2 The process of system modeling.

proach by choosing a certain type of model to represent the actual system. The black belt's choice will be based on several criteria such as system complexity, the objectives of system modeling, and the cost of modeling. As shown in Figure 14.3, we can classify different types of models into four major categories: physical models, graphical models, mathematical models, and computer models. The following is a summary of those types of models.

1. Physical models. Physical models are tangible prototypes of the actual products or processes, created by using a one-to-one scale or any other feasible scale of choice. Such models provide a close-to-reality, direct representation of the actual system and can be used for demonstrating the system's structure, the role of each system element, and the actual functionality of system of interest in a physical manner. They help designers achieve a deeper understanding of system structure and details and allow them to try out different configurations of design elements before the actual build of the product and deployment of the process. They are used primarily in large-scale projects to examine a limited set of behavioral characteristics in the system. For example, in the transactional world, an extra customer service clerk can be added to check the reduction of waiting time.
2. Graphical models. Graphical models are abstracts of the actual products or processes created by using graphical tools. This type of modeling starts from paper and pencil sketches, and proceeds to engineering drawings, pictures, and movies of the system of interest. Most common graphical representations of systems include the process and functional mapping techniques presented in Chapter 8.
3. Mathematical models. Mathematical modeling is the process of representing system behavior with formulas, mathematical equations, or transfer functions. Such models are symbolic representations of systems functionality, decision (control) variables, responses, and constraints. They assist in formulating the system design problem in a form that is solvable using graphical and calculus-based methods. Mathematical models use mathematical equations (transfer functions), probabilistic models, and statistical methods to represent the key relationships among system components. A transfer function equation can be derived in a number of ways. Many of them come from DOE

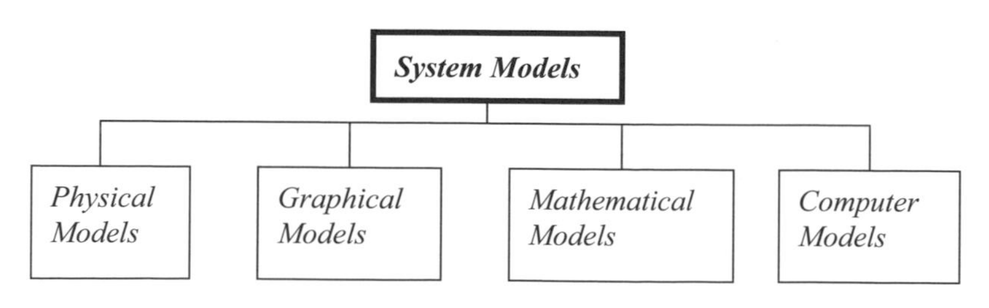

Figure 14.3 Types of models.

studies or are from relationships and are then tested against real data. Typically, a transfer function formula is a closed-form relationship between a dependent variable (Y), a design parameter, one or more independent variables (X), and process variables, with the form of $Y = f(X)$.

4. Computer models. Computer models are numerical, graphical, and logical representation of a system (a product or a process) that utilizes the capabilities of a computer: fast computations, large capacity, consistency, ability to do animation, and accuracy. Computer simulation models, which represent the middleware of modeling, are virtual representations of real-world systems on the computer. They assist in capturing the dynamics and logic of system processes and estimating the system's long-term performance. Simulation models are models that allow the user to ask "what if" questions about the system, present changes that are made in the physical conditions or their mathematical representation, and run the system many times for long periods to simulate the impacts of such changes and to evaluate different "what if" scenarios. The model results are then compared to gain insight into the behavior of the system. Computer simulations of products and processes using different application programs and software tools are typical examples of computer models. Mathematical models that represent complex mathematical operations, control systems, computer algorithms, and other systems can be built with computer tools. A discrete event simulation (DES) model that mimics the operation of an automobile's final assembly line is another example of a process computer model. This type of computer model is the major focus of this book. A DES process model can be used to estimate throughput by running the model dynamically and tracking its throughput hour by hour or shift by shift as desired. Accurate and well-built computer models compensate for the limitations of other types of models. They are built using software tools, which is easier, faster, and cheaper than building physical models. In addition, the flexibility of computer models allows for quick changes, easy testing of what-ifs, and quick evaluation of system performance for experimental design and optimization studies. Computer models also provide the benefits of graphical models with modern animation and logical presentation tools.

14.3 SYSTEM MODELING WITH DISCRETE EVENT SIMULATION (DES)

By utilizing computer capabilities in logical programming, random number generation, fast computation, and animation, discrete event simulation (DES) modeling is capable of capturing the characteristics of the real-world process and estimating a system's performance measures at different settings of its design parameters. To measure such performance, DES imitates the stochastic and complex operation of a real-world system as it evolves over time and seeks to describe and predict the system's actual behavior. The discrete event approach is powerful enough to provide the analyst with key information that quantifies system characteristics such as design/decision parameters and performance measures. Such information makes a DES model an effective decision support system that estimates how many resources the process/project owner needs in the system (such as how many operators, machines, trucks, etc.) and how to arrange and balance the resources to avoid bottlenecks, cross traffic, backtracking, and excessive waiting lines, waiting times, or inventories.

Towards this end, system modeling with DES includes mimicking the structure, layout, data, logic, and statistics of the real-world system and representing them in a DES model. An abstraction of system modeling with DES is illustrated in Figure 14.4. Abstracting the real-world system with a DES model can be approached by specifying the details and extent of the five system modeling elements shown in the figure.

14.3.1 System Structure

A system DES model is expected to include the structure of the simulated actual system. This structure is basically the set of system elements (E) in terms of physical components, pieces of equipment, resources, materials, flow lines, and infrastructure. Modeling such elements thoroughly is what makes the model realistic and representative. However, the level of detail and specification of model structural elements is mainly dependent on the objective and purpose of building the model. For example, the details of a machine's components and kinematics may not be helpful in estimating its utilization and effectiveness. Hence, basic graphical representations such as process mapping of structural elements are often used in DES models for animation purposes. On the other hand, physical characteristics such as dimensions, space, distances, and pathways that often impact flow routes, cycle times, and capacity should be part of the model structure. Table 14.1 shows examples of structural elements that impact or do not impact model performance of a plant or a manufacturing system.

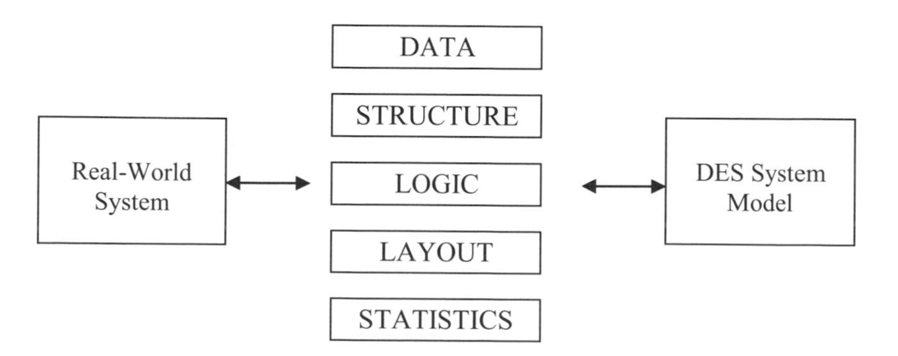

Figure 14.4 System modeling with DES.

Table 14.1 Examples of modeled structural elements

Modeled structural elements	Impacted model factor
Conveyor length	Conveyor capacity
Unit load dimensions	Number of stacked units
Buffer size	Buffer capacity
Length of aisles and walkways	Walking distance and time
Size of AGV	Number of AGV carriers
Length of monorail	Carrier traveling time
Dock spacing of power and free throughput	Power and free throughput
Dimensions of storage units	AS/RS storage and retrieval time

14.3.2 System Layout

A system layout is simply a way of configuring system structural elements. The layout specifies where to locate pieces of equipments, aisles, repair units, material handling systems, storage units, offices, loading/unloading docks, and so on. Similar to system structure, model elements are placed and sized according to the specified layout results in a more representative DES model. Sticking to layout specifications helps capture the flow path of material or entities within the system. Hence, flow diagrams are often developed using system layouts. When designing new systems or expanding existing ones, the layout often plays an important role in assessing design alternatives. What does the system look like? How close is the packaging department to the shipping dock? Those are just examples of questions answered using a system layout.

Facility planning is used to design the layout of a plant or a facility. Department area and activity relationship charts (see Figure 14.5) are often used to provide a design for a facility layout. Locations of departments, distances between them, and interdepartmental flow need to be captured in the DES model to provide accurate system representation. An example of a system layout of a manufacturing facility is shown in Figure 14.6.

14.3.3 System Data

Real-world systems often involve tremendous amounts of data while functioning. Data collection systems (manual or automatic) are often used to collect critical data for various purposes such as monitoring operations, process control, and generating summary reports. DES models are data-driven; hence, pertinent system data should be collected and represented in the model. Of course, not all system data are necessary to build and run the DES model. Deciding on what kind of data are necessary for DES is highly dependent on the model structure and the goal of simulation. Generally speaking, all system elements are defined using system data. Examples include parameter settings of machines, material handling systems, and storage systems. Such parameters include interarrival times, cycle times, transfer times, con-

Reasons for Closeness

Code	Reason
1	Type of customer
2	Ease of supervision
3	Common personnel
4	Contact necessary
5	Share same price
6	Psychology

Importance of Closeness

Value	Closeness
A	Absolutely necessary
E	Especially important
I	Important
O	Ordinary closeness OK
U	Unimportant
X	Undesirable

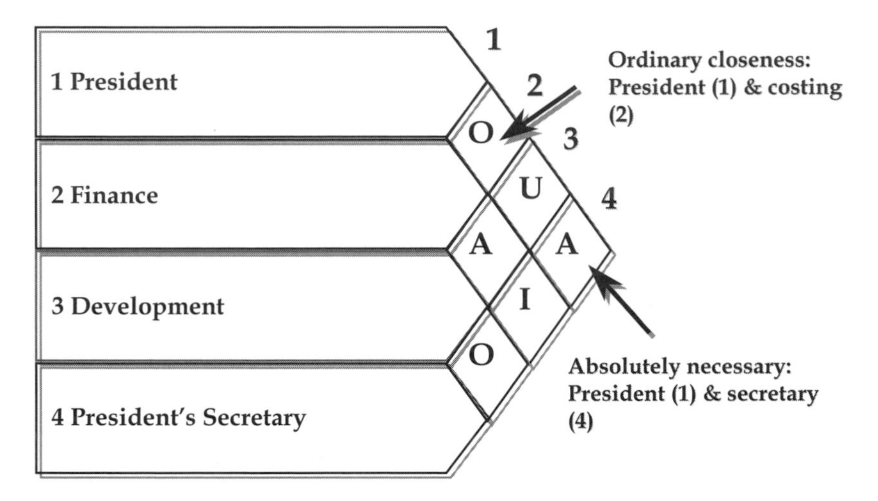

Figure 14.5 Activity relationship chart.

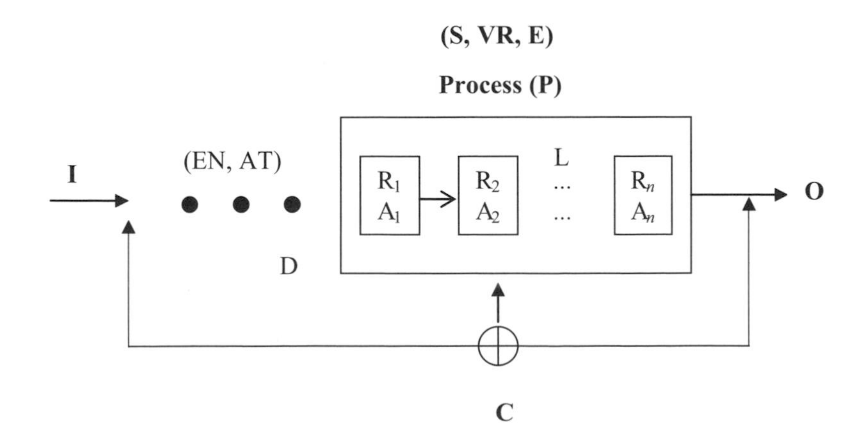

Figure 14.6 DES system elements.

veying speed, time to failure, time to repair, product mix, scrap rates, and so on. Model performance and results are highly dependent on the quality and accuracy of such data. As the saying goes, "garbage in, garbage out."

Thus, developing a DES model requires the analyst to precisely define data elements needed, the method of collecting such data, and how it will be represented and used in the DES model.

14.3.4 System Logic

System logic is the rules and procedures that govern the behavior and the interaction of different model elements. It defines the relationships among model elements and how entities flow within the system. Programming capability of simulation languages is often utilized to implement the designed system logic in the developed DES model. Similarly, real-world systems often involve a set of simple or complex logical designs that control system performance and direct its behavior. Abstracting needed logic into the DES model is a critical modeling task. It is probably the most difficult modeling challenge that faces simulation analysts, especially because of limitations of some simulation software packages.

A lot of decision points at model operation and merging points, scheduling rules, routing schemes, and operational sequences need to be built into the DES model in order to reflect the actual performance of the underlying system. Table 14.2 shows examples of such logical designs.

14.3.5 System Statistics

System statistics are means of collecting run time information and data from the system. Such statistics are necessary to control operation and flow of system activities and elements. In the system DES model, therefore, statistics are collected and accumulated to provide a summary of results at the end of the run time. Such statistics are used to model the real-time monitoring gauges and clocks in the real-world system. Because of model flexibility, however, some statistics that are used in the model may not actually be in the real-world system. This is because statistics do not

Table 14.2 Examples of model logical designs

Model activity	Logical design
Parts arriving at loading dock	Sorting and inspecting scheme
Requesting components	Model mix rules
Producing an order	Machines scheduling rules
Material handling	Carrier routing rules
SPC quality control	Decision rules
Machining a part	Sequence of operation
Forklift floor operation	Driver dispatching rules
AS/RS storage system	AS/RS vehicle movement rules

impact the model performance. Therefore, we can define statistics that are necessary to the system operation and other statistics that may provide useful information during run time, and summarize the results at the end of run time. Table 14.3 shows examples of model statistics.

14.4 ELEMENTS OF DISCRETE EVENT SIMULATION

A DES model is built using a set of model components (building blocks) and executed based on a set of DES mechanisms. The nature and functionality of those components and mechanisms may vary from one simulation package to another. However, components of DES models can be defined using a certain set of DES elements. These elements are related to the definition and the structure of a system as defined in the previous sections in terms of system inputs, processes (elements and relationships), outputs, and controls. Hence, DES elements include system entities (EN) characterized by their attributes (AT), system state (S), variables (VR) that describe the system state, events (E) that occur within the system and change its state, activities (A) or tasks performed in the system, resources (R) or means by which activities are performed, delays (D) that take place in the system, and the logic (L) that governs the system operation. System state, variables, and events are directly related. These elements are shown in Figure 14.6. This section discusses those elements and Section 14.5 presents key mechanisms used to execute DES models.

14.4.1 System Entities

Entities (EN) are the items that enter the system as inputs (I), are processed through the system resources (R) and activities (A), and depart the system as outputs (O). As DES's dynamic objects, they often represent traceable elements that are of interest to system designers and managers. Examples include parts or products in manufacturing, customers in banking, calls in a customer-service center, patients in health systems, letters and packages in postal services, documents in an

Table 14.3 Examples of model statistics

Model statistic	Measured value
Jobs produced per hour	System throughput
Percent of machine busy time	Machine utilization
Number of units in system	Work-in-progress level
Time units spent in system	Manufacturing lead time
Number of defectives	Process quality
Number of machine failures	Maintenance plan
Number of units on a conveyor	Conveyor utilization
Number of units on a buffer	Buffer utilization

office, insurance policies in an insurance company, data in information systems, and so on.

In DES, entities are characterized by attributes (AT) such as price, type, class, color, shape, ID number, origin, destination, priority, due date, and so on. Specific values of such attributes are tied to entities and can differ from one entity to another. Hence, attributes can be considered as *local* variables that are tied to individual entities. Those attributes can be used at different locations and instances within the DES model to make various decisions such as directing the flow of entities, assigning them to storage locations, activating resources, and so on.

Also, the type of entity is the basis for classifying DES systems as discrete or continuous. Discrete entities are modeled with discrete systems. Continuous entities, such as flow of bulk materials, fluids, and gases, are modeled with continuous systems. Our focus in this text is on discrete entities since they represent the majority of process and service systems.

14.4.2 System State

The system state (S) is the description of system status at any point in time. It describes the condition of each system component in the DES model. As shown in Figure 14.6, those elements include system inputs (I), processes (P), outputs (O), and controls (C). Hence, at any point in time, the system state defines the state of system inputs (types, amount, mix, specifications, arrival process, source, attributes, and so on). It also defines the system process in terms of type of activities (A), number of active resources (R), number of units in the flow, utilization of resources, time in state, delay time, time in system, and so on. Similarly, the system state defines the state of model outputs (types, amount, mix, specifications, departure process, destination, attributes, and so on).

A DES model records changes in the system state as it evolves over time at discrete points in time to provide a representation of the system behavior and to collect statistics and performance measures that are essential to system design and analysis. System state variables are used to quantify the description of the system state.

14.4.3 State Variables

DES models include a collection of variables that describe the system state at any specific point in time. Such variables contain and represent the information needed to describe system status and understand its performance. Examples include number of units in the system and each resource's status (idle, busy, broken, blocked).

In addition to overall system-level variables, state variables also include input factors such as system design or control parameters ($x_1, x_2, \ldots, x_n$). They represent the set of independent variables since changing their values affects system behavior. Hence, different settings of such controllable factors often lead to different sets of model outcomes. Recall the transfer function in Chapter 6. Examples include interarrival times, number of operators and resources, service time, cycle time, capacity of buffers, speeds of conveyors, and so on.

They also include system response (output) variables ($y_1, y_2, \ldots, y_n$) that are measured from system performance. Such variables are dependent variables that represent the set of system performance measures. Examples include system throughput per hour, average utilization of resources, manufacturing lead time, number of defectives, and so on. Figure 14.7 shows a schematic representation of state variables.

14.4.4 System Events

An event (E) is an instantaneous occurrence that changes the system state. As discussed earlier, the system state is the description of system status at any time, which is defined by a set of state variables. Events are the key element in DES models, which are characterized as being event-driven; updating the system state, collecting system statistics, and advancing the simulation clock take place at the event's occurrence. The set of events (E_1, E_2, E_3, . . . , E_n) that occur at their corresponding times (T_1, T_2, T_3, . . . , T_n) are stored chronologically in an event list (EL), as will be discussed later in the context of DES mechanisms. System states (S_1, S_2, S_3, . . . , S_n) are changed based on the events and their implications; hence, model behavior is referred to as event-driven. Figure 14.8 shows how the system state is updated at event occurrences.

An event's occurrence in a DES model can be the arrival of an entity to the system, the start of a coffee break, the end of a shift, the failure of a resource, a change in batch size, the start of a new production schedule, the departure of an entity, and so on. Since DES model elements are interrelated and the DES environment is dynamic, the occurrence of such events often leads to a series of changes in the system state. For example, the arrival of a customer at a bank increases the waiting line if the bank teller is busy or changes the teller status from idle to busy if the teller is available. Similarly, when the customer departs the bank, the server status is changed back to idle, another customer is requested from the waiting line, and number of served customers is increased. Hence, affected state variables are updated, relevant statistics are accumulated, and the simulation clock is advanced to the next event in the event list.

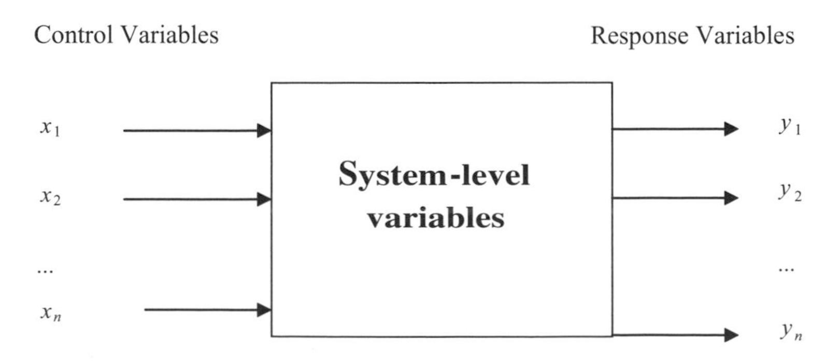

Figure 14.7 System state variables.

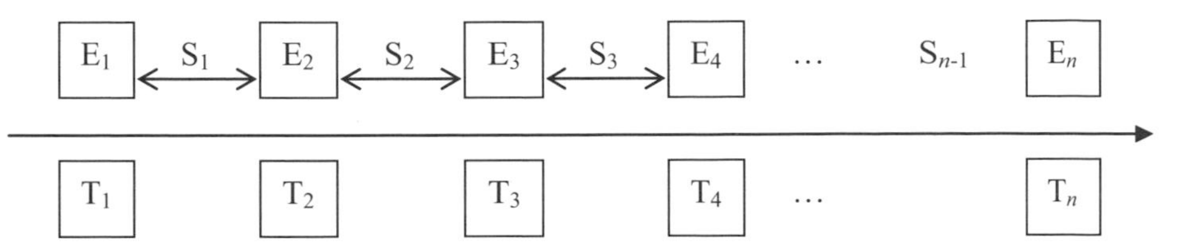

Figure 14.8 Events driving the system state.

14.4.5 System Activities

A process is a sum of activities (A) and tasks that are performed in the model for a specified time. Processes are simply determined by establishing a sequence for the operations needed to process the entity. Examples of such operations include receiving, directing, transferring, cleaning, machining, packaging, and shipping. Such operations either directly process the entity, such as cutting and serving activities, or are indirectly involved in processing system entities, such as material handling, inspection, and storage activities.

A process map often includes a sequence of all activities or operations required to process an entity, a classification of different types of those operations, and specifications of such operations. A process map often classifies operations into processes (circles), transport (arrows), storage (triangles), delay (D), and inspection (squares), as shown in Table 14.4. Similarly, activities in a DES model can be classified in order to provide a better understanding of the process flow and a deeper knowledge of the process operations.

System processes and operations (activities) can be also classified into value-added and non-value-added operations:

- Value-added activities. The customer defines value. The activities transform the characteristics of an entity from one form to another, producing attributes the customer is willing to pay for. By making changes to entities, such activities increase their value.
- Non-value-added activities. Many operations performed on entities may not add value to them. Some are needed to complement the system functionality

Table 14.4 Activities in process operations

Symbol	Process meaning	Example
○	Processing operation	Drilling operation
➔	Transporting operation	Forklift transfers unit load to storage
△	Storage operation	Finished goods (FG) are stored before shipping
D	Delay operation	Work-in-process (WIP) units are delayed before final assembly
□	Inspection operation	One of every 100 units are inspected at the QC station

(e.g., transporting and transferring operations, storing and delaying materials, and quality control inspections) but many are not. The job of a Six Sigma belt is to identify these unneeded non-value added activities and eliminate them. They cause variation and consume resources unnecessarily. Examples include unnecessary inspections or the measurement of noncritical and nonsignificant transfer function variables.

- Time duration of a system process step is specified in three ways, depending on the nature of the activity. These are:

  Fixed time duration. The time assigned to a process step has a fixed value and possesses very minimal variability. Typical examples include fixed cycle times in automatic operations, planned stoppage times, and timed indexing operations.

  Probabilistic time duration. The time assigned in this case incorporates randomness and variability; hence, activity time changes from one entity to another. Sampling from theoretical or empirical statistical distributions is often used to represent process step time duration.

  Formula-based time duration. In this case, activity time is calculated using an expression of certain system variables. For example, loading time is determined based on the number of parts loaded, whereas walking time is determined as a function of load weight.

14.4.6 System Resources

A system resource (R) represents the tool or the means by which model process steps are carried out. Examples include pieces of equipment, operators, personnel, doctors, repairmen, machines, specialized tools, and other means that facilitate the processing of entities. An entity is the object that moves from one process step to another based on a predescribed value-added route or stream. It is the part in a manufacturing system, the patient in the hospital system, the case paperwork in a law office, and so on. Allocation and scheduling scheme are often implemented to provide the best task assignments to resources. Table 14.5 shows examples of system resources in both manufacturing and service applications.

Key factors that impact the performance of resources include capacity, speed, and reliability. All are explained below.

Table 14.5 Examples of system resources

Manufacturing systems	Service systems
Machines and machine centers	Doctors
Operators and general labor	Bank tellers
Inspectors and quality controllers	Hospital beds
Repairmen and maintenance crew	Drivers
Assembly stations and tools	Drive-through windows

Capacity impacts resource utilization, which measures the percentage of resource usage. Since resources consume capital, design teams and their management prefer to increase the utilization of resources through better scheduling and resource allocation. Resource capacity is often measured by how many entities (transactions) are able to access the resource. Most resources such as machines, bank tellers, and doctors often treat one entity at a time. Some, however, perform batch processing of identical or different entities; examples are filling machines, assembly stations, testing centers, and group-based activities. Capacity limitations often impact the utilization of resources, so designers often strive to strike balance between both measures. Therefore,

$$\% \text{ Utilization} = \text{Actual Capacity}/\text{Designed Capacity}$$

Resource speed determines the productivity of the resource, often measured as throughput or yield. Similarly, a resource throughput is a higher-the-better measure. Eliminating waste using lean thinking and Six Sigma deployment over time and addressing non-value-added inefficiencies in operation increases throughput.

Resource speed is the factor that determines the throughput or yield of a resource. Fast resources often run with short processing times and result in processing more entities per time unit. Examples include speed of a transfer line, cycle time of a machining center, service time at a bank teller, and diagnostic time at a clinic. Units produced per hour (UPH) or per shift (UPS) are common throughput measures. Cycle time is often the term that is used to indicate the speed of machines and resources in general. Process throughput, therefore, is the reciprocal of cycle time. That is,

$$\text{Resource Throughput} = 1/\text{Cycle Time}$$

For example, a process with a cycle time of 60 seconds produces 1/60 units per second or one unit per minute (on average).

Finally, resource reliability determines the percent of resource uptime (availability). It is always desirable to increase the uptime percentage of resources through better maintenance and workload balancing. Table 14.6 summarizes the three resource factors of a M_2, a hypothetical machine resource.

Resource reliability is a resource factor that determines the resource uptime or availability. It is often measured in terms of mean time between failures (MTBF),

Table 14.6 How factors impact system resources

Resource factor	Performance measure	Units metric	Example
Capacity	Utilization	Percent Busy	M_2 is 85% utilized
Speed	Throughput	Units per hour	M_2 produces 60 UPH
Reliability	Uptime	Mean time between failures	M_2 has MTBF of 200 minutes

where a resource failure is expected to occur each MTBF time unit. Repair time, after a failure occurs, is measured by mean time to repair (MTTR). Resource uptime percentage is determined by dividing MTBF by available time (both MTBF and MTTR) as follows:

$$\text{Resource Uptime \%} = \text{MTBF}/(\text{MTBF} + \text{MTTR})$$

14.4.7 System Delay

A system delay (D) is a step that takes place within the system processes but has no specified time duration. The time duration is determined during run time based on the dynamic and logical interactions among system elements. Examples include customer's waiting time, delays in units sequencing, and delays caused by logical relationships. Measuring delays by accumulating the time of delay occurrences is a key capability of DES models. Statistics important to decision making can be collected on such capability, such as average waiting time for bank customers, manufacturing lead time, and the time span from the start to the end of parts manufacturing, including delays. These are rough order of magnitude measures since no customer actually feels the average of a process.

14.4.8 System Logic

System logic (L) controls the performance of processing steps and dictates the time, location, and method of their execution. DES models run by executing logical designs such as rules of resources allocation, parts sequencing, flow routing, task prioritization, and work scheduling. Route sheets, work schedules, production plans, and work instructions are examples of methods used to implement the system logical design. Table 14.7 summarizes examples of system logic.

14.5 DES MECHANISMS

As discussed earlier, DES models are dynamic, event-driven, discrete in time, stochastic (randomized and probabilistic), and graphically represented. Such characteristics are established in the model based on different mechanisms, including

Table 14.7 Examples of system logic

Logical design	Implementation method	Example
Material flow	Route sheet	Machining sequence
Workforce scheduling	Work schedule	Nurses' shifts in hospitals
Product-mix schedule	Production plan	Production lot size
Prioritizing tasks	Shop floor controls	Expedited processing
Standard operations	Work instructions	Assembling products

events list (EL) creation, sampling from probability distributions with random number generation (RNG), accumulating statistics, and animation.

DES functionality includes the creation of an event list (EL), advancing the simulation clock (event-driven advancement), updating the event list, updating pertinent statistics, and checking for termination. Figure 14.9 shows the architecture of DES functionality.

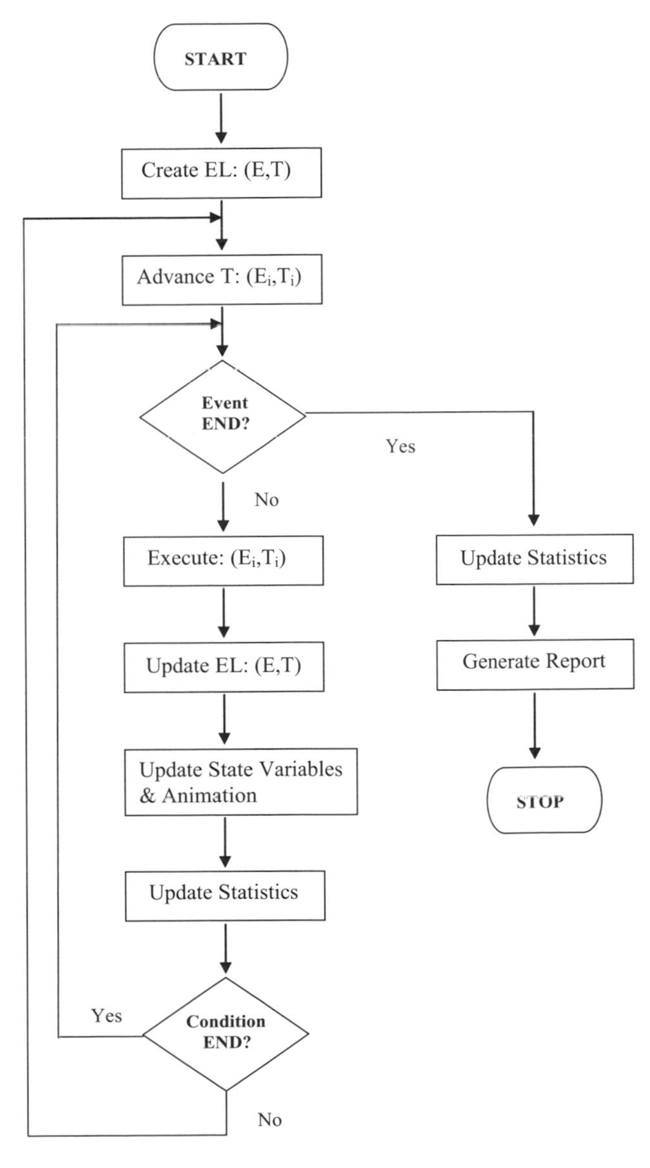

Figure 14.9 DES functionality.

14.5.1 Discrete-Event Mechanism

Events in DES are discrete since they take place at discrete points in time. DES functionality is, therefore, based on creating an event list (EL) and executing the events of the list chronologically. Such functionality is based on a discrete-event and a clock-advancement mechanism. The discrete-event mechanism is the most distinctive aspect of DES simulation models. As discussed earlier, an event (E) is defined as an occurrence, a situation, or a condition that results in changing the state (S) of the system model instantaneously. The state of a system is represented by a collection of state variables that describe the system status.

Arrival and departure of entities such as parts and customers are events that change the state of a production process or a service station. Information about event types and their scheduled occurrence times are stored in an event list (EL). An event list comprises a set of events and event–time combinations. That is:

$$EL = \{(E_1, T_1), (E_2, T_2), (E_3, T_3), \ldots, (E_n, T_n)\}$$

At each event occurrence, two actions take place: the EL is updated in terms of content and time and the collected statistics are updated. Upon the occurrence of an event (E_1), the executed event will be removed from the list and the next most imminent event will go to the top of the list. Other events may enter or leave the list accordingly. Updating the event list includes two main operations, as shown in Figure 14.10:

1. Event removal. Upon the occurrence of an event (E), the list may be updated by removing one or more events from the list, resulting in change in the chronological order of events execution. An event may be cancelled from the EL for many reasons; for example, when the occurrence of an event (E_i) precludes its occurrence. Event removal can take place in two locations in the EL:

   a. From the top of the EL. When an event is removed from the top of the EL, it is practically done or executed. For example, event E_1 in Figure 14.10 was processed; hence, the event is removed from the top of EL_1 when updating the list to become EL_2.

EL_1

Event (E)	Time (T)
E_1	T_1
E_2	T_2
E_3	T_3
E_4	T_4
E_5	T_5

← Event Removal (at E_1)

Event Addition (→ E_7, E_6 in EL_2)

← Event Cancellation (at E_4)

EL_2

Event (E)	Time (T)
E_7	T_7
E_2	T_2
E_6	T_6
E_3	T_3
E_5	T_5

Figure 14.10 Event list (EL) operations.

   b. From any other location within the EL. When an event is removed from any other location in the EL besides the top of the list, it is practically cancelled. For example, event E_4 in Figure 14.10 was cancelled; hence, the event is removed from EL_1 and did not show up in EL_2.
2. Event addition. Upon the occurrence of an event (E), the list may also be updated by adding one or more events to the list, resulting in changing the chronological order of event execution. Event additions can take place in two locations in the EL:
   a. To the top of EL. When an event is added to the top of the EL, it is considered the most imminent event to take place. For example, Event E_7 in Figure 14.10 was added to top of EL_2 and will be processed first. E_7 was not a member of EL_1 when the list was updated to become EL_2.
   b. To any other location within the EL. When an event is added to any other location in the EL besides the top of the list, it is just being added as a future event. For example, Event E_6 in Figure 14.10 was added to EL_2 and will be processed right after E_2.

In a discrete-event mechanism, we view a system as progressing through time from one event to another, rather than changing continuously. Since many real systems (plants, banks, hospitals, etc.) can be viewed as queuing networks, discrete events occur in the simulation model in a manner similar to a queuing system (arrival and departure of events). For instance, a customer joins a queue at a discrete instant of time and at a later discrete instant leaves the bank. There is no continuously varying quantity that says a customer is 65% in the queue. Time is accumulated upon the occurrence of events at discrete points in time. The customer is either in or out the queue. This view results in a computationally efficient way of representing time. The ability to represent time in an event-driven manner is the greatest strength of DES since it captures the dynamic behavior of real-world production and business systems.

14.5.2 Time-Advancement Mechanism

The association of time with events (T_1, T_2, T_3, . . . , T_n) in a simulation model is maintained using a *simulation clock* time variable. This variable is updated through the *next-event time-advancement* mechanism that advances the simulation clock to the time of the most imminent event in the event list. For example, Figure 14.11 shows an event list of 10 events (E_1, E_2, E_3, . . . , E_{10}) that are chronologically ordered as T_1, T_2, T_3, . . . , T_{10}. In DES, the time periods between events (E_1 and E_2, for example) are skipped when executing the model, resulting in a compressed computation time. Hence, the total time required to process the 10 events in a DES computer model is much shorter than the actual clock time. The mechanism simply states that if no event is occurring in a time span, there is no need to observe the model and update any statistics, so the time is simply skipped to the next event scheduled in the EL. This event-driven time-advancement process continues until some specified stopping condition is satisfied.

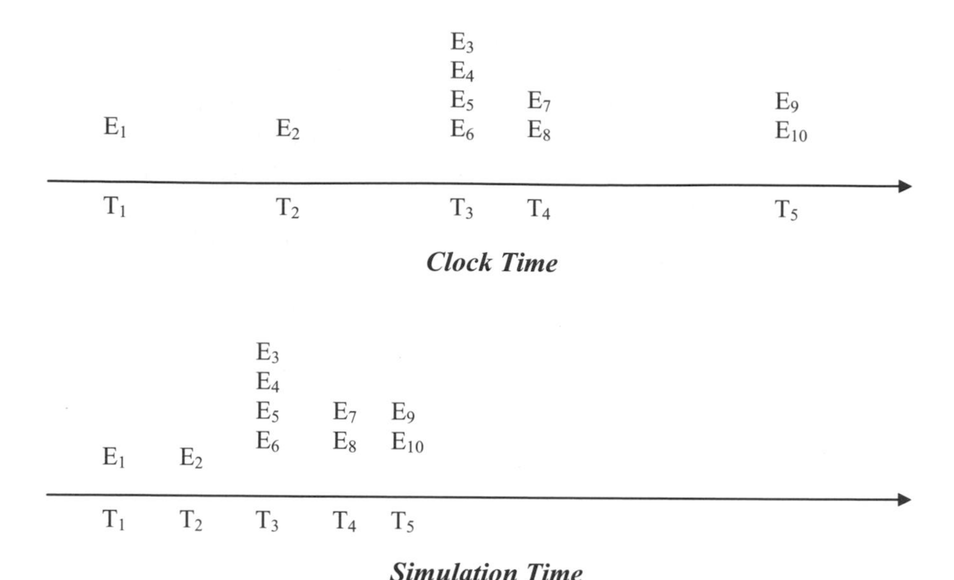

Figure 14.11 Time advancement and compression in DES.

14.5.3 Random Sampling Mechanism

The stochastic (random) nature of simulation models is established by means of sampling from probability distributions using random generation techniques—mathematical methods used to generate streams of pseudo-random numbers. A random number generation (RNG) mechanism allows one to capturing the random variations in real-world systems in DES models. The ability to induce randomness is one of the important aspects of DES.

From a practical point of view, random numbers are the basic ingredient in the simulation models of most real-world systems. A machine's cycle times, customers service times, equipment failures, and quality control tests are some examples of random elements in DES models. Random model inputs often lead to random model outputs. Thus, simulation outcomes, representing real-world systems, are often stochastic and lead to inconsistent performance levels.

We can refer to the factors that contribute to variation in simulation outputs as *noise factors*.[1] Individual values of such factors change with time due to the actual behavior of the real-world process. Examples include an entity's interarrival time, service or processing time, equipment time between failure (TBF) and time to repair (TTR), and percentages of scrap and rework. Such factors are often presented in simulation models by a random generation from a certain probability distribution. Key distribution parameters are often estimated through statistical approxima-

[1]These factors, in general, are responsible for causing functional characteristics of a process to deviate from target values. Controlling noise factors is very costly or difficult, if not impossible. Noise factor treatment strategies are usually addressed by robust design methodology.

tions based on fitting a collected set of empirical data to one of the commonly used distributions.

The behavior of a DES model is, therefore, driven by the stochastic nature of real-world processes such as an entity's arrival process, in-system service process, and output departure process. Such variability in model processes results in stochastic variability in model response (system performance measures). Therefore, the throughput of a production system that is subject to variability in material delivery, tool failure, and operator efficiency is represented by the DES model in a stochastic manner. Based on the inherent fluctuations, the throughput (yield) could be a low value in one shift and then higher in another. The amount of variability in model outputs depends on the amount of variability in model inputs, along with the dynamics caused by the logical interactions among different model elements and processes. Thus, simulation runs yield estimates of such performance measures in terms of means and variances to assess such variability. Figure 14.12 presents examples of variation factors in a DES model.

The probabilistic nature of the time period (T) that separates successive events when executing a DES model represents the merit of randomness in simulation models. Events (such as customer arrival, machine failure, and product departure) can occur at any point in time, stimulating a stochastic simulation environment. To generate successive random time samples ($T = T_1, T_2, \ldots$) between events, *sampling* from probability distributions is used. Hence, most simulation languages, as well as generic programming languages, include a RNG engine to generate event times and other modeled random variables. For example, interarrival times of certain entities to the system are generated randomly by sampling from the exponential distribution, whereas the number of arriving entities per unit time is randomly generated from the Poisson distribution.

14.5.4 Statistical Accumulation Mechanism

Statistics are the set of performance measures and monitoring variables that are defined within the DES model to quantify its performance and collect observations about its behavior. These statistics are defined in the model for various reasons, such as:

- Monitoring simulation progress. Values of the set of model statistics can be defined to be part of model animation. Such values are updated discretely throughout model run time. The simulation analyst can observe the progress of simulation by monitoring the changes that occur in the values of the defined statistics. For example, the Six Sigma team can continuously observe the changes that occur in the number of units/customers in the system.
- Conducting scheduled simulation reviews. It is often required to review the system state at certain points in the simulation run time. These could be the end of one week of operation, the completion of a certain order, or at a model-mix change. With model statistics defined, the Six Sigma team can halt simulation at any point in time, determined by the simulation clock, and review the progress of the simulation by checking the values of such statistics. For exam-

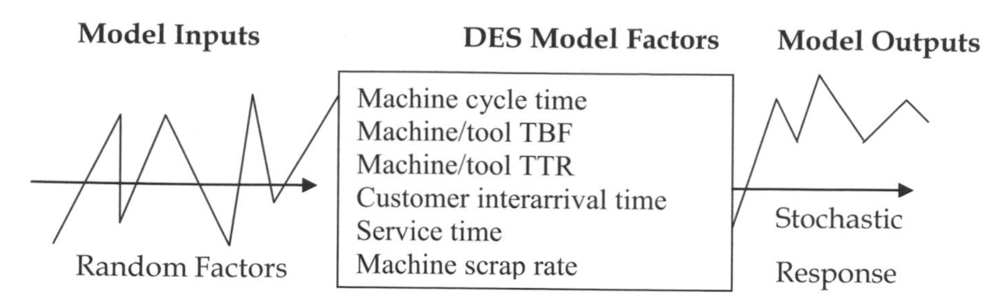

Figure 14.12 Examples of variation factors in a DES model.

ple, the team can stop the model at the end of the day shift at a fast food restaurant and review the number customers who waited more than 10 minutes at the restaurant drive-through window.

- Summarizing model performance. At the end of the simulation run time, the model generates a report that includes averages of the defined statistics along with variability measures such as variance and confidence intervals. These statistics can be used as a summary of the overall system performance. For example, the average throughput for a production line is found to be 55.6 units per hour (UPH).

Collecting statistics from a DES model during run time and at the end of simulation is achieved using a statistical accumulation mechanism. Through this mechanism, the model keeps track of the defined statistics, accumulates their values, and provides averages and other statistics of their performance. This mechanism is activated at the occurrence of model events. As shown in Figure 14.9, model statistics are updated after the execution of each event as well as at the end of simulation.

For example, when defining customers' waiting time as a statistic in a DES model of a bank, the waiting time statistical accumulator updates the overall customers' waiting time upon the arrival or departure of customers. As the simulation run ends, the total accumulated waiting time is divided by the number of customers (processed in the bank during run time) to obtain the average waiting time for bank customers. In manufacturing, a plant statistic can be the average units produced per hour. Other statistics in the bank example and a plant example are shown in Table 14.8.

14.5.5 Animation Mechanism

Animation in systems simulation is a useful tool for the Six Sigma team, the their champion, and process owner. Most graphically based simulation software packages have default animation elements for system resources, buffers, operations, labor, and so on. This is quite useful for model debugging, validation, and verification. This type of animation comes with little or no additional effort and gives the Six Sigma team additional insight into how the model works and how to test different scenarios. Further, the more realistic the animation, the more useful the model becomes to the team in testing scenarios and implementing solutions.

Table 14.8 Examples of bank and plant statistics

Statistics in a bank	Statistics in a plant
Average customer waiting time	Average throughput per hour
Average time spent in system	Number of defectives per shift
Percent of time the bank teller is idle	Average manufacturing lead time
Number of customers served per day	Percentage of machine utilization
Maximum length of the waiting line	Average number of units in buffer

Simulation models can run and produce results without animation. The model logic is executed, model statistics are accumulated, and a summary report is produced at the end of model run time. Simulation analysts used to rely on the results of the model program in order to validate the model and correct errors in logic programming and model parameters. Only the programmer was able to verify model correctness and usefulness. The process owner and champion used to review the model results and suggest what-if scenarios and experiments to be performed. Today, however, with animation included in most simulation packages, owners and champions can watch the model run (in two- or three-dimensional graphical representation), track model entities, observe the impact of changes, and test model behavior at any selected point in the process flow. Such capability has helped teams sell new concepts, compare alternatives, and optimize performance. Finally, an animated model can be used for training purposes and as a demonstration tool in various kinds of industries.

Along with animation capability, DES models are often combined with good model management tools. A user-friendly graphical user interface (GUI) for graphics editing and libraries of graphical tools have made model animation an easy and enjoyable process. Tools have been developed that combine a database with simulation to store models, data, results, and animations. Modules graphical and parametric representation systems for basic model components such as parts, machines, resources, labor, counters, queues, conveyors, and many others are available in most simulation packages.

14.6 MANUAL SIMULATION EXAMPLE

Discrete event simulation models are driven by the mechanisms of discrete events, time-advancement, random sampling, statistical data accumulation, and animation, as discussed in the previous section. Such mechanics can be better understood by analyzing a queuing system. Queuing systems have a close relationship to simulation models. Queuing models are simple and can be used analytically to compare the results to simulation model results. Analytical models that can be analyzed with simulation include inventory models, financial models, and reliability models.

Since we have close-form formulas to analyze simple queuing models, the need to approach such models with simulation is often questionable. Simulation is often used in such cases to clarify the application of DES mechanisms. Also, when assumptions that are required for developing analytical queuing systems do not apply, simulation is used to analyze such systems. Event arrival and departure, determining each event's time of occurrence, and updating the event list are examples of such mechanisms. Based on the mechanics of modeling discrete events, the model state variables and collected statistics are determined at different instants when executing the model.

To clarify the mechanics that take place within the DES model, an example of a discrete event simulation model that represents a simple single-server queuing model is presented. In this model, cars are assumed to arrive to a single-bay oil-change service station based on an exponentially distributed interarrival time (t) with a mean of 20 minutes. Two types of oil changes take place in the station bay: regular oil changes that take about 15 minutes on average, and full-service changes that take about 20 minutes on average. Station history shows that only 20% of customers ask for a full-service oil change for their cars. A simulation model is built to mimic the operation of the service station. Model assumptions include:

- There are no cars in the oil-change station initially (the queue is empty and the oil-change bay is empty).
- The first car arrives at the beginning of simulation (clock time $T = 0$).
- Cars' interarrival times are exponentially distributed (this assumption is essential in analyzing the model as a queuing system), whereas oil-change time is discretely distributed: a 15 minute service time has a probability of 80%, and 20 minute service time has a probability 20%.
- The move time from the queue to the oil-change bay is negligible.
- Cars are pulled from the queue based on first in/first out (FIFO) discipline.
- No failures are expected to occur in the oil-change bay.

The logic of the simulation model is performed based on the following mechanics. In the discrete event mechanism, entities (cars) arrive and depart the service station at certain points in time. Each arrival/departure is an event and each event is stored chronologically in an event list (EL) according to the following formulas:

$$\textit{Next arrival time} = \textit{Current simulation clock time} + \textit{generated interarrival time}$$

$$\textit{Next departure time} = \textit{Current simulation clock time} + \textit{generated service time}$$

For each car arrival, a certain logic is executed based on the discrete event and time advancement mechanisms. If the bay is idle (empty), the arriving car at time T enters the bay for service, and RNG is used to randomly sample a service time (s) and schedule the car departure to be at time $t = T + s$. If the bay is busy, the car enters the station queue and waits for the bay to be empty, based on the FIFO discipline. Once the car finishes service, an interarrival time (a) is randomly sampled using RNG and a new arrival is scheduled at time $t = T + a$. Statistics are accumulated and collected and simulation continues for another arrival similarly. The execution of event arrival is shown in Figure 14.13.

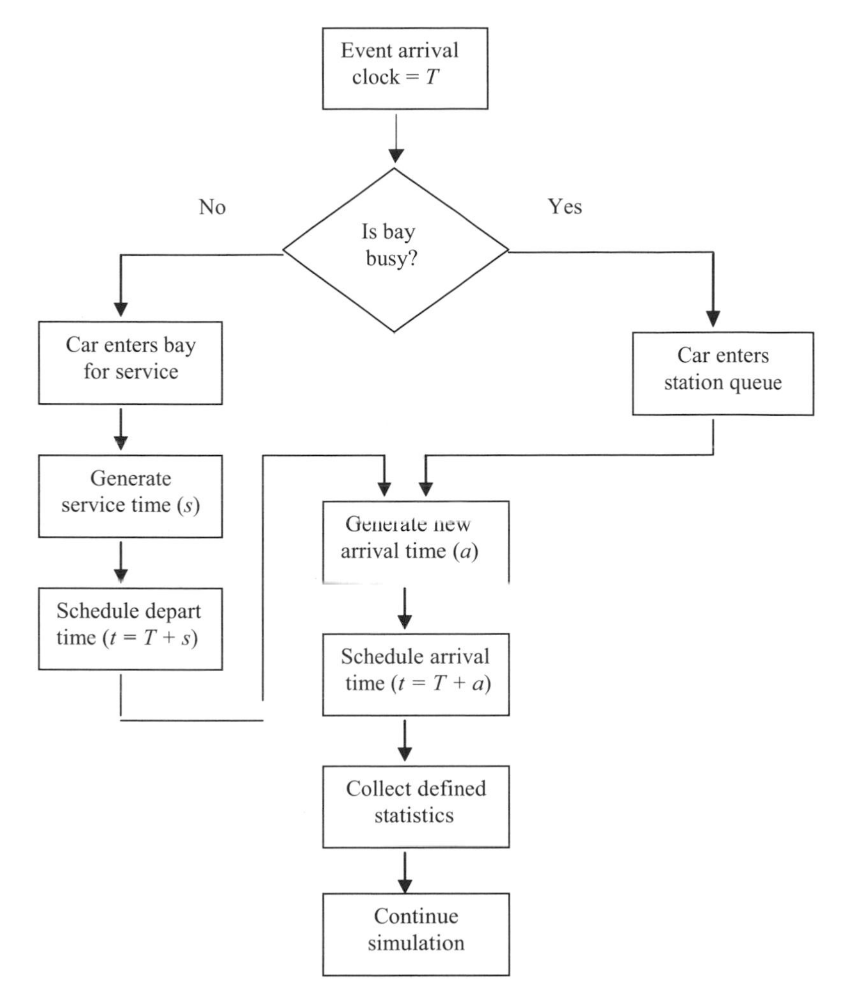

Figure 14.13 Execution of car arrival process.

Cars' interarrival time is generated randomly from a continuous exponential distribution (mean = 20 minutes), whereas the service time is generated randomly from a discrete distribution (15 minutes with probability 80%, and 20 minutes with probability 20%). Sampling from the exponential distribution with a cumulative distribution function of $F(t) = 1 - e^{-\lambda t}$, $t > 0$, results in $t = -(1/\lambda) \ln (1 - R)$, where $R = F(t)$. Values of R used to obtain successive random time samples (t) are selected from a uniform (0, 1) distribution using the LCG method. Similarly, R is generated for sampling from the discrete distribution to determine the service time (s), where s = 15 minutes if $0 \ R \leq 0.8$ and $s = 20$ minutes if $0.8 < R \leq 1.0$.

Therefore, when starting the simulation ($T = 0$, the time on the simulation clock), we assume that the first car (Customer 1) arrives at $T = 0$. The arrival of Customer 1 is the event that changes the state of the service station from idle to busy. The service time assigned for Customer 1, using $R = 0.6574$, is $s = 15.00$ minutes. Thus, Customer 1 departure time is scheduled as $0 + 15.00 = 15.00$ minutes.

While the car of Customer 1 receives the service of regular oil change, another car (Customer 2) may drive into the service station. The arrival time of Customer 2 is determined using $R = 0.8523$ as $t = -(20) \ln (1 - 0.8523) = 38.24$ minutes. Hence, the arrival time for Customer 2 is scheduled as $0 + 38.24 = 38.24$ minutes.

Based on this mechanism, the first event list (EL) is formed with two scheduled events on the simulation clock:

(E_1, t_1): Departure of Customer 1. Departure time is $T = 15.00$ minutes.
(E_2, t_2): Arrival of Customer 2. Arrival time is $T = 38.24$ minutes.

Apparently, Customer 1 will depart the service station before Customer 2 arrives. E_1 will be removed from top of the EL and the list will be reduced to E_2 only. The simulation clock will be advanced to next event (E_2) time ($T = 15.00$ minutes) in order to execute E_2. The departure of Customer 1 is the event that changes the state of the service station from busy to idle.

The service time assigned for Customer 2 using $R = 0.8867$ is $s = 20.00$ minutes. Hence, the departure time of Customer 2 is scheduled as $38.24 + 20.00 = 58.24$ minutes. After the arrival of Customer 2, another car (Customer 3) is scheduled to arrive at $t = -(20) \ln (1 - 0.2563) = 5.92$ minutes, using $R = 0.2563$. The arrival of Customer 3 is scheduled as $38.24 + 5.92 = 44.16$ minutes.

Based on this mechanism, the event list is updated with the two scheduled events (E_3 = departure of Customer 2, and E_4 = arrival of Customer 3). The two events are scheduled on the simulation clock as follows:

E_3: $T = 58.24$ minutes
E_4: $T = 44.16$ minutes

Apparently, Customer 3 will arrive at the service station before the departure of Customer 2, resulting in a waiting time of $58.24 - 44.16 = 14.08$ minutes.

Simulation run time continues using the described mechanism until a certain terminating condition is met at a certain number of time units (for example, three shifts of production, one year of service, 2000 units are produced, and so on). A simulation table that summarizes the dynamics of the first 20 cars arriving at the service station is shown in Table 14.9, in which the following variables are used:

A: Customer number (1, 2, . . . , 20)
B: Randomly generated interarrival time in minutes for each customer
C: Arrival time of each customer
D: Randomly generated service time in minutes for each customer
E: Time at which service (oil change) begins
F: Time each customer waits in the queue for service to begin

Table 14.9 Simulation table of single-bay oil-change example

A	B	C	D	E	F	G	H	I
1	0.00	0.00	15.00	0	0.00	15.00	15.00	0.00
2	38.24	38.24	20.00	38.24	0.00	58.24	20.00	23.24
3	5.92	44.16	20.00	58.24	14.08	78.24	34.08	0.00
4	31.40	75.56	15.00	78.24	2.68	93.24	17.68	0.00
5	13.25	88.81	20.00	93.24	4.43	113.24	24.43	0.00
6	28.12	116.93	15.00	116.93	0.00	131.93	15.00	3.69
7	14.35	131.28	15.00	131.93	0.65	146.93	15.65	0.00
8	15.22	146.50	15.00	146.93	0.43	161.93	15.43	0.00
9	21.87	168.37	20.00	168.37	0.00	188.37	20.00	6.44
10	13.98	182.35	15.00	188.37	6.02	203.37	21.02	0.00
11	36.54	218.89	15.00	218.89	0.00	233.89	15.00	15.52
12	9.95	228.84	15.00	233.89	5.05	248.89	20.05	0.00
13	10.54	239.38	15.00	248.89	9.51	263.89	24.51	0.00
14	23.64	263.02	20.00	263.89	0.87	283.89	20.87	0.00
15	11.70	274.72	15.00	283.89	9.17	298.89	24.17	0.00
16	15.90	290.62	15.00	298.89	8.27	313.89	23.27	0.00
17	28.70	319.32	15.00	319.32	0.00	334.32	15.00	5.43
18	25.65	344.97	15.00	344.97	0.00	359.97	15.00	10.65
19	22.45	367.42	20.00	367.42	0.00	387.42	20.00	7.45
20	12.50	379.92	15.00	387.42	7.50	402.42	22.50	0.00
Total	379.92		330.00		68.66		398.66	72.42

G: Time at which service (oil change) ends

H: Total time a customer spends in system (waiting plus service)

F: Idle time for the station bay

Statistics are collected from the model by accumulating state variables and averaging them over the run time. Examples of such statistics in the service station example are:

- Average customer waiting time = Total waiting time in queue/Number of customers, which is determined to be 68.66/20 = 3.43 minutes.
- Average car service time = Total service time/Number of customers, which is determined to be 330.00/20 = 16.50 minutes.
- Average interarrival time = Sum of inter-arrival times/Number of customers – 1, which is determined to be 379.92/(20 – 1) = 19.99 minutes.
- Average waiting time for customers = Total waiting time/Number of customers who wait, which is determined to be 68.66/12 = 5.72 minutes.
- Average time in system (TIS) = Total time-in-system/Number of customers, which is determined to be 398.66/20 = 19.93 minutes. This is identical to the sum of the average waiting time and average service time, 3.43 + 16.5 = 19.93.

- The probability that a customer has to wait = Number of customers who wait/Number of customers, which is determined to be 12/20 = 60%.
- The utilization of the station bay = [1.00 – (Total idle time/Total run time)] × 100%, which is determined to be 100% × [1.00 – (72.42/402.42)] = 82.00%

14.7 VARIABILITY IN SIMULATION OUTPUTS

In designing, analyzing, and operating complex production and business systems with DES, one is interested not only in performance evaluation but also in sensitivity analysis, experimentation, and optimization. However, simulation outcomes, representing real-world systems, are often stochastic and lead to inconsistent performance levels.

We can classify the factors that contribute to variation in simulation outputs as controllable and uncontrollable (noise) factors. Model design factors such as buffer sizes, number of production resources, cycle times of automatic operations, and speeds of conveyance systems are usually considered controllable factors. Different settings of system controllable factors may result in different sets of model outputs. However, design teams running a simulation study can control such factors and change their parameters to the benefit of the process performance. Once set to their optimum levels, control factors do not change with time or changes in running conditions.

On the other hand, random factors are those uncontrollable factors whose individual values change over time due to the actual behavior of the real-world process; hence, they are often presented in simulation models by random generation from a certain probability distribution. Key distribution parameters are often estimated through statistical approximations based on fitting a collected set of empirical data to one of the commonly used distributions (see Chapter 2). The behavior of DES models is, therefore, driven by the stochastic nature of real-world processes such as an entity's arrival process, in-system service process, and output departure process. Variability in these processes is caused by random factors such as arrival rate, service/processing rate, equipment time between failure (TBF) and time to repair (TTR), percentages of scrap and rework (measured by defects per opportunity and defects per million opportunities) and so on.

Such variability in model processes results in stochastic variability in model response (system performance measures). Therefore, the throughput of a production system, for example, is subject to variability in material delivery, tool failure, and operator efficiency and is represented in a DES model by stochastic behavior. Based on the inherent fluctuations, the throughput could be a low value in one shift and a higher one in another. The amount of variability in model outputs depends on the amount of variability in model inputs along with the dynamics caused by the logical interactions among different model elements and processes. Thus, simulation runs yield estimates of such performance measures (throughput in this case) in terms of means and variances to reflect such variability.

Noise factors can also arise from uncontrollable internal sources of variation within the simulated system. For example, in a plant, such factors can be the plant climate, team dynamics, union relations, and worker strikes. They also represent external factors such as the variation in the plant logistics, supply chain dynamics, and environmental changes. Such factors also cause variations in model outcomes. Although DES can include certain assumptions that represent model noise factors, most practical simulation studies are focused on modeling controllable and random design factors. Hence, by determining optimum settings to model controllable factors, statistical and simulation-based optimization methods aim at obtaining a system response that meets design expectations with minimal sensitivity to variations (it is robust) in modeling random factors.

In addition to controllable, random, and noise factors, artificial factors are a source of variation in simulation outputs. Artificial factors are those simulation-specific factors such as system initialization state, warm-up period,[2] run length,[3] termination condition, and random number streams. Changing the settings of such factors from one simulation run to another often results in changes in model outcomes. Hence, testing such factors in the model and providing proper settings for different model run controls to obtain a steady-state response is basically a prerequisite for applying statistical and simulation-based optimization methods. Figure 14.14 presents a summary of different sources of variation in simulation models.

14.7.1 Variance Reduction Techniques

It has been shown in the previous section that the stochastic nature of simulation often produces random outputs. Random and probabilistic model inputs result in random model outputs. As the amount of input variability increases, the observed model response tends to be more random, with high variance and wide confidence intervals around the response mean. As a result, less precise performance estimates are obtained from the simulation model. Thus, the objective of using variance reduction techniques (VRTs) is primarily to make better decisions based on simulation results by reducing the variability in model outcomes. The response mean and confidence interval are typically used to describe the performance measure of a simulation model. Hence, reducing the response variance results in reducing the width of the confidence interval around the mean, which increases the precision of simulation estimation.

[2]Warm-up period is a transient simulation state. A steady-state or nonterminating simulation applies to systems whose operation continues indefinitely or for a long time period with no statistical change in behavior. Such periods include a relatively short transient (warm-up) period and a relatively long steady-state period. The warm-up period is excluded from results analysis and simulation statistics are collected from the steady-state period.

[3]The simulation run length represents the time period in which model behavior is observed, statistics are accumulated and collected, and performance measures are recorded. It is essential to set the model to run for enough time past the warm-up period in order to collect representative data and arrive at correct conclusions about the model behavior.

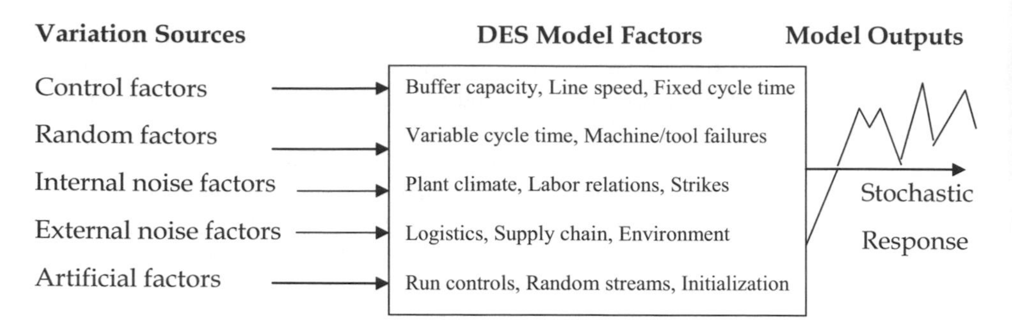

Figure 14.14 Sources of variation in a simulation model.

Several methods are used to reduce the variance in simulation model response. The simplest way is to eliminate the sources of variations in model inputs by simply using expected values of model parameters rather than empirical or theoretical statistical distributions. This results in a stable model response that is simply not correct and does not represent the actual behavior of the simulated process. Therefore, a VRT is needed that reduces the amount of variation in model response without impacting model representation of the simulated system.

Another method of reducing response variance is to increase the run length for steady-state simulation so that the sample size used in determining the response confidence interval is very large. Increasing run length is particularly beneficial when using the batch-means, steady-state method. When using the independent replication method, the number of replications must be increased. For terminating simulation when the run length is limited by the nature of the system pattern of operation, the number of simulation replications can be increased. The following formula can be used to approximate the number of simulation replications (n') that are needed to bring the confidence interval half-width (hw) down to a certain specified error amount at a certain significance level (α):

$$n' = \left[\frac{(Z_{a/2})s}{hw} \right]^2$$

where s is the sample standard deviation. In simulation, however, we can benefit from our control of model variability through random number generation (RNG). Although there is a class of methods that can work on RNG to reduce the model-produced variability, using common random numbers (CRNs) is one of the most popular VRTs. CRN is based on using one exact source (stream) of RNG for all simulation replicas instead of using a RNG for each independent simulation replica. The objective is to induce certain kinds of correlations (synchronization) in the produced random numbers so that some variation is reduced.

To maintain the independence of simulation replicas, which is essential in random sampling, the CRN uses different random number seeds that are assigned at the beginning of each independent simulation replica. Care should be taken in picking the seed so that the segments of the random stream are not shared between replicas. Simulation software tools provide the CRN option that automatically assigns seed values to each replica so that the possibility of sharing segments of the random stream is minimized. Each stochastic element in the model, such as cycle times, arrival times, and failure times, can be assigned a different random number stream to keep random numbers synchronized across the simulated system.

Using CRN is particularly useful when comparing the performance of two or more simulated systems (strategies or design alternatives) because using the CRN results in positive correlation among compared strategies, which actually reduces the variance of sample differences collected from the model but does not reduce the variance of the simulation output. If no correlation is created between compared σ^2 systems or a negative correlation was created, using CRN will result in either no impact or an increase in the variance obtained from comparing simulated systems. The variance (σ^2) of two independent systems (x_1 and x_2) is expressed as follows:

$$\sigma^2_{x_1-x_2} = \sigma^2_{x_1} + \sigma^2_{x_2}$$

When using a synchronized CRN, however, a positive correlation is induced between x_1 and x_2, which results in less variance (reduced by the covariance "COV" component) and is expressed as follows:

$$\sigma^2_{x_1-x_2} = \sigma^2_{x_1} + \sigma^2_{x_2} - 2\,\text{cov}(x_1, x_2)$$

Several other VRTs can be used in addition to the CRN method. Examples include antithetic variates, control variates, indirect estimation, and conditioning.

14.8 SERVICE PROCESSES SIMULATION IMPLICATIONS

Several considerations should prepare the design team for planning a business process simulation study and for addressing the suitability of their choice of simulation package. First, a clear classification of their project business process is needed. It could be product development; a production process;[4] problem solving, including distribution[5] processes; or a service-based process.[6] A team usually designs the product development processes. These processes are analyzed using DFSS tools. However, process cycle times and resource requirements are usually handled by simulation, which produces more accurate results. Many DES simulation packages are capable of representing the highly variable steps in tightly coupled resources. Modeling shifts, downtime, overtime, and learning curves with multiple replicas of

[4]Typical examples are order fulfillment, accounts payable, and claims processing activities.

[5]Includes transportation and delivery processes, in which products or people are carried between locations via a distribution network.

[6]Example of service processes are telephonic services (call centers), restaurants, copy centers, hospitals, and retail stores.

resources are some of the important considerations in building a valid simulation of a product development process.

In the context of Six Sigma, discrete event simulation is quite suitable for production process problem solving when outputs are produced in a batch or continuous flow mode in relatively high volumes. Tasks such as assembly, disassembly, setup, inspection, and rework are typical steps in production processes with queuing rules and downtime modeling. Such processes are usually modeled to obtain steady-state behavior past the warm-up period. In modeling distribution processes in a production environment, it is important to define attributes for flow entities in order to keep track of unique characteristics such as value-adds, cost, and distance traveled. Due to the transient nature of distribution processes, the simulation model warm-up period is usually longer than other production processes.

In a service-based industry, processes present a major area for employment of simulation studies. These processes are typically characterized by a non-value-added time that exceeds the value-added (processing time). The simulation of service processes represents a challenge because both the entities and processing resources are typically people. Modeling human behavior is complex and unpredictable. For example, customers calling a service center may hold or hang up. Modeling adaptation is required to model such situations. Also, processing times are highly variable and customer arrivals are random and cyclical, adding to the model complexity. Model accuracy demands representative probability distributions. Otherwise, a steady state may not be reached.

14.9 SUMMARY

The discrete event simulation (DES) techniques presented in this chapter are most capable and powerful tools for business process simulation of transactional nature within DFSS and Six Sigma projects. DES provides modeling of entity flows with capabilities that allow the design team to see how flow objects are routed through the system. DES capability is increasing due to new software tools and a wide spectrum of real-world applications.

Basic concepts of simulation modeling include the system concept, model concept, and simulation concept. Systems include inputs, entities, relationships, controls, and outputs. System elements should be tuned toward attaining an overall system goal. A system model is a representation or an approximation of a real-world system. Models can be physical, graphical, mathematical, or computer models. The goal of modeling is to provide a cheaper, faster, and easier-to-understand tool for process analysis, design, and improvement. Simulation is the art and science of mimicking the operation of a real-world system on a computer. It is aimed at capturing complex, dynamic, and stochastic characteristics of real-world processes. Other types of models fall short in this regard. Based on type of state variables, computer simulation models can be discrete, continuous, or combined. They can also be deterministic or stochastic based on randomness modeling. Finally, they can be static or dynamic based on the changes of system state.

15

DESIGN VALIDATION

15.1 INTRODUCTION

The final aspect of DFSS methodology that differentiates it from the prevalent "launch and learn" method is design validation. This chapter covers in detail the verify phase of the DFSS (ICOV) project road map (see Figure 5.1). Design validation helps identify unintended consequences and effects of design, aids in developing plans, and helps reduce risk for full-scale commercialization.

At this final stage before the full-scale production stage (see Chapter 1, Section 1.3.6), we want to verify that the design's performance is capable of achieving the expectations of customers and stakeholders at Six Sigma performance levels. We want to accomplish this assessment in a low-risk, cost-effective manner. This chapter will cover the service-relevant aspects of design validation, including prototyping, testing, and assessment.

15.2 DESIGN VALIDATION STEPS

Design validation can take on many different forms depending on the magnitude of the design change and the complexity of the environment in which the design will operate. Typically, it is desirable to understand the performance of the new design as early as possible in the design cycle with the highest confidence achievable. In order to accomplish this, we can rely on the design of experiments methods detailed in Chapters 12 and 13 but often these DOEs are specialized around a subset of input-significant factors and do not provide a full functional test. Product-specific validation often means building prototypes and testing them in a "beta" environment. The beta environment is one in which real end users utilize the new design in their actual application and provide feedback. In the context of writing this book, people proofread the draft (prototype) chapters in a beta environment and provided valuable feedback. In the beta environment, end users are willing to test the new design because they are offered something of value in return, such as free products or early access to new technologies. In order for this beta testing to provide useful feedback to the design team, frequent debriefing on the experience and detailed notes on the operating environment are necessary.

As we move into a service validation environment, many concepts from the product validation environment are directly applicable but often it is just as easy to run the new service or process in parallel with the legacy design.

We introduce the design validation flowchart in Figure 15.1. In this flowchart, we see that the first thing to be done is to reassess the makeup of the design team. Often at this stage, the team needs to be expanded by adding members who have practical operations experience or who can mentor customers as they use the new process or service. The next step is to ensure that all team members are up to date with the design intent and objectives. Next, the team defines the evaluation criteria and the associated budgets and schedules related to this. Step 4 consists of estab-

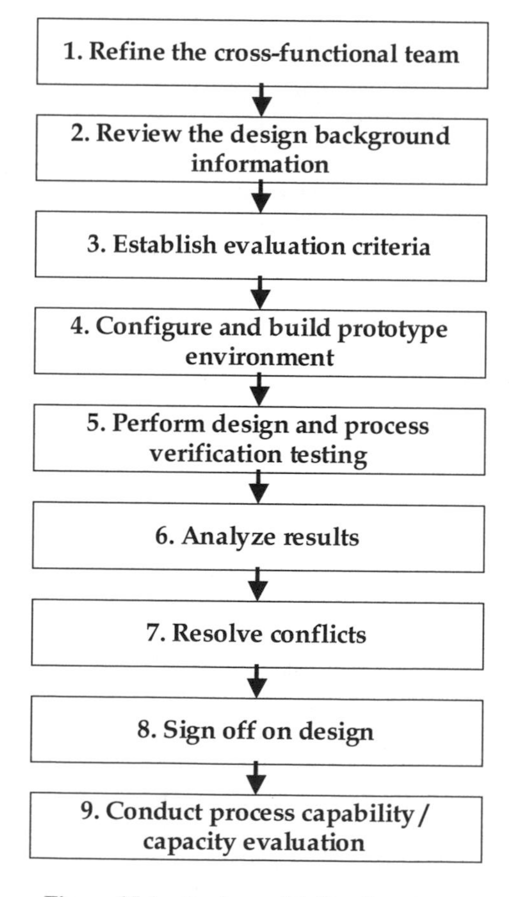

Figure 15.1 Design validation flowchart.

lishing the prototype environment so that is as close to the production environment as possible, considering cost and risk factors. In step 5, the process/service is run and observed based on the evaluation criteria and environmental factors are determined. Once the verification testing results are available, they are analyzed. It is important to note that analysis should not be left until the end of the testing as the team will want to be sensitive to any initial defects as soon as possible and make go/no-go decisions after each completed cycle. After the results have been analyzed, the team makes decisions as to whether conflicts exist that require design modifications. If design modifications are made, a decision must be made whether to loop back to Step 4 again. Once the team is satisfied that the new design is correct, the documentation should be formally signed off. The last step is to conduct process capability analysis and any capacity planning.

The key elements of validation, then, are to create a prototype, subject it to a realistic test environment, summarize the observed performance, and feed back any required changes into the design. We will next cover the details of each of these steps as well as the need for statistical process control and control plans.

15.3 PROTOTYPE BUILDING

In order to produce a useful prototype of the new design that is sufficient to allow for evaluation, consideration must be given to the completeness and stability of the elements that make up the design. These elements can be classified as design parameters that are manifested as components and subsystems, and production process variables with all steps and subprocesses. We can look at components, subsystems, and systems to determine the lack of knowledge of performance and the interaction of preexisting (carried over) subsystems in a new system. In order to assist in determining if the prototype is complete enough and relevant for testing, it is helpful to assess the elements as being a:

- Current design
- Analogous design
- New design, low risk
- New design, high risk

Risk can be assessed from FMEA and technology road maps or other sources (e.g., history) and/or DFSS tools usage. If any of the elements are in the new design, high-risk category, they should not be compromised and must be included in the prototype.

The prototype in a product environment may not have been produced on actual production equipment or with the actual production process and people. The same can happen in a service environment. New software or organization structures may be months away and you still want to know if the new design will perform properly. In an enterprise resource planning (ERP) or customer relationship management

(CRM) system deployment, the configured software is often built into a conference room pilot (CRP) on nonproduction hardware and tested under test script scenarios. The built CRP is a prototype.

Prototypes usually involve some trade-off between the final completed design and some elements that will not be available until later. Prototype designs are tested off-line under controlled conditions. Some examples are:

- Initial mock-up hardware that is created in a lab environment.
- New software to prioritize/route calls for a call center is tested off-line.
- A modified version of the current month-end close process in accounting is tested in parallel with an actual system prior to pilot testing.
- An Excel model is used to conduct an off-line sensitivity analysis of sales commission by salesperson prior to real application of customers.

Prototypes should be evaluated using objective and robust analytical techniques to objectively demonstrate design performance for a product/service pilot. The testing method is critical to accomplishing objective decision making and obtaining information in the most cost-effective manner.

15.4 TESTING

When testing prototypes, it is vital to follow a rigorous process in order to assess the performance and learn from any unusual observations. Many times, unexpected results may occur, and without the correct approach these will have to be written off as unusual observations when, in fact, they may be significant events that can either differentiate your new design or plague it. The key elemets of a rigorous test plan are:

- Assign individual names to each role.
- Confirm that the test environment has been established and is ready for use.
- Finalize/confirm testing estimates.
- Estimate hours to be spent by each individual.
- Plan calendar duration of all testing.
- Set testing milestones (target dates).
- Finalize test cases.
- Finalize test scenarios.
- Describe inputs and expected results.
- Implement defect tracking mechanisms.
- Confirm issue resolution procedures that should include decisions to defer, fix, or withdraw.
- Finalize acceptance test criteria/comparison tests.

- Perform cycle tests.
- Do performance test.
- Do integration test.
- Do regression test.

Some of the specific types of testing include pilot tests and pre-production (low-rate production) tests. A pilot tests prototype design under limited real-world conditions to verify original design assumptions versus actual performance. Preproduction testing takes a new product or process through concept and pilot evaluation to full production or roll out.

In validation, the DFSS team conducts a progressive sequence of verification tests to quantify the ability of the service to satisfy FRs under conditions that approximate the conditions of actual use. The DFSS team needs to conduct specification and verification tests and make initial adjustments to testing and measurement equipment. Initial conditions are verified and recorded according to a test plan. The team then begins collecting data for those tests that are designed to analyze performance and records any abnormalities that may occur during test and may facilitate subsequent concern resolution. This involves tiered testing (testing that parallels hierarchy) as part of the test strategy, to enable functional verification at progressive levels of process mappings.

The first hierarchical level is associated with subprocess-step-level testing. The second hierarchical level is associated with subprocesses. The third level of testing is for the process, which includes the functional subprocesses. Functional insensitivity to the testing environment (lack of reliability) will be most quantifiable at the step and subprocess hierarchical levels. This is due in part to the number of tests that can be effectively performed, and to the greater ability to measure functional requirements performance at lower levels of design complexity. The issue just noted is not a hard-and-fast rule but is certainly observable in the greater number of tests that are run (hence, the improved quantification) at the component and subprocess levels.

Testing costs at the subprocess level are typically one-tenth of those at the next level. This also includes finalizing packaging, shipping, and delivery processes (if any). In addition, the teams need to verify time studies and process capabilities. The validation staff includes seasoned personnel with direct operating and regulatory compliance experience in addition to the DFSS team. The extended DFSS team should identify critical process parameters and production limitations and manage the launch through process efficiency studies using simulation, design of experiments for prevalidation investigations, writing the process validation protocols, and overseeing the work.

Validation planning should begin early in the design process, starting with the *characterize* DFSS phase. It is most important to integrate validation activities early in design rather than as an afterthought. Significant design issues must be reviewed, understood, and accepted by the extended DFSS team before progressing too far into detailed design.

15.5 CONFIDENCE INTERVAL OF SMALL SAMPLE VALIDATION

When testing in a preproduction environment, the scope of the available samples or resources is often very small due to cost or time factors. Proof of concept is usually determined after a single successful transaction but robust performance can only be determined after a significant number of cycles have been experienced. It is recommended that the legacy system be baselined for statistical parameters and the new design validated to a degree of statistical significance. Power and sample size calculations can be made based on the underlying variation assumptions and the change in performance expected. If a process or service will only experience a single cycle in the period of a year, then it is very unlikely that you will be able to validate the process statistically. Many new processes or services experience many cycles per week and it is reasonable to validate these with 95% confidence intervals for evaluating the results as statistically valid and different.

The statistical factors that affect the sample size are the confidence interval desired, the system noise (standard deviation), and the delta change between the signals (averages). The sample size (n) is given by

$$n = \left(\frac{Z_{\alpha/2}\sigma}{\delta} \right)^2 \tag{15.1}$$

where $\delta = \text{error} = |\bar{y} - \mu|$. For example, take a legacy process that "approves" 25 transactions per day with an average approval time of 98 hours and standard deviation of 13.7 hours. How many observations would be needed in order to have 95% confidence that there was a statistically significant difference? If we take the cases of differences in average approval time of 2, 12, and 94 hours, we find the following required sample sizes:

2 hour difference—612 samples
12 hour difference—19 samples
94 hour difference—3 samples.

The factors that affect the sample size are the confidence interval desired, the system noise (standard deviation), and the delta change between the signals (averages).

If in your validation process you are unable to achieve such a level of test confidence, it is recommended that you roll out your deployment gradually and incrementally. Start out with typical users who are less sensitive to any issues and pulse them for performance. As you feel more comfortable with the performance, increase the scope and continue to ramp up the new process/service until the old process/service is retired.

15.6 CONTROL PLANS AND STATISTICAL PROCESS CONTROL

As we build upon the DOE transfer functions, finalize the process capability, and update the design scorecard, we become cognizant of the key process input variables (KPIV) and the key process output characteristics (KPOC) that are critical for monitoring and controlling in the production environment. One of the best tools to start the process of implementing process control is the control plan depicted in Figure 15.2. In this control plan, take the KPIVs and the KPOCs and list them in the column labeled Specification Characteristic. In the columns labeled Sub Process and Sub Process Step, input the corresponding labels.

The next required element is the Specification Requirement. In this cell, the upper and lower specification limits (USL and LSL) are typically entered. In some processes, only one limit may be required or just the target value for attribute type requirements.

The next cell is for the measurement method. How will the measurement be acquired? Many times this is the result of a calculation from several fields, as in the case of processes and services. For instance, process time might be calculated from the field of start time and end time. Other times, the measurement may require surveys or audits.

The cell labeled Sample Size is for how many measurements will be required at the specified frequency. This refers back to the underlying statistical significance that is required to capture the variation.

In the frequency cell, enter the required frequency. This can be, for example, every cycle, every 20 cycles, quarterly, or annually, depending on the stability of the feeder process and the nature of the measured characteristic.

The column labeled Who is for the assigned person(s) who will collect the measurements. The Where Recorded is the location where the collected measurement is documented for referral or audit purposes. The Decision Rule/Corrective Action column is for the rules for what action to take if the measurement is outside of the specification requirement. This can either be a set of actions or reference to a policy or procedure. The column labeled SOP Reference is for the document name or number that is the standard operating procedure for the subprocess or subprocess step.

It is important to note that if the design team has followed all of the optimization and validation steps and the design is operating at a very low defect rate (approaching Six Sigma capability) then there should be relatively few KPIVs and the KPOCs should be the prioritized CTQs from the HOQ1.

15.7 STATISTICAL PROCESS CONTROL

Once the data measurements are collected as prescribed in the control plan, the method for charting the data should be determined. Since statistical process control (SPC) is a well-developed subject and there are many textbooks on the subject, we will not be covering the methodology in any detail in this book. We do feel, however, that in the transactional world of services and business processes, there is little evidence of the use of SPC, so brief coverage of the subject will be provided. We will cover the selection of the charting method and some of the basic rules of interpretation. For specific books on the subject, see Wheeler (1995) and Wheeler and Chambers (1992).

Process Control Plan

Process Name:	______	Prepared by:	______	Page:	____ of ____
Customer	______ Int/Ext ____	Approved by:	______	Document No:	______
Location:	______	Approved by:	______	Revision Date:	______
Area:	______	Approved by:	______	Supercedes:	______

Sub Process	Sub Process Step	CTQ		Specification Characteristic	Specification/ Requirement	Measurement Method	Sample Size	Frequency	Who Measures	Where Recorded	Decision Rule/ Corrective Action	SOP Reference
		KPOC	KPIV		USL LSL							

Figure 15.2 Process control plan.

15.7.1 Choosing the Control Chart

Figure 15.3 shows the decision tree for what type of SPC method to use. To follow this chart to the correct method, you enter at Start and determine if the data you are collecting is of the attribute or variable type. Attribute data is counting or categorical (pass/fail, go/no-go) in nature, and variable data are measures that typically can have many values of a continuous nature. For example, if we measure day of the week, then this is attribute data, whereas if we measure time to complete in minutes, this is variable data. If we follow the Variables side of the chart, the next decision point is to determine if we are measuring individual values or rational subgroups. The answer to this comes directly from the Sample Size cell in the control plan. If we are measuring individuals, then the most common chart method is the individual moving range chart (IMR), which is useful when the measured value is changing slowly. If the process is changing rapidly then the moving average (MA) chart is used. If the measure is rational subgroups (sample size greater than one), then we enter another decision point to determine whether the subgroup size is less than six. Again, the subgroup size question is answered by the Sample Size cell in the control plan. For subgroups less than six, the chart of choice is called X-bar, moving range (Xbar, MR), whereas if the subgroup size is greater than six, the recommended chart is the X-bar, standard deviation chart (Xbar, S).

Moving back to the top decision point and following the Attribute side to the left, the first decision point is whether we are measuring defects or defectives. We are counting defects when we count the number of defects in a purchase order or invoice. We are counting defectives when we count a purchase order or invoice as defective regardless of the number of defects that the item contains. For example on a purchase order, the fields of Deliver To Address, Deliver Date, Item Quantity, Item Description, and Unit Price could all be defects, but we count the purchase order as a defective. First, let us assume that we are measuring defects. The first check is to see if the Poisson assumptions are satisfied. You have Poisson data when the data meet the following conditions:

- The data are counts of discrete events (defects) that occur within a finite area of opportunity.
- The defects occur independently of each other.
- There is equal opportunity for the occurrence of defects.
- The defects occur rarely (when compared to what could be).

Assuming that the assumptions are satisfied, then the next decision point is whether the area of opportunity is constant from sample to sample. In our example of the purchase order, the opportunities are those data elements that we defined, and these are constant. A changing opportunity would be one in which we were measure the number of line items per purchase order; this would definitely be nonconstant. With constant opportunity, the C chart is the correct method, whereas if it is nonconstant, the U chart is the proper method.

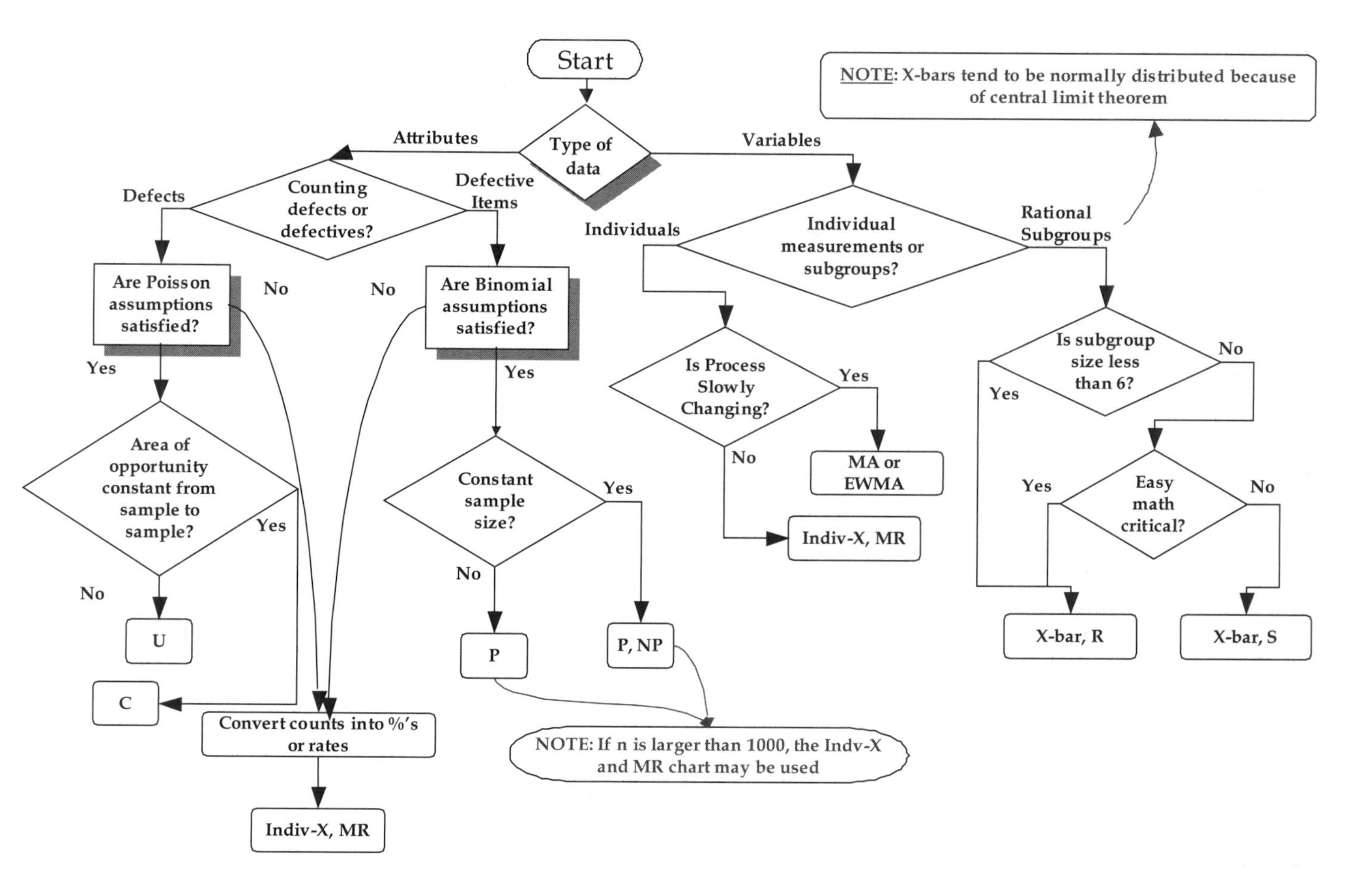

Figure 15.3 SPC charting method selection.

Returning to the first decision point for attribute data, we move to the Defective Items path on the righthand side. The first check is to see if the binomial assumptions are satisfied. You have binomial data when it meets the following conditions:

- Each item is the result of identical conditions.
- Each item results in one of two possible outcomes (pass/fail or go/no-go).
- The probability of success (or failure) is constant for each item.
- The outcomes of the items are independent.

Assuming that these assumptions are met, we then need to determine if the sample size is constant or not. With a constant sample size, the P chart is the correct method, but for nonconstant sample size the NP chart or the P chart are the correct methods. You may note from Figure 15.3 that the most robust chart is the I-MR chart, which can be used for individual variable measures and for the defects or defectives if the number of data points is large or the results are converted to percentages.

15.7.2 Interpreting the Control Chart

Once the proper charting method is selected, the data is plotted and the charts are interpreted for stability. A process that is in control exhibits only random variation. In other words, the process is stable or consistent. There is no evidence of special-cause variation.

Control charts assess statistical control by determining whether the process output falls within statistically calculated control limits and exhibits only random variation. A process is classified as in control if the data demonstrate stability on the control chart. The typical indicators that a process is unstable or out of control is when there are definite trends upward or downward (usually seven consecutive points). This is evidence of cyclical behavior, either by shift, supplier, or even season. The chart in Figure 15.4 depicts the distribution that occurs on most control charts. The diagram is symmetrical about the centerline which is the mean. The zone defined by the two “C”s is called Zone C and is the ± 1 standard deviation about the mean. The area encompassed by the two “B”s (and the previous “C”s) is called Zone B and encompasses the ± 2 standard deviations. The final Zone is called Zone A and encompasses the ± 3 standard deviations.

The specific tests used are (Nelson tests):

- Any point outside the control limits (outside of Zone A)
- Nine points in a row on same side of the center line
- Six points in a row, all increasing or all decreasing
- Fourteen points in a row, alternating up and down
- Two out of three points in the same Zone A or beyond
- Four out of five in same Zone B or beyond

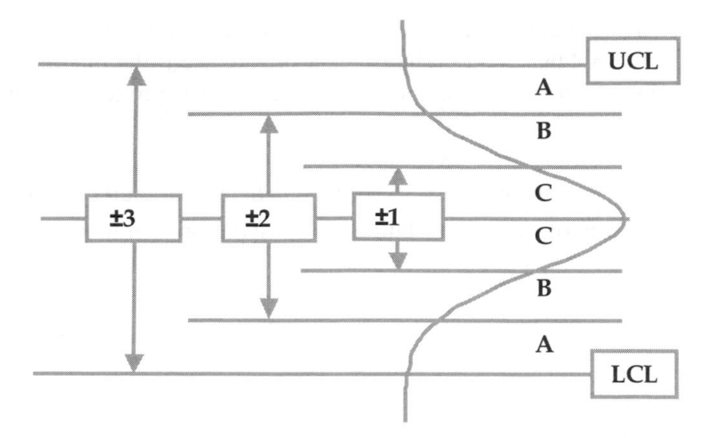

Figure 15.4 Control chart zones and distribution.

- Fifteen points in a row in either Zone C
- Eight points in a row or more outside of Zone C on the same side of the center line

15.7.3 Taking Action

When interpreting control charts, if they are in control, everything is fine and processing should continue. If one of the stability rules are violated, then a decision must be made. These rules indicate that a special cause situation has occurred. Every attempt must be made to discover the root cause of the special cause situation. This does not mean that the process is not producing acceptable products.

15.8 PROCESS CAPABILITY

A process can be in control and still have poor capability. SPC on a process with poor capability will only assure us that the process will not degrade further; it will not prevent defects. It is also necessary to determine if the process is meeting the customer's requirements. The control chart has no relationship to the voice of the customer (VOC); it is strictly the voice of the process (VOP). The last step of validation is to ensure process capability. Before we get to this point, we must ensure that we have adequate measurement capability and the process is in a state of stability as indicated by the control chart.

Process capability can be assessed in either the short-term frame or long-term frame. In the short term, it is the condition without any sources of variation between the rational subgroups. This is the best the process can perform under ideal situations. The customer can get glimpses of the short-term capability but, more than likely they, will experience the full variation and, therefore, the long-term capability is a better gauge of what the customer will experience. Capability can be measured with several indices. There are the traditional methods defined by Cp and Cpk and there are the sigma values that correspond to the part per million defects. Each of these methods utilizes specification limits as defined by the customer and can be defined as the USL – LSL. We are interested in how many times the process variations can fit with these limits. Obviously, the more times the process spread can fit in the specification limits, the less is the chance that a defect will be experienced. Simply put, the capability is $(USL - LSL)/6\sigma_{LT}$, where σ_{LT} can be calculated with the following formula for an overall long-term sample of size n:

$$\sigma_{LT} = \sqrt{\frac{\sum_{i=1}^{n}(x_i - \bar{x})^2}{n - 1}}$$

We can use the Cp measure, which is, $(USL - LSL)/6\sigma_{LT}$, or the Cpk:

$$C_{pk} = \min\left[\left(\frac{USL - \bar{x}}{3\sigma_{LT}}\right), \left(\frac{\bar{x} - LSL}{3\sigma_{LT}}\right)\right]$$

15.9 SUMMARY

Validation is a critical step in the DFSS road map (see Chapter 5) and needs to be thought out well in advance of final deployment of the new design. Validation often requires prototypes that need to be near "final design" but are often subject to trade-offs in scope and completeness due to cost or availability. Assessing the components and subsystems of any new design against the type of design risk assists in determining when a prototype is required and what configuration must be available. Once prototypes are available, a comprehensive test plan should be followed in order to capture any special event and to populate the design scorecard. Any go/no-go decision should be based on statistically significant criteria and all lessons learned should be incorporated into the final design if appropriate.

Before hand-off to the production environment, the process stability and capability should be assured.

16

SUPPLY CHAIN PROCESS

16.1 INTRODUCTION

In order to appreciate the design activity that will be covered in Chapter 17, we must first describe the aspects of the supply chain.

16.2 SUPPLY CHAIN DEFINITION

The supply chain has many definitions and covers many traditional functions as well as life cycle aspects of acquiring resources for use in the production of value for end users. The traditional definition of the supply chain (Figure 16.1) is the collection of activities that connect suppliers to end users, forming a chain of events. Value-added operations supply end customers, and this relationship is repeated with suppliers and their suppliers.

In the life cycle of acquiring resources, processing them into value-added solutions, and providing postdelivery support there are seven main process-focused requirements that must be met to varying degrees. These processes are:

1. Demand forecasting
2. Planning and scheduling
3. Strategic sourcing
4. Processing
5. Asset and inventory management
6. Distribution and logistics
7. Postservice support

We shall briefly describe what is performed in each of these process steps.

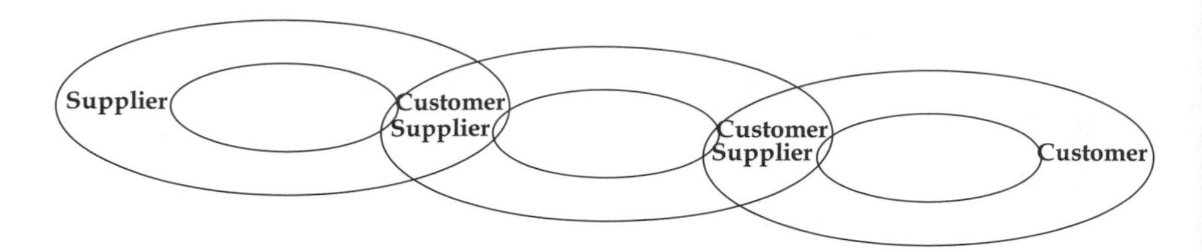

Figure 16.1 Supply chain model.

16.2.1 Demand Forecasting

Demand forecasting is the process in which future needs are projected in order to understand the landscape and horizon. Will the future demand be increasing? By how much? Understanding this future landscape allows for strategic planning to occur; this applies to all of the life cycle elements—acquire, process, distribute, and support. Will a single supplier be adequate? Will more storage be required? What method of distribution will be cost-effective? Typical questions include the general magnitude of demand as well as any lumpiness or seasonality.

16.2.2 Planning and Scheduling

Once the future demand is understood well enough to take action, the commensurate planning and scheduling of demand on suppliers, processes, and distribution can be undertaken. Some of these areas will have adequate or excess capacity (ability to fulfill requirements), whereas others will have insufficient capacity and require scaling up or augmentation. New levels of demand also bring into play other resources or solutions that previously were uneconomical or required switching. The supply chain is a dynamic environment with new suppliers entering the fray as well as customer demand and needs evolving over time. Planning and scheduling must be done at the correct level of detail and frequency, and on the correct elements. Understanding options and constraints in the supply chain is critical to maintain responsiveness and cost-effectiveness.

16.2.3 Strategic Sourcing

Strategic sourcing is the dynamic management of the external supply chain to align value and introduce competitive behavior. It is about performance measurement and supplier development.

16.2.4 Processing

Processing is the internal value stream that you manage and provide resources for. Aligning and optimizing this value stream is the predominant focus of internal supply chain management.

16.2.5 Asset and Inventory Management

Assets are the capital resources that you apply to produce the value you deliver to customers. Minimizing the capital you have tied up while optimizing the fulfillment of your customer's needs is necessary for profitability and longevity. Inventory can be supply chain resources or intellectual property (data-based information). These inventory items also need to be managed so that you have the minimum amount on hand to be responsive to the customer's needs. If you have too little, you fail to satisfy customers, whereas if you have too much, you lose profitability.

16.2.6 Distribution and Logistics

Once you produce a solution for your customer, you need to determine how to get it to them. If you are providing information, then you need to decide if an electronic format is sufficient or if hard copy is required. This may require delivery of the hard copy by land mail, express mail, or courier. The same aspects can be explored with hardware. Do we ship form the plant or from regional warehouses?

16.2.7 Postservice Support

After a service or solution is provided, there will often be some type of service required. This may be a field upgrade such as updating virus definition files, or it may be postservice follow-up such as that performed after placing an employment candidate and surveying what the level of satisfaction is from both parties. The activities may range from standard maintenance items to standard wear-and-tear replacement activities. All of these elements have a bearing on the first two steps of demand forecasting and planning and scheduling, as well as on the process design and provisioning (resourcing) that are required to fulfill the postservice support.

16.3 SUMMARY

The supply chain process covers the life cycle of understanding needs, producing, distributing, and servicing the value chains from customers to suppliers. It applies to all contexts of acquiring resources and transforming them into value for customers. This can be product based, information based, or even resource based, such as capital equipment, employees, raw materials, components, information, expense items, or services. Because of its broad applicability to all aspects of consumption and fulfillment, the supply chain is the ultimate "service" for design consideration.

17

DFSS IN THE SUPPLY CHAIN

17.1 INTRODUCTION

In this chapter, we will cover specific elements of the DFSS road map (see Chapter 5) that lead to the successful deployment of a full supply chain redesign in a legacy environment in which every function was performing its own sourcing and buying and having incestuous relationships with the suppliers. This redesign is more than a reorganization as it represents a major change in strategy and structures, with greater active management of $4 billion spent in 12 countries, including $3.9 billion spent on materials resold to customers and $113 million in expense-related purchases, including consultants, information technology, travel-related services, and office supplies.

17.2 THE PROCESS

In the context of Figure 3.2, we will demonstrate the progression from Stage 1—Idea Creation through Stage 7—Launch Readiness, that is, the implementation of the transformation from a decentralized, relation-based procurement model to a full supply chain model delivering double-digit productivity on the total purchase base. Figure 17.1 (a replica of Figure 5.1) will provide the road map that we will follow.

17.3 STAGE 1—IDEA CREATION

In this stage, we learn why the project was conceived and what its scope and resources are. Based on the existing model of sourcing and integrated supply chain management in its manufacturing businesses, this company wanted to instill the same rigor in its nonmanufacturing businesses to leverage and control spending on all purchased items, including office supplies, rent, information technology purchases and licenses, and travel.

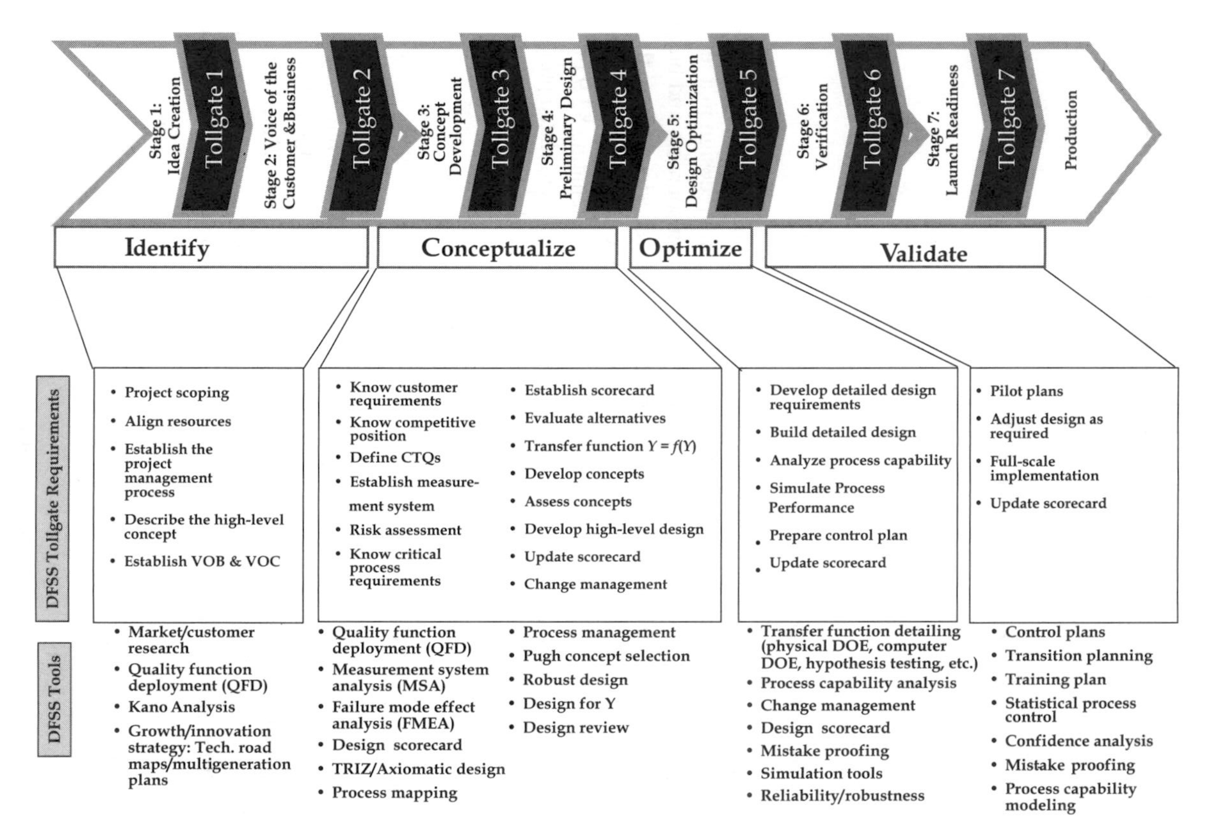

Figure 17.1 DFSS project road map.

17.4 STAGE 2—VOICE OF THE CUSTOMER AND THE BUSINESS

17.4.1 The Voices

The design team realized that the actual customer base would include the corporate parent looking for results, a local management team that wanted no increased organizational cost, and the operational people who needed materials and services. Through interviews laced with a dose of common sense, the following customer wants and needs were derived:

- Price deflation
- Conforming materials
- Fast process
- Affordable process
- Compliant
- Ease of use

Figure 17.2 shows the affinity diagram that was used to determine these high-level needs. You will notice that there are several comments under each high-level heading representing the various voices of the business and customers of the process. It is also of interest that there was one "conforming" element that was "on the frame" and considered nice to have but not a mandate. Also, there were two items that were out of the frame—"take care of my people" and "same suppliers." These items came about because some of the people and suppliers in the existing

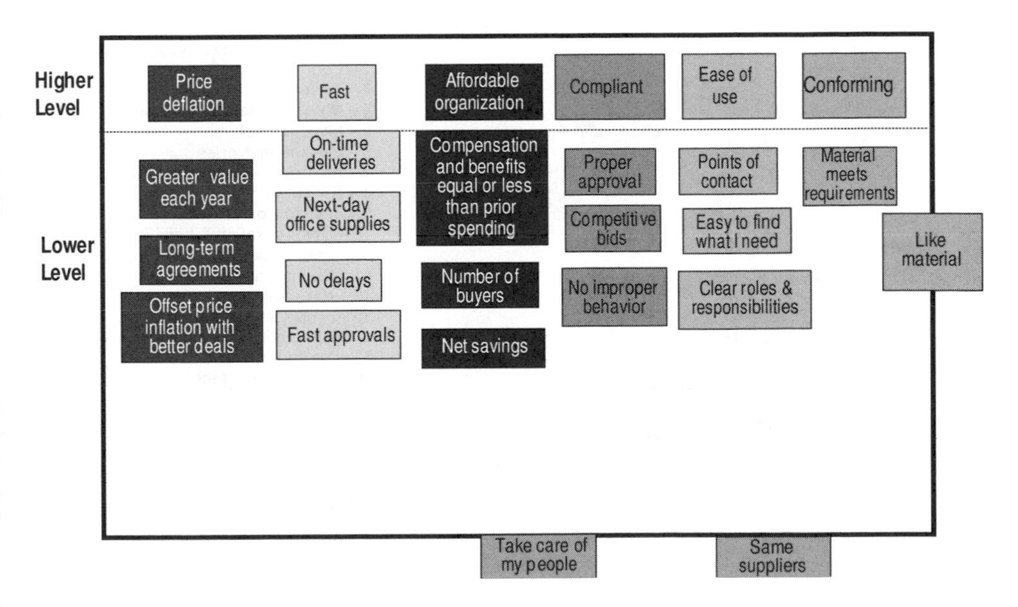

Figure 17.2 Full supply chain affinity diagram.

process were at risk of being displaced. These items were also processed through a Kano analysis in Figure 17.3 to determine the priority of the items. Based on the fact that items 16, 17, and 18 were all delighters and of medium importance, the design team decided to take these out of scope. The existing process did not address these items and it was felt that by centralizing the control of all purchases, they would be improved by the new process without specific design requirements.

We take the remaining elements and begin to build our HOQ 1 as we did in Chapter 7. Figure 17.4 shows the completed HOQ 1 that was built in Chapter 7, Section 7.11.1. The critical-to-satisfaction measures were determined to be:

- Percent year-over-year price deflation
- Percent defective deliveries
- Percent late deliveries
- Number of compliance violations
- Dollar cost per transaction
- Cost to organization/Cost of orders
- Speed of order (hours)
- Number of process steps

17.4.2 Risk Assessment

The team next thought about what program risks would be associated with the transformation to a full supply chain organization structure. Besides the typical

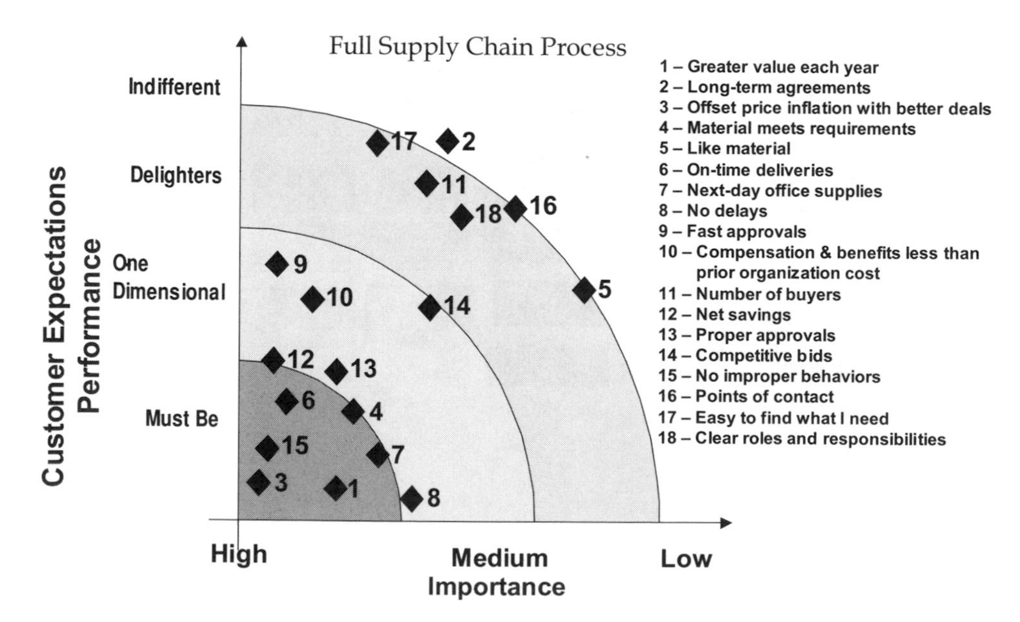

Figure 17.3 Full supply chain Kano analysis.

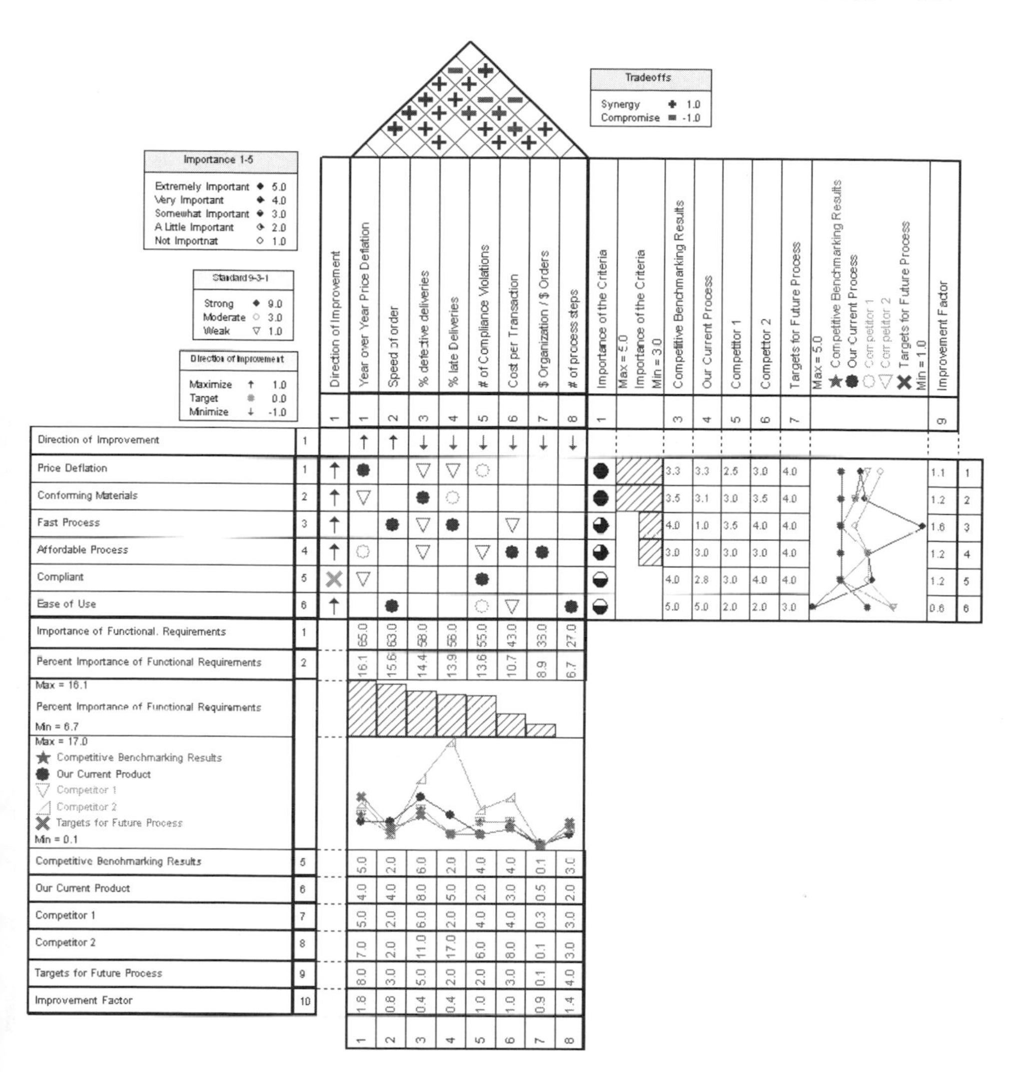

Figure 17.4 HOQ 1 full supply chain process.

risks associated with schedule and resources, the team identified risks in the categories of unstable playing field due to acquisitions, changing corporate strategy, and change in customer base affecting spend. The complete set of risks is shown in the FMEA of Figure 17.5.

Following the normal method of treating all RPNs greater than 125 and all severity of occurrence of 9 or higher, the following became the initial focal point for taking actions:

- Unsatisfied clients due to poor supplier performance resulting in negative support.

Risk Category	Potential Failure Mode	Potential Failure Effects	SEV	Potential Cause	OCC	Current Controls	DET	RPN
Management	Unstable environment	Lack of savings	7	Corporate directions	6	Strong linkage to corporate	2	84
Management	Unstable environment	Lack of resources	9	Loss of customers	2	Monthly operations review	2	36
Management	Unstable environment	Lack of savings	7	Acquisition	8	Deal review board	2	112
Management	Unstable environment	Lack of resources	9	Acquisition	8	Deal review board	2	144
Client	Lack of buy-in	Continued leakage	7	Unwilling to give up control	8	Monthly A/P report	4	224
Client	Unsatisfied	Negative support	9	Poor supplier performance	8	None	8	576
Suppliers	Lack of qualified suppliers	Lack of savings	7	Unable to qualify suppliers	3	None	4	84
Suppliers	Lack of qualified suppliers	Diminshed performance	6	Too small at new supplier	4	Detailed analysis of supplier	2	48
Suppliers	Switching costs	Diluted net savings	5	Long-term agreements	5	Supplier notification	6	150
Technology	Availability	Lack of measurements	5	IT Resources	5	IT project pipeline	3	75
Technology	Affordable	Diluted net savings	5	Conversion/implementation	2	Appropriation process	2	20
Project	Lack of qualified resources	Delayed benefits	7	Unwilling to release individuals	6	Manpower review	3	126
Project	Lack of qualified resources	Continued legacy behavior	8	Unable to find qualified	3	Recruiters	2	48

Figure 17.5 Full supply chain project FMEA.

- Lack of buy-in from clients due to their unwillingness to give up control that results in leakage in spending.
- High switching costs to change suppliers due to long-term contracts or unique requirements that results in diluted savings.
- Lack of critical resources due to a major acquisition, which deflects assigned resources from leveraging supplier performance.
- Lack of qualified resources for new competency model.
- Lack of resources due to change in customer base.

The team needed to ensure that any new suppliers or suppliers that lowered their pricing maintained the level of performance that the business was accustomed to. In regard to the leakage, the team needed to review the accounts payable data for supplier and cost centers to see if spending could be accounted for. The other items were deemed to require watching but no further action at this time.

17.5 STAGE 3—CONCEPT DESIGN

The current organizational structure had purchasing resources basically doing transactional work, with ad hoc reporting capability. This structure was replicated in each country. When the team completed the activity of HOQ 2, the output depicted in Figure 17.6 was obtained.

The prioritized functional requirements with the corresponding CTS in parentheses are:

- Resources to achieve deflation (percent year-over-year price deflation)
- Resources to achieve delivery quality (percent defective deliveries)
- Resources to achieve on time delivery (percent late deliveries)
- Organizational design (spend/organizational cost)
- Supplier support (spend/organizational cost)
- Process design (number of process steps)
- Approval speed (speed of order)
- Measures of delivery quality (percent defective deliveries)
- Delivered quality improvement methods (percent defective deliveries)
- Amount of spending controlled (spending/organizational cost)
- Staffing methods (number of compliance violations)
- Organizational cost (cost per transaction)
- Systems process cost (cost per transaction)
- Approval levels (number of process steps)
- Methods for improving on-time delivery (percent late deliveries)
- Measures of on-time delivery (percent late deliveries)
- System availability (speed of order)

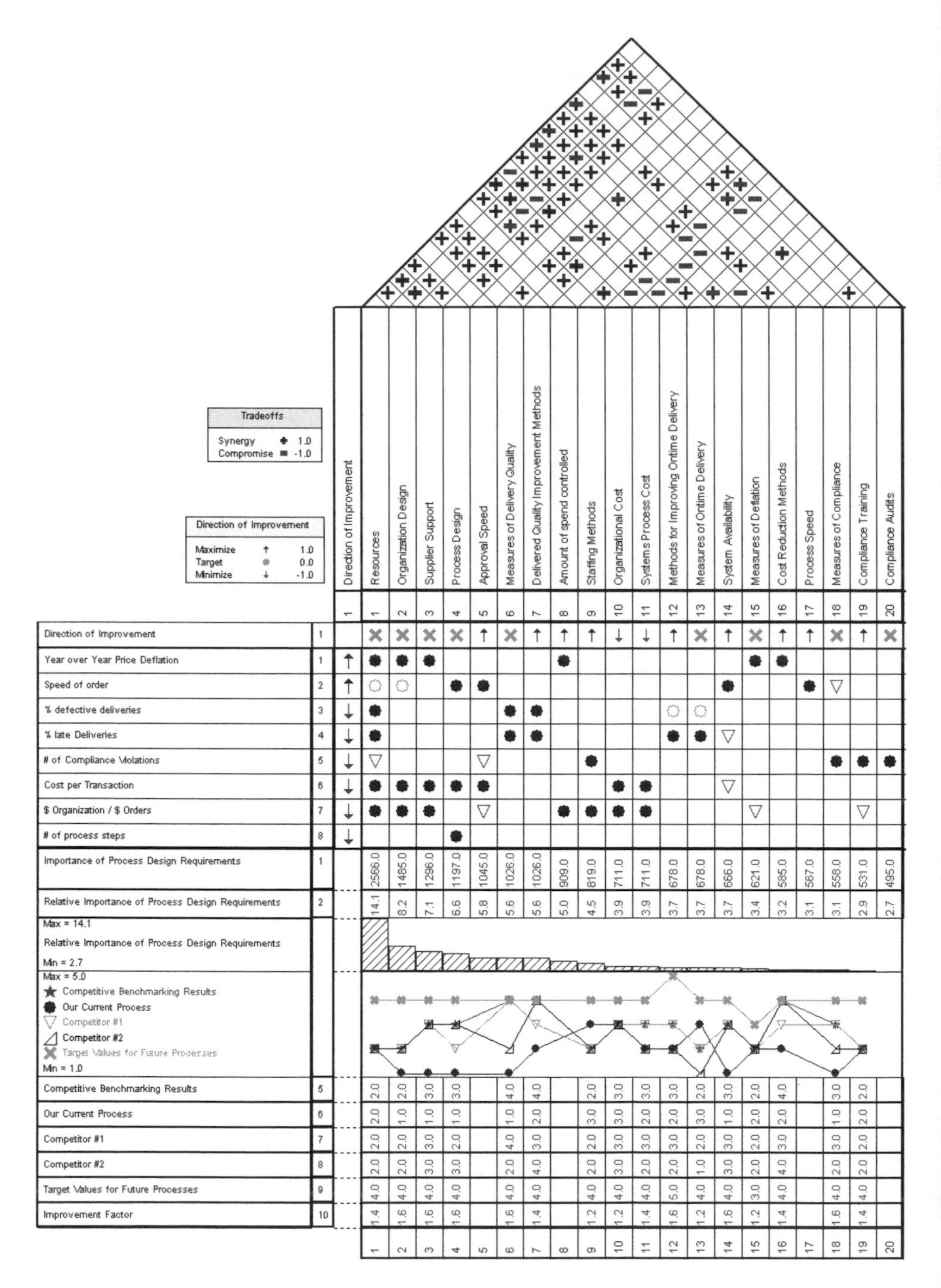

Figure 17.6 Full supply chain HOQ2.

		1	2	3	4	5	6	7	8
1	Importance of Functional. Requirements	65.0	63.0	58.0	56.0	55.0	43.0	36.0	27.0
2	Percent Importance of Functional Requirements	16.1	15.6	14.4	13.9	13.6	10.7	8.9	6.7
	Max = 16.1 Percent Importance of Functional Requirements Min = 6.7								
	Max = 17.0 Competitive Benchmarking Results Our Current Product Competitor 1 Competitor 2 Targets for Future Process Min = 0.1								
5	Competitive Benchmarking Results	5.0	2.0	6.0	2.0	4.0	4.0	0.1	3.0
6	Our Current Product	4.0	4.0	8.0	5.0	2.0	3.0	0.5	2.0
7	Competitor 1	5.0	2.0	6.0	2.0	4.0	4.0	0.3	3.0
8	Competitor 2	7.0	2.0	11.0	17.0	6.0	8.0	0.1	3.0
9	Targets for Future Process	8.0	3.0	5.0	2.0	2.0	3.0	0.1	4.0
10	Improvement Factor	1.8	0.8	0.4	0.4	1.0	1.0	0.9	1.4

Standard 9-3-1		
Strong	◆	9.0
Moderate	◇	3.0
Weak	▽	1.0

Figure 17.6 (*continued*).

- Measures of deflation (percent year-over-year price deflation)
- Cost reduction methods (percent year-over-year price deflation)
- Process speed (speed of order)
- Measures of compliance (number of compliance violations)
- Compliance training (number of compliance violations)
- Compliance audits (number of compliance violations)

Also notice that what we would like to see in the correlation room (room 4 in the HOQ) matrix is as close as possible to the diagonal of strong relationships shown in Figure 17. 7. This will facilitate reduction or elimination of cascading the coupling

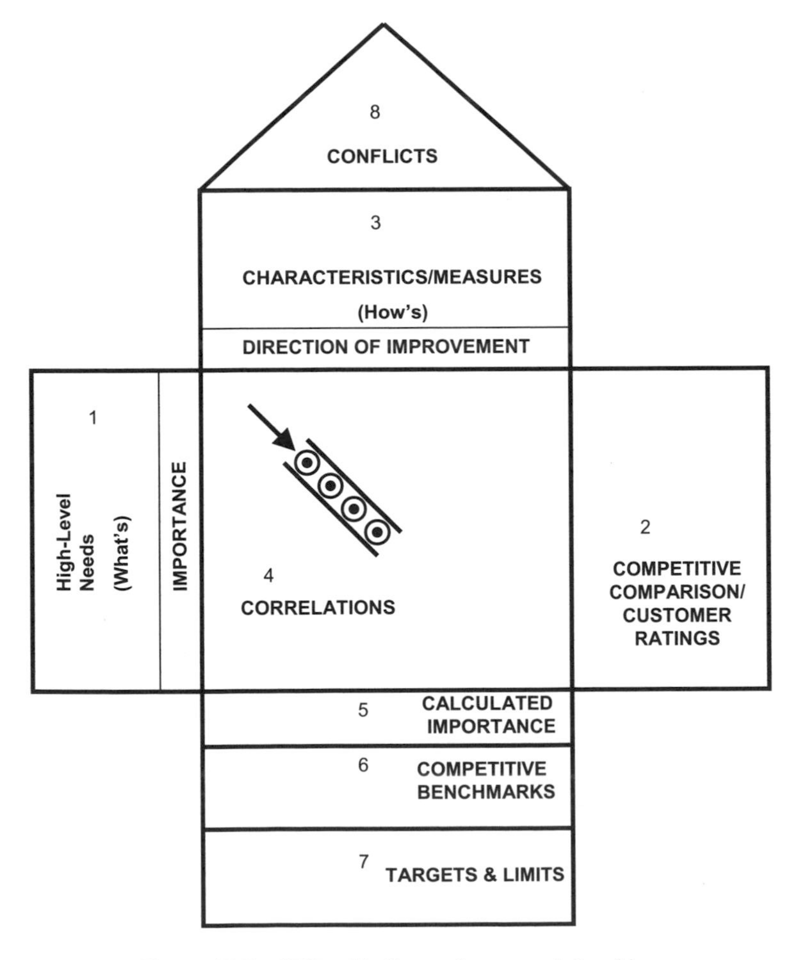

Figure 17.7 QFD with diagonal strong relationships.

vulnerability to HOQ 2 and beyond (recall Section 8.5.2 and Figure 8.13). At this point, we will highlight the following submatrices of room 4.

Case 1: Uncoupled Case
$FR1$ = "year-over-year price deflation" and $FR2$ = "speed of order" are both uncoupled in $DP3$ = "supplier support" and $DP4$ = "process design." The design matrix is given as

$$\begin{Bmatrix} FR1 \\ FR2 \end{Bmatrix} = \begin{bmatrix} X & 0 \\ 0 & X \end{bmatrix} \begin{Bmatrix} DP3 \\ DP4 \end{Bmatrix} \tag{17.1}$$

That is, the design team is free to decide in which way they utilize the design parameters to satisfy the functional requirements. There is no design sequence to follow under the assumption that only these two DPs will be used for the two FRs mentioned above.

Case 2: Decoupled Case
A prime example is when $FR6$ = "cost per transaction" and $FR7$ = "$ organization/$ order" are decoupled in $DP2$ = "organization design" and $DP9$ = "staffing methods." The design matrix is given as

$$\begin{Bmatrix} FR6 \\ FR7 \end{Bmatrix} = \begin{bmatrix} X & 0 \\ X & X \end{bmatrix} \begin{Bmatrix} DP2 \\ DP9 \end{Bmatrix} \tag{17.2}$$

That, is the design team has to satisfy $FR6$ first, using $DP2$, fix $DP2$ when they satisfy $FR7$, and then use $DP9$ *only* to satisfy $FR7$, as was discussed in the context of Figure 8.13.

Case 3: Coupled Case
In this case, we find that $FR3$ = "percent defective deliveries" and $FR4$ = "percent late deliveries" are coupled in $DP7$ = "delivered quality improvement methods" and $DP6$ = "measures of delivery quality." The design matrix is given by

$$\begin{Bmatrix} FR3 \\ FR4 \end{Bmatrix} = \begin{bmatrix} X & X \\ X & X \end{bmatrix} \begin{Bmatrix} DP7 \\ DP6 \end{Bmatrix} \tag{17.3}$$

That is, the best that the team can do is to compromise by satisfying the two FRs if they want to live with that. The other option is to use the TRIZ contradiction table (Appendix 9.A) to make either or both of the off-diagonal elements zero or approximately zero. Which TRIZ solution principle is applicable here? The answer is left as an exercise.

The above three cases are just examples of what the team may face in resolving design vulnerabilities. There are many other examples of the case of 2×2 matrices. Notice that we concentrate only on strong relationships and we did not examine 3×3, 4×4, or other size submatrices. For more details on such handling, see El-Haik (2005).

17.5.1 Concept Development

From the list of functional requirements (FRs), concepts for the fulfillment of the requirements can be developed. We also have trade-off compromises for several of the FRs; for example, the number of resources is a trade-off with the organizational cost. These trade-offs are couplings that are design vulnerabilities that will affect the next HOQ. It is best if the team can solve them at this stage with some creative thinking. Because we are dealing with multiple countries, it was determined that a four-tier approach should be pursued. First, any commodity that could be controlled at the corporate level would be. The next tier would be the items that were business specific and could be controlled at the business unit level. The third tier would be by major geographical area. There were three major areas defined: North America, Europe, and the Far East. The fourth tier would be the country-specific spending that could not be consolidated.

Through benchmarking, it was determined that a commodity team approach would provide the best solution for achieving the delivery, quality, and cost-reduction objectives while staying close to the needs of the client and be able to competitively leverage the supply chain.

At the time this project was active, there were limited options for some of the technological solutions. The morphological matrix of possible solutions is summarized in Figure 17.8. For each functional requirement, possible solutions are brainstormed and placed into the matrix.

The next step was to look at these options and choose the ones that seemed to be a best fit. Figure 17.9 depicts what will be labeled Concepts 1 and 2. The way to interpret this is that the functional requirement of organizational design will be fulfilled by a centralized commodity team, supplier support will be provided by internal company resources, and so on. For concept 1 the team plans to rely on the existing systems while for concept 2 they would rely on an Ariba Catalogue system for expense item management

Figure 17.10 depicts what will be called Concept 3. The major difference between Concept 1 and Concept 2 is the use of contracted resources instead of company resources. This means that you can acquire experienced talent but at the cost of affordability and use of company personnel.

Once a set of concepts has been generated, they can be evaluated through the Pugh selection method (see Chapter 8). The initial datum is generally the existing system. The legend on the right-hand side outlines what the other concepts are. The CTSs are taken from the HOQ 1 and are used to evaluate how the concepts fulfill the CTSs. When a concept provides more functionality to the CTS compared to the datum, it is scored as "+"; if it is the same, it is scored as "S"; and if it offers diminished functionality, it is scored as "–." The team evaluated the four concepts and found that Concept 1 provided the best solution (Figure 17.11).

Concept 2 provided a good solution, as evidenced by the number of +'s and S's. However, it was not as strong as Concept 1, as there are five +'s in Concept 1, is 5 versus three in Concept 2. Concept 3 was definitely inferior, as it contains five minuses and only a single S.

Functional Requirements	Option A	Option B	Option C	Option D
Organization Design	By location	By commodity team		
Organization Design	Decentralized	Centralized	Contracted	
Supplier Support	Company resources	Contracted resources	Supplier self-development	
Delivered Quality Improvement Methods	Problem solving	DMAIC	Lean	Contractual requirement
Measures of Delivery Quality	Manual	Automatic		
Measures of On-time Delivery	Manual	Automatic		
Methods for Improving On-time Delivery	Traditional problem solving	DMAIC	Lean	Contractual requirement
Measures of Deflation	Manual	Automatic		
Cost-Reduction Methods	Competitive bidding	Consolidation	DMAIC/Lean	Contractual requirement
Measures of Compliance	Failure reporting	Audit of activity		
Compliance Training	Formal class room	Web-based learning		
Process Design	N/A			
Number of resources	N/A			
Staffing Methods	N/A			
Approval Policies	N/A			
Levels of Approval	N/A			
Systems Cost	N/A			
Staffing Cost	N/A			
System Availability	N/A			

Figure 17.8 Full supply chain morphological matrix.

Functional Requirements	Option A	Option B	Option C	Option D
Organization Design	By location	By commodity team		
Organization Design	Decentralized	Centralized	Contracted	
Supplier Support	Company resources	Contracted resources	Supplier self-development	
Delivered Quality Improvement Methods	Problem solving	DMAIC	Lean	Contractual requirement
Measures of Delivery Quality	Manual	Automatic		
Measures of On-time Delivery	Manual	Automatic		
Methods for Improving Ontime Delivery	Traditional problem solving	DMAIC	Lean	Contractual requirement
Measures of Deflation	Manual	Automatic		
Cost-Reduction Methods	Competitive Bidding	Consolidation	DMAIC/Lean	Contractual requirement
Measures of Compliance	Failure reporting	audit of activity		
Compliance Training	Formal classroom	Web-based learning		
Process Design	N/A			
Number of Resources	N/A			
Staffing Methods	N/A			
Approval Policies	N/A			
Levels of Approval	N/A			
Systems Cost	N/A			
Staffing Cost	N/A			
System Availability	N/A			

Figure 17.9 Full supply chain morphological matrix Concept 1.

Functional Requirements	Option A	Option B	Option C	Option D
Organization Design	By location	By commodity team		
Organization Design	Decentralized	Centralized	Contracted	
Supplier Support	Company resources	Contracted resources	Supplier self-development	
Delivered Quality Improvement Methods	Problem solving	DMAIC	Lean	Contractual requirement
Measures of Delivery Quality	Manual	Automatic		
Measures of On-time Delivery	Manual	Automatic		
Methods for Improving On-time Delivery	Traditional problem solving	DMAIC	Lean	Contractual requirement
Measures of Deflation	Manual	Automatic		
Cost Reduction Methods	Competitive bidding	Consolidation	DMAIC/Lean	Contractual requirement
Measures of Compliance	Failure reporting	Audit of activity		
Compliance Training	Formal classroom	Web-based learning		
Process Design	N/A			
Number of resources	N/A			
Staffing Methods	N/A			
Approval Policies	N/A			
Levels of Approval	N/A			
Systems Cost	N/A			
Staffing Cost	N/A			
System Availability	N/A			

Figure 17.10 Full supply chain morphological matrix Concept 3.

Put +, -, or S in each cell to represent whether concept is SIGNIFICANTLY better, worse, or same as the datum concept.

Work this way --->

CTSs \ Concept	Import. Rating	0	1	2	3
Price Deflation	H		+	+	+
Conforming Materials	H	D	+	S	-
Fast Process	M	A	S	S	-
Affordable Process	M	T	+	S	-
Compliant	H	U	+	+	-
Ease of Use	M	M	+	+	-
			5	3	1
			0	0	5
S			1	3	0

Datum	Concept Summary
1	Commodity Team with existing systems
2	Commodity Teams with Ariba Catalog Systems
3	Commodity team with contratced resources

Figure 17.11 Full supply chain morphological Pugh matrix.

17.6 STAGE 4—PRELIMINARY DESIGN

Once the conceptual design is selected, the next phase entails detailed design to the level of design parameters that is done in HOQ 3.

The design parameters from HOQ 3 were developed in Chapter 7 and are repeated here. The functional requirements are in parentheses:

- Type of resources (resources)
- Number of resources (resources)
- Type of resources (organization design)
- Quantity of resources (organization design)
- Location of resources (organization design)
- Type of support (supplier support)
- Frequency of support (supplier support)
- Magnitude of support (supplier support)
- Intuitive (process design)
- Workflow (process design)
- Approval levels (approval speed)
- Speed of approval levels (approval speed)
- By supplier (measures of delivery quality)
- By commodity (measures of delivery quality)
- By part (measures of delivery quality)
- By buyer (measures of delivery quality)
- Lean methods (methods for improving delivery quality)
- Variation reduction methods (methods for improving delivery quality)
- Contract delivery quality terms (methods for improving delivery quality)
- Consolidated buying (amount of spend controlled)
- Background checks (staffing methods)
- Screening methods (staffing methods)
- Compensation and benefits (organizational cost)
- Travel and living (organizational cost)
- Depreciation cost (systems process cost)
- Telecommunication cost (systems process cost)
- Mail cost (systems process cost)
- Lean methods (methods for improving on-time delivery)
- Variation reduction methods (methods for improving on-time delivery)
- Contract on time delivery terms (methods for improving on time delivery)
- By supplier (measures of on-time delivery)
- By commodity (measures of on-time delivery)
- By part (measures of on time delivery)

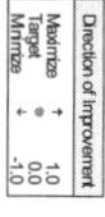

Direction of Improvement
Maximize 1.0
Target 0.0
Minimize -1.0

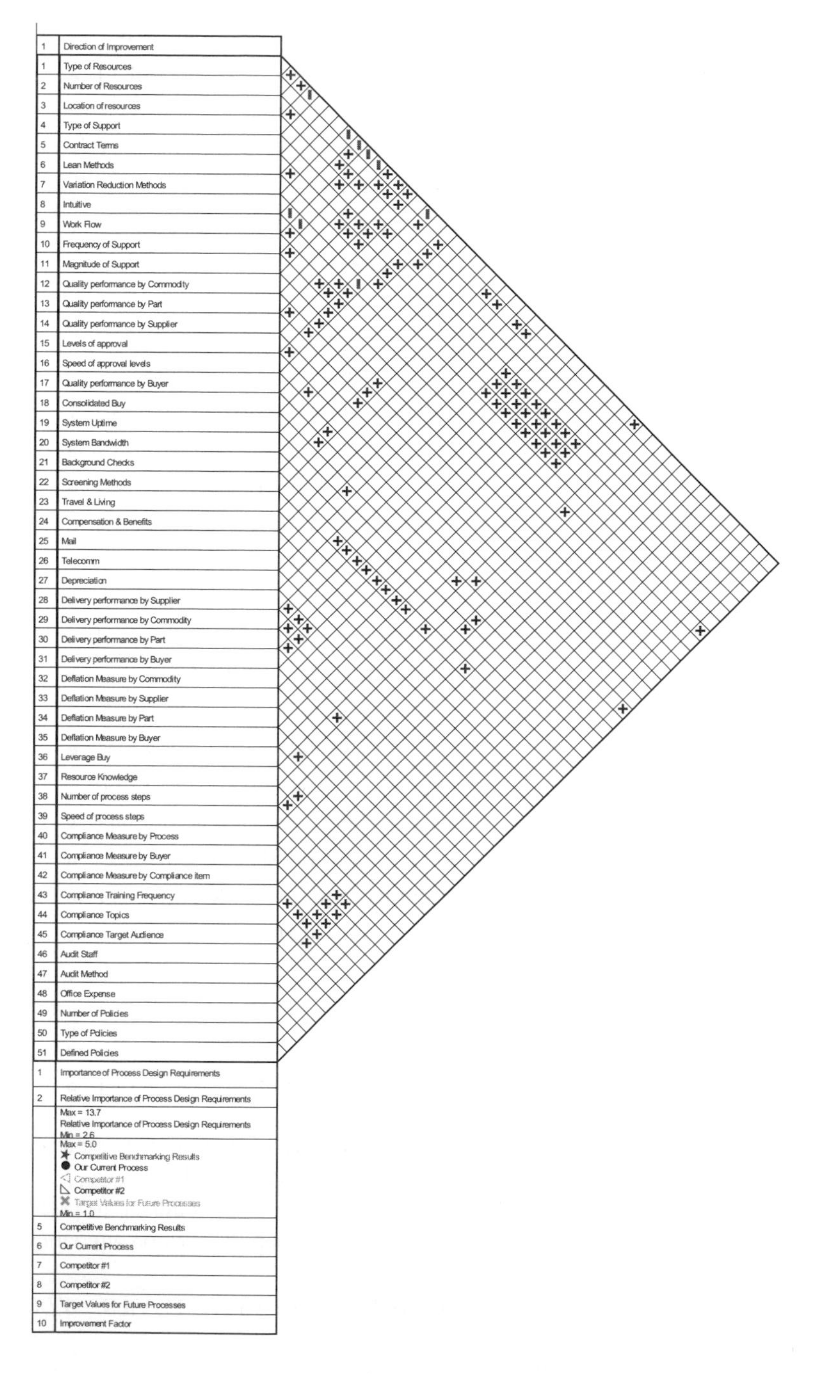

1 Direction of Improvement
1 Type of Resources
2 Number of Resources
3 Location of resources
4 Type of Support
5 Contract Terms
6 Lean Methods
7 Variation Reduction Methods
8 Intuitive
9 Work Flow
10 Frequency of Support
11 Magnitude of Support
12 Quality performance by Commodity
13 Quality performance by Part
14 Quality performance by Supplier
15 Levels of approval
16 Speed of approval levels
17 Quality performance by Buyer
18 Consolidated Buy
19 System Uptime
20 System Bandwidth
21 Background Checks
22 Screening Methods
23 Travel & Living
24 Compensation & Benefits
25 Mail
26 Telecomm
27 Depreciation
28 Delivery performance by Supplier
29 Delivery performance by Commodity
30 Delivery performance by Part
31 Delivery performance by Buyer
32 Deflation Measure by Commodity
33 Deflation Measure by Supplier
34 Deflation Measure by Part
35 Deflation Measure by Buyer
36 Leverage Buy
37 Resource Knowledge
38 Number of process steps
39 Speed of process steps
40 Compliance Measure by Process
41 Compliance Measure by Buyer
42 Compliance Measure by Compliance Item
43 Compliance Training Frequency
44 Compliance Topics
45 Compliance Target Audience
46 Audit Staff
47 Audit Method
48 Office Expense
49 Number of Policies
50 Type of Policies
51 Defined Policies
1 Importance of Process Design Requirements
2 Relative Importance of Process Design Requirements
Max = 13.7
Relative Importance of Process Design Requirements
Min = 2.6
Max = 5.0
Competitive Benchmarking Results
Our Current Process
Competitor #1
Competitor #2
Target Values for Future Processes
Min = 1.0
5 Competitive Benchmarking Results
6 Our Current Process
7 Competitor #1
8 Competitor #2
9 Target Values for Future Processes
10 Improvement Factor

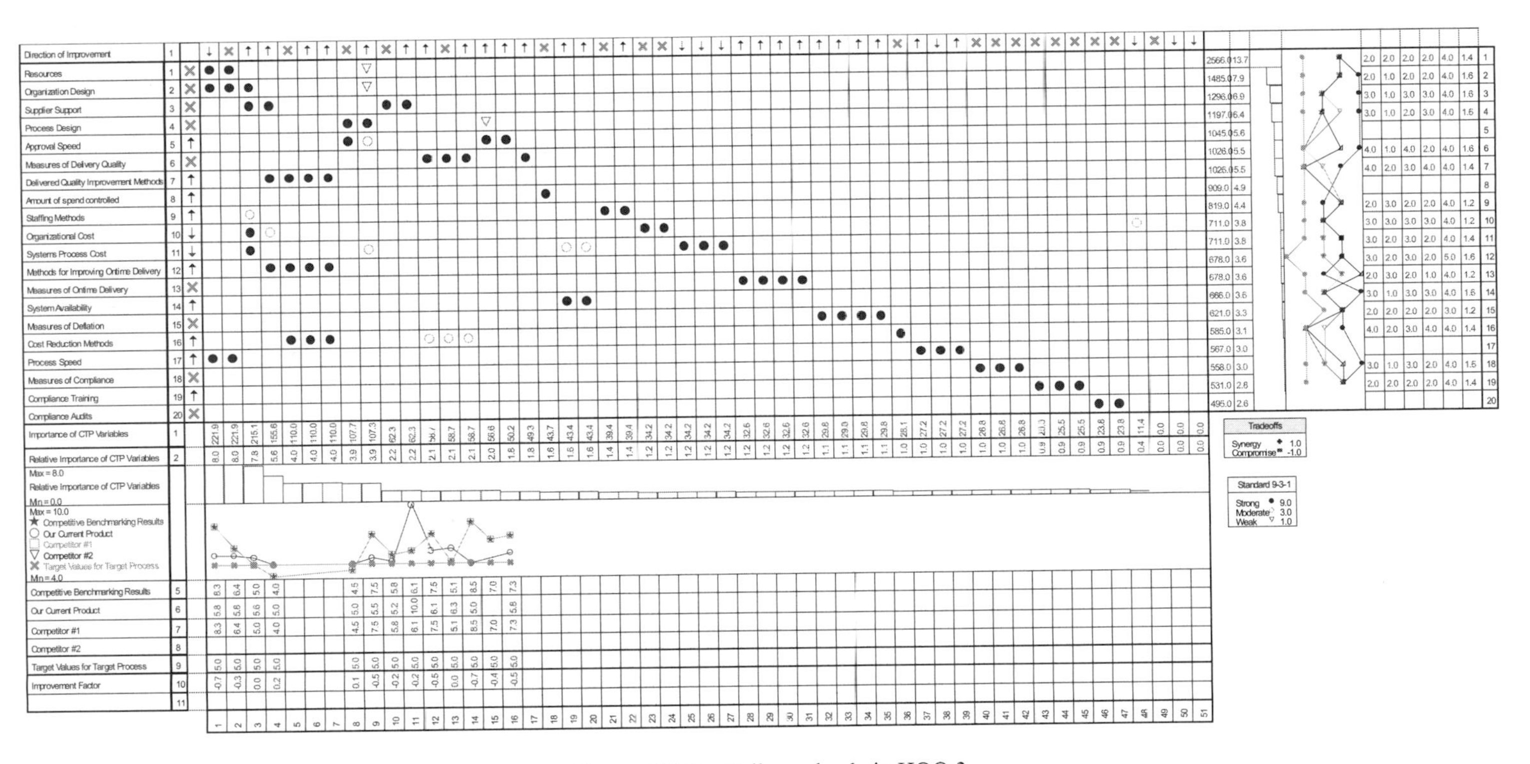

Figure 17.12 Full supply chain HOQ 3.

- By buyer (measures of on time delivery)
- Uptime (system availability)
- Bandwidth (system availability)
- By supplier (measures of deflation)
- By commodity (measures of deflation)
- By part (measures of deflation)
- By buyer (measures of deflation)
- Leverage buy (cost-reduction methods)
- Supplier deflation performance (cost-reduction methods)
- Lean methods (cost-reduction methods)
- Variation-reduction methods (cost-reduction methods)
- Contract cost-reduction terms (cost-reduction methods)
- Number of process steps (process speed)
- Speed of each process step (process speed)
- Resource quantity (process speed)
- Resource skill (process speed)
- By compliance category (measures of compliance)
- By buyer (measures of compliance)
- By process (measures of compliance)
- Target audience (compliance training)
- Content (compliance training)
- Frequency (compliance training)
- Auditors (compliance audits)
- Audit method (compliance audits)

Most of these parameters come directly from the morphological matrix; however, the ones in the matrix that are denoted N/A require another level of decomposition. For example, the sixth DP is "Type of Support," aligned with the FR "Supplier Support." The types of support that may be required are order status, invoice status, productivity support, ramp-up support, and so on. Ramp-up support refers to the situation in which you are ordering a new part from an existing supplier or a new supplier ramps up to full production. The decomposition complexity of some of these can have a major impact on the type and quantity of resources that will be required. In a Six Sigma deployment, the tactical support elements of order status and invoice status should not be required, since all of the elements of the process work the first time. This puts more focus on the strategic support items such as productivity support and ramp-up support. It is at this level that the design scorecard could provide valuable guidance in terms of the capacity planning for the support elements. Figure 17.13 shows a top-level design scorecard. If you look at the rolled throughput yield (RTY) column, you can quickly see that the two items that are not performing at high levels are productivity and the on-time delivery. These two

Full Supply Chain Processes	Performance	
	DPM	RTY
On-Time Delivery	45,500	95.4%
Delivered Quality	10,050	99.0%
Productivity	85,716	91.4%
Invoice Efficiency	9,950	99.0%
Processing Cost per Order	3	100.0%

Figure 17.13 Top-level design scorecard.

items will require the support of additional resources, based on their performance levels.

17.7 STAGE 5—DESIGN OPTIMIZATION

Once the basic design parameters were decided on, the team began fitting the solutions to the commodities and locations, and performing trade-offs between scope and coverage. As mentioned earlier, following the deployment strategy of the hierarchal assignment of responsibility to the highest level of consolidation requires a massive reshuffling of responsibilities. The other transfer functions that had to be considered in this realignment were the number of suppliers to be managed by any one individual and the number of transactions that would be performed on a weekly basis per supplier. When all of the trade-offs and capacity planning were complete, the clean-sheet organization structure depicted in Figure 17.14 was created.

In addition to the organizational structure there was also a process flow created for the sourcing aspect. Figure 17.15 depicts the macro-level flow of this process. The process flow is generic enough to apply to any purchased commodity, yet it has many of the details that align with the design parameters from the HOQ 3.

Another element was the selection of supplier strategy. Figure 17.16 shows a methodology that was prevalent in the parent company and was adopted for use in this deployment. Based on the criticality of the items being procured and the type of supplier base, it is simple to pick the most effective strategy.

What is meant by value of product or service in Figure 17.16 is the relevant value these items have in the value delivery stream. A personal computer would have a higher value than, say, a paper clip, and an active component in a product would most likely have a higher value than a label on the product. Once the value of the product or service is selected, it becomes a simple task to determine how many "qualified suppliers" exist to provide this product or service. Qualified means that the supplier meets the screening criteria that the business sets for doing business with reputable, reliable suppliers capable of providing conforming products and services.

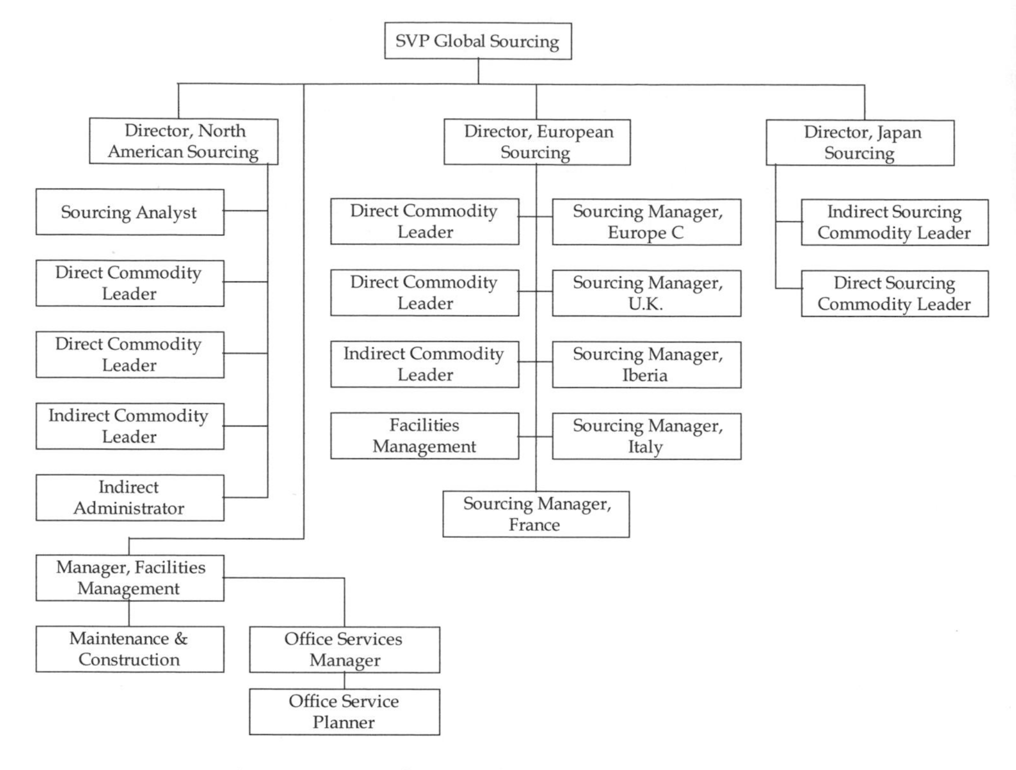

Figure 17.14 Full supply chain new organization structure.

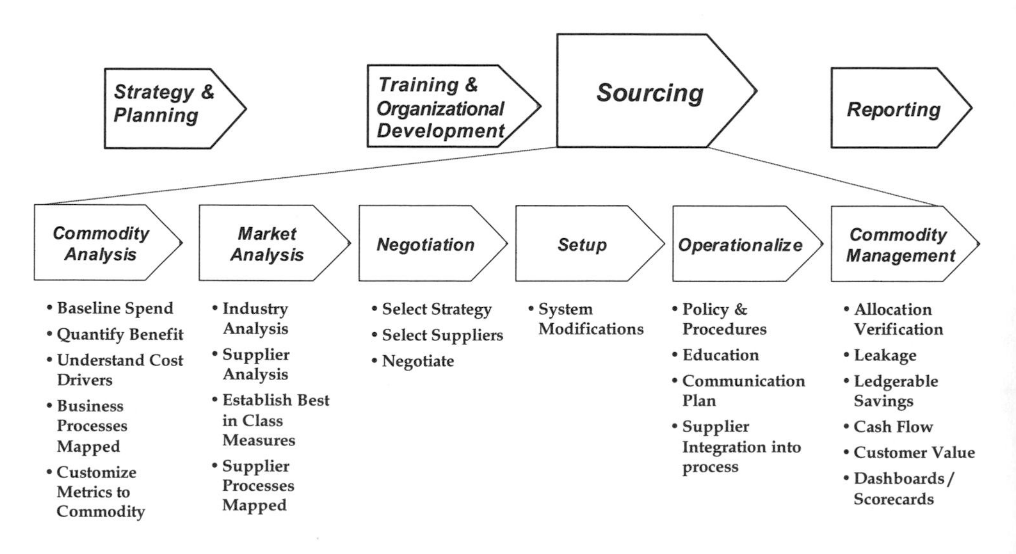

Figure 17.15 Sourcing macro process.

Figure 17.16 Supplier segmentation and strategy.

So completing the optimization phase provided the organizational structure, the capacity loading and alignment, the process flow, and specific solutions for the tools used to drive the speed, cost, and quality performance.

17.8 STAGE 6—VERIFICATION

This deployment was not piloted on a small scale because it had already been demonstrated at other business locations. The risk was that the commodities in this business would not respond in a similar fashion. The new organization was launched at the beginning of the fiscal year, following a ramp-up of the resources and some orientation and preliminary setup (commodity plans) by a small task force. Each of the positions had a competency model defined for them and existing employees where allowed to apply for these positions. All positions were filled based on screening for the following:

- Productivity/expense reduction history
- Negotiation ability
- Integrity
- Strategic sourcing experience
- Knowledge of competitive bidding strategies
- Knowledge of purchasing law

- Risk assessment ability
- Six Sigma knowledge and experience

Each of the above criteria can be seen in the flowdown of requirements such as "compliant," "price deflation," and other wants and needs. As a result of this screening based on the new requirements, only 40% of the positions were filled with existing employees. The nonselected individuals were incorporated into other open positions since this was a high-growth business (+20% per year).

17.9 STAGE 7—LAUNCH READINESS

With only 40% of the existing employees being placed into the new positions, the transformation to the new organization structure was going to meet with some resistance and there might even be a few individuals motivated to find fault with it. In order to manage this conversion, the company's change management process was utilized to prepare a compelling reason to believe a change was required, and during the first several months and at the end of one year, the dramatic results were shared broadly. Soon, even the doubters became believers. Figure 17.17 shows a neutralized version of the change message.

17.10 PRODUCTION

The new organization was deployed and monthly management reviews were held to help the team understand the activities and planned benefits. Later, the reporting be-

Based on today s market realities of deflation, we are seeing increasing compression of our margins from customers for services. In reaction to these market pressures we must radically re-engineer the Sourcing activities to drive balance between purchase costs and revenues.

The framework for this re-engineering exists in many of the other business units who have faced this reality earlier than the services businesses. Their efforts indicate that there will be no correlation between prior sourcing activities and new activities.

As we redefine the activities, new position guides will be created and posted that define the new work activities as well as competencies for the individuals. Every incumbent will be given consideration for the jobs that are close in fit to skills, background and career goals.

In the case that you are no longer interested in the type of work the new Sourcing objectives present, you will be given every opportunity to place into the open needs of the company.

Figure 17.17 Change management message.

came strictly report based, without any formal meetings. During the first year, the group costs were increased by 14%, even though the head count was reduced by 19%, and the benefits exceeded $32,000,000 ($12,000,000 on the expense items and $20,000,000 on the rebilling), a 1000% increase in benefits. Another way of looking at this is that savings per head went from $80,000 to over $1,000,000.

This increased performance also translated into improved customer loyalty, as performance from their perspective improved dramatically, and the company was better able to maintain profit margins while being price sensitive to the paying customer.

17.11 SUMMARY

Following the DFSS road map presented in Chapter 5 helps accelerate a new process introduction and aligns the benefits for customers and stakeholders. The tools and tollgates allow for risk management, creativity, and a logical documented flow that is superior to the "launch and learn" mode that many new organizations, processes, or services are deployed with. Not all projects will use all of the DFSS tools and methodology, and some will use some to a greater extent than others. In this case, the design scorecard was shown at only the highest level. In some service processes, it is required to show the process and infrastructure levels as well as the top level. The transfer functions in this example were simple dollar savings per head and dollar cost per head.

REFERENCES

Abbatiello, N. (1995), *Development of Design for Service Strategy,* MS Thesis, University of Rhode Island.

Akao, Y. (1972), "New Product Deployment and Quality Assurance—Quality Deployment System" (in Japanese), *Standardization and Quality Control, 25,* 14, 7–14.

Akao, Y. (1997), "QFD: Past, Present, and Future," www.qfdi.org/qfd_history.pdf, Proceedings of the the 3rd Annual International QFD Conference at Linkoping, October 1997.

Alexander, C. (1964), *Notes on the Synthesis of Form,* Harvard University Press, Cambridge, MA.

Al-Aomar, R. (2002), "A Methodology for Determining System and Process-level Manufacturing Performance Metrics," SAE 2002, *Transactions Journal of Materials & Manufacturing, V111-5,* pp. 1043–1050.

Al-Aomar, Raid (2000), "Model-Mix Analyses with DES," in *Proceedings of the 2000 Winter Simulation Conference,* pp. 1385–1392, Orlando, FL.

Almquist, W. (2001), "Boost Your Marketing ROI with Experimental Design," *Harvard Business Review,* October.

Altshuler, G. S. (1988), *Creativity as Exact Science,* Gordon & Breach, New York.

Altshuler, G. S. (1990), "On the Theory of Solving Inventive Problems," *Design Methods and Theories, 24,* 2, 1216–1222.

Anderson, D. M. (1990), *Design for Manufacturability,* CIM Press, Lafayette, CA.

Arciszewsky, T. (1988), "ARIZ 77: An Innovative Design Method," *Design Methods and Theories, 22,* 2, 796–820.

Arimoto, S., Ohashi, T., Ikeda, M., and Miyakawa, S. (1993), "Development of Machining Productivity Evaluation Method (MEM)," *Annals of CIRP, 42,* 1, 119–1222.

Ashby, W. R. (1973), "Some Peculiarities of Complex Systems," *Cybernetic Medicine,* 9, 1–7.

Automotive Industry Action Group (AIAG) (2001), *Potential Failure Mode and Effects Analysis (FMEA) Reference Manual,* 3rd ed., AIAG, Southfield, MI.

Beitz, W. (1990), "Design for Ease of Recycling (Guidelines VDI 2243)," in *ICED Proceedings 90,* Dubrovnik, Heurista, Zurich.

Boothroyd, G. and Dewhurst, P. (1983), *Product Design for Assembly Handbook,* Boothroyd-Dewhurst Inc., Wakefield, RI.

Boothroyd, G., Dewhurst, P., and Knight, W. (1994), *Product Design of Manufacture and Assembly,* Marcel Dekker, New York.

Bothe, D. R. (1997), *Measuring Process Capabilities,* McGraw-Hill, New York.

Box, G. E. P. and Draper, N. R. (1987), *Empirical Building and Response Surfaces,* Wiley, New York.

Box, G. E. P., Hunter, W. G., and Hunter, J. S. (1978), *Statistics for Experimenters,* Wiley, New York.

Brejcha, M. F. (1982), *Automatic Transmission,* 2nd ed., Prentice-Hall, Englewood Cliffs, NJ.

Bulent, D. M., Kulak, O. and Tufekci, S. (2002), "An Implementation Methodology for Transition from Traditional Manufacturing to Cellular Manufacturing Using Axiomatic Design," in *Proceedings of the International Conference on Axiomatic Design,* Cambridge, MA.

Burchill, G. and Brodie, C. H. (1997), *Voices Into Choices: Acting on the Voice of the Customer,* Joiner Associates, Inc., Madison WI.

Bussey, L. E. (1998), *The Economic Analysis of Industrial Projects,* Prentice-Hall, Upper Saddle River, NJ.

Carnap, R. (1977), *Two Essays on Entropy,* University of California Press, Berkeley.

Cha, J. Z. and Mayne, R. W. (1987), "Optimization with Discrete Variables via Recursive Quadratic Programming," in *Advances in Design Automation—1987,* American Society of Mechanical Engineers, September, 1987.

Chase, K. W. and Greenwood, W. H. (1988), "Design Issues In Mechanical Tolerance Analysis," *Manufacturing Review, 1,* 1, 50–59.

Chen, E. (2002), "Potential Failure Mode and Effect Analysis: Enhancing Design and Manufacturing Reliability," A George Washington Center for Professional Development Course.

Chen, G. and Kapur, K. C. (1989), "Quality Evaluation System Using Loss Function," in *Proceedings of International Industrial Engineering Conference,* Toronto, Canada, 1989.

Clausing, D. P. and Ragsdell, K. M. (1984), "The Efficient Design and Development of Medium and Light Machinery Employing State-of-The-Art Technology," in *International Symposium on Design and Synthesis,* Tokyo, July 11–13.

Clausing, D. P. (1994), *Total Quality Development: A Step by Step Guide to World-Class Concurrent Engineering,* ASME Press, New York.

Cohen, L. (1988), "Quality Function Deployment and Application Perspective from Digital Equipment Corporation," *National Productivity Review, 7,* 3, 197–208.

Cohen, L. (1995), *Quality Function Deployment: How to Make QFD Work for You,* Addison-Wesley, Reading, MA.

Cochran, W. and Cox, G. (1957), *Experimental Design,* Wiley, New York.

Crew, K. (1998), "Design for Manufacturability—Assembly Guidelines," DRM Associates, http://www/npd-solutions.com/dfmguidelines.html.

Crosby, P. (1979), *Quality is Free,* McGraw-Hill, New York.

Dadd, P. and Woodford, D., http://www.gamedev.net/reference/articles/article1832.asp.

Deming, W. E. (1986), *Out of the Crisis,* MIT Center for Advanced Engineering Studies, Cambridge, MA.

Dewhurst, P. (1989), “Cutting Assembly Costs with Molded Parts,” *Machine Design,* July.

Dewhurst, P. and Blum, C. (1989), Supporting Analyses for the Economic Assessment of Die Casting in Product Design, *Annals of CIRP, 28,* 1, 161.

Dixon, J. R. (1966), *Design Engineering: Inventiveness, Analysis, and Decision Making,* McGraw-Hill, New York.

Domb, E. and Dettner, W. (1999), “Breakthrough Innovation in Conflict Resolution: Marrying Triz and the Thinking Process,” www.triz-journal.com/archives/1999.

Domb, E. (1997), “Finding the Zones of Conflict: Tutorial,” www.triz-journal.com/archives/1997.

Dorf, R. C. and Bishop, R. H. (2000), *Modern Control Systems,* 9th ed., Prentice Hall, Upper Saddle River, NJ.

Dovoino, I. (1993), “Forecasting Additional Functions in Technical Systems,” in *Proceeding of ICED—93,* vol. 1, pp. 247–277, The Hague.

Drury, C. G. (1992), “Inspection Performance,” in *Handbook of Industrial Engineers,* 2nd ed., pp. 2282–2314, edited by G. Salvendy, Wiley, New York.

Drury, C. G. and Kleiner, B. M. (1984), “A Comparison of Blink Aided and Manual Inspection Using Laboratory and Plant Subjects,” in *Proceedings of International Conference on Occupational Ergonomics,* Rexdale, Ontario, Human Factors Association of Canada, pp. 670–676.

Drury, C. G. and Kleiner, B. M. (1993), “Design and Evaluation of an Inspection Training Program,” *Applied Ergonomics, 24,* 2, 75–82.

El-Haik, B. (1996), *Vulnerability Reduction Techniques in Engineering Design,* Ph.D. Dissertation, Wayne State University, 1996.

El-Haik, B. (2005), *Axiomatic Quality: Integrating Axiomatic Design with Six Sigma Reliability and Quality Engineering,* Wiley, Hoboken, NJ.

El-Haik, B. and Yang, K. (1999), “The Components of Complexity in Engineering Design,” *IIE Transactions, 31,* 10, 925–934.

El-Haik, B. and Yang, K. (2000a), “An Integer Programming Formulation for the Concept Selection Problem with an Axiomatic Perspective (Part I): Crisp Formulation,” in Proceedings of the 1st International Conference on Axiomatic Design, MIT, Cambridge, MA, pp. 56–61.

El-Haik, B. and Yang, K. (2000b), “An Integer Programming Formulation for the Concept Selection Problem with an Axiomatic Perspective: Fuzzy Formulation,” Proceedings of the 1st international Conference on Axiomatic Design, MIT, Cambridge, MA, pp. 62–69.

Feigenbaum, A. (1983), *Total Quality Control,* 3rd ed., McGraw-Hill, New York.

Fisher, R. (1925), *The Design of Experiments,* Oliver & Boyd, London and Endinburgh.

Foo, G., Clancy, J. P., Kinney, L. E., and Lindemudler, C. R., “Design for Material Logistics,” *AT&T Technical Journal, 69,* 3, 61–67.

Fowlkes, W. Y. and Creveling, C. M. (1995), *Engineering Methods for Robust Product Design,* Addison-Wesley, Reading, MA.

Fredriksson, B. (1994), *Holistic Systems Engineering in Product Development,* Saab-Scania Griffin, Linkoping, Sweden.

Galbraith, J. R. (1973), *Designing Complex Organizations,* Addison-Wesley, Reading, MA.

Gardner, S., and Sheldon, D. F. (1995), “Maintainability as an Issue for Design,” *Journal of Engineering Design, 6,* 2, 75–89.

Gershenson, J. and Ishii, K. (1991), "Life Cycle Serviceability Design," in *Proceedings of ASME Conference on Design Theory and Methodology,* 1991.

Grossley, E. (1980), "Make Science a Partner," *Machine Design,* April 24.

Handerson, R. and Clark, K. B. (1990), "Architectural Innovation: The Reconfiguration of Existing Product Technologies and the Failure of Established Firms," *Administrative Science Quarterly, 35,* 1, 9–30.

Harry, M. J. (1994), *The Vision of 6-Sigma: A Roadmap for Breakthrough,* Sigma Publishing Company, Phoenix, AZ.

Harry, M. J. (1998), "Six Sigma: A Breakthrough Strategy for Profitability," *Quality Progress,* May, 60–64.

Harry, M. and Schroeder, R. (2000). *Six Sigma: The Breakthrough Management Strategy Revolutionizing the World's Top Corporations.* Doubleday, New York.

Hauser, J. R. and Clausing, D. (1988), "The House of Quality," *Harvard Business Review, 66,* 3, 63–73.

Hines, W. M. and Mongomery, D. C. (1990), *Probability and Statistics in Engineering and Management,* 3rd ed., Wiley, New York.

Hintersteiner, J. D. (1999a), "A Fractal Representation for Systems," in *International CIRP Design Seminar,* Enschede, the Netherlands, March 24–26.

Hintersteiner, J. D. and Nain, A. S. (1999b), "Integrating Software into Systems: An Axiomatic design Approach," in *3rd International Conference on Engineering Design and Automation,* Vancouver, B.C., Canada, August 1–4.

Holland, P. and Carvens, A. (1973), "Fractional Factorial Experimental Designs in Marketing Research," *Journal of Marketing Research,* 270–276.

Hornbeck, R. W. (1975), *Numerical Methods,* pp. 16–23, Quantum Publishers, New York.

Hubka, V. (1980), *Principles of Engineering Design,* Butterworth Scientific, London.

Hubka, V. and Eder, W. (1984), *Theory of Technical Systems,* Springer-Verlag, New York.

Kacker, R. N. (1985), "Off-line Quality Control, Parameter Design, and the Taguchi Method," *Journal of Quality Technology, 17,* 176–188.

Huang, G. Q. (Ed.) (1996), *Design for X: Concurrent Engineering Imperatives,* Chapman & Hall, London.

Hutchins, S. (2004), http://216.239.57.104/search?q=cache:WTPP0iD4WTAJ:cipm.ncsu.edu/symposium/docs/Hutchins_text.doc+product+multi-generation+plan&hl=en.

Ishikawa, K. (1968), *Guide to Quality Control,*

Johnson, R. A., and Wichern, D. W. (1982), *Applied Multivariate Statistical Analysis,* Prentice-Hall, Englwood Cliffs, NJ.

Juran, J. (1951), *Quality Control Handbook,* McGraw-Hill, New York.

Kacker, R. N. (1986), "Taguchi's Quality Philosophy: Analysis and Comment," *Quality Progress, 19,* 12, 21–29.

Kapur, K. C. (1988), "An Approach for the Development of Specifications for Quality Improvement," *Quality Engineering, 1,* 1, 63–77.

Kapur, K. C. (1991), "Quality Engineering and Tolerance Design," *Concurrent Engineering: Automation, Tools and Techniques,* 287–306.

Kapur, K. C. and Lamberson, L. R. (1977), *Reliability In Engineering Design,* Wiley, New York.

Kim, S., J., Suh, N. P., and Kim, S. G. (1991), "Design of Software System Based on Axiomatic Design," *Annals of the CIRP, 40,* 1, 165–70.

Knight, W. A. (1991), Design for Manufacture Analysis: Early Estimates of Tool Costs for Sintered Parts," *Annals of CIRP, 40,* 1, 131.

Kota, S. (1994), "*Conceptual Design of Mechanisms Using Kinematics Building Blocks—A Computational Approach,*" Final report, NSF Design Engineering Program, Grant # DDM 9103008, October, 1994.

Ku, H. H. (1966), "Notes on the Use of Propagation of Error Formulas," *Journal of Research of The National Bureau of Standards: C. Engineering and Instrumentation, 70,* 4, 263–273.

Kusiak, A. and Szczerbicki, E. (1993), "Transformation from Conceptual to Embodiment Design," *IIE Transactions, 25,* 4, 1993.

Lee, T. S. (1999), *The System Architecture Concept in Axiomatic Design Theory: Hypotheses Generation and Case-Study Validation,* M.S. Thesis, Department of Mechanical Engineering, MIT, Cambridge, MA.

Lerner, (1991), "Genrich Altshuller: Father of Triz," *Ogonek.* (Russian magazine), www.aitriz.org/downloads/ga40p.pdf.

Magrab, E. B. (1997), *Integrated Product and Process Design and Development,* CRC Press.

Magner, N., Welker, R. and Campbell, T. (1995), "The Interactive Effect of Budgetary Participation and Budget Favorability on Attitudes Toward Budgetary Decision Makers: A Research Note," *Accounting, Organization and Society, 20,* 7/8, 611–618.

Mann, D. and Domb, E. (1999), "40 Inventive (Business) Principles with Examples," www.triz-journal.com/archives/1999.

Maskell, B. H. (1991), *Performance Measurement for World Class Manufacturing,* Productivity Press.

Matousek, R. (1957), *Engineering Design: A Systematic Approach,* Lackie & Son, London.

McClave, J., Benson, P. G., and Sincich, T. (1998), *Statistics for Business and Economics,* 7th ed., Prentice-Hall, Upper Saddle River, NJ.

McCord, K. R. and Eppinger, S. D. (1993), "Managing the Integration Problem in Concurrent Engineering," MIT Sloan School of Management working paper 359-93-MSA.

Miles, B. L. (1989), "Design for Assembly: A Key Element Within Design for Manufacture," *Proceedings of IMechE, Part D, Journal of Automobile Engineering,* 203, 29–38.

Mizuno, S. and Akao, Y. (1978), Quality Function Deployment: A Company Wide Approach (in Japanese), JUSE Press.

Mizuno, S. and Akao, Y. (1994), *QFD: The Customer-Driven Approach to Quality Planning and Deployment* (trans. by G. H. Mazur), Asian Productivity Organization.

Montgomery, D. (1991), *Design and Analysis of Experiments,* 3rd ed., Wiley, New York.

Montgomery, D. (1997), *Design and Analysis of Experiments,* 4th ed., Wiley, New York.

Mostow, J. (1985), "Toward Better Models of the Design Process," *AI Magazine,* 44–57.

Mughal, H., and Osborne, R. (1995), "Design for Profit," *World Class Design to Manufacture, 2,* 5, 160–26.

Myres, J. D. (1984), *Solar Applications In Industry and Commerce,* Prentice-Hall, Englewood Cliffs, NJ.

Nair, V. N. (1992), "Taguchi's Parameter Design: A Panel Discussion," *Technometrics, 34,* 127–161.

Navichandra, D. (1991), Design for Environmentality," in *Proceedings of ASME Conference on Design Theory and Methodology,* New York.

Neibel, B. W. and Baldwin, E. N. (1957), *Designing for Production,* Irwin, Homewood, IL.

Nordlund, M. (1996), *An Information Framework for Engineering Design Based on Axiomatic Design,* Doctoral Thesis, The Royal Institute of Technology (KTH), Department of Manufacturing Systems, Stockholm, Sweden.

Nordlund, M., Tate, D., and Suh, N. P. (1996), "Growth of Axiomatic Design through Industrial Practice," in *3rd CIRP Workshop on Design and Implementation of Intelligent Manufacturing Systems,* Tokyo, Japan, June 19–21, pp. 77–84.

O'Grady, P. and Oh, J. (1991), "A Review of Approaches to Design for Assembly," *Concurrent Engineering, 1,* 5–11.

Owens, D. A. (2004), http://216.239.57.104/search?q=cache:mG3a-1wdz4sJ:mba.vanderbilt.edu/david.owens/Papers/2000%2520Owens%2520Status%2520in%2520Design%2520Teams%2520DMJ.pdf+motivating+design+teams&hl=en.

Papalambros, P. Y. and Wilde, D. J. (1988), *Principles of Optimal Design,* Cambridge University Press, Cambridge, UK.

Park, R. J. (1992), *Value Engineering,* R.J. Park & Associates, Inc., Birmingham, MI.

Pech, H. (1973), *Designing for Manufacture,* Pitman & Sons, London.

Penny, R. K. (1970), "Principles of Engineering Design," *Postgraduate,* 46, 344–349.

Phadke, M. S. (1989), *Quality Engineering Using Robust Design,* Prentice-Hall, Englewood Cliffs, NJ.

Phal, G. and Beitz, W. (1988), *Engineering Design: A Systematic Approach,* Springier-Verlag, New York.

Pimmler, T. U. and Eppinger, S. D. (1994), "Integration Analysis of Product Decomposition," *Design Theory and Methodology, 68,* 343–351.

Plackett, R. L. and Burman, J. P. (1946), "The Design of Optimum Multifactorial Experiments," *Biometrika, 33,* 305–325.

Pugh, S. (1991), *Total Design: Integrated Methods for Successful Product Engineering,* Addison-Wesley, Reading, MA.

Pugh, S. (1996) in *Creating Innovative Products Using Total Design,* edited by Clausing and Andrade, Addison-Wesley, Reading, MA.

Ramachandran, et. al. (1992), "Initial Design Strategies for Iterative Design," *Research in Engineering Design, 4,* 3.

Ross (1988)

Ruchti, B. and Livotov, P. (2001), "TRIZ-based Innovation Principles and a Process for Problem Solving in Business and Management," in *Proceedings of the European TRIZ Association,* November.

Rantanen, K. (1988), "Altshuler's Methodology in Solving Inventive Problems," in *Proceedings of ICED—88,* Budapest, 23–25 August.

Reinderle, J. R. (1982), *Measures of Functional Coupling in Design,* Ph.D. Dissertation, Massachusetts Institute of Technology, June, 1982.

Reklaitis, J. R., Ravindran, A., and Ragsdell, K. M. (1983), *Engineering Design Optimization,* Wiley, New York.

Rinderle, J. R. (1982), *Measures of Functional Coupling in Design,* Ph.D. Dissertation, Massachusetts Institute of Technology.

Rother M. and Shook, J. (2000), *Learning to See: Value Stream Mapping to Create Value and Eliminate Muda,* The Lean Enterprise Institute.

Sackett, P. and Holbrook, A. (1988), "DFA as a Primary Process Decreases Design Deficiencies," *Assembly Automation, 12,* 2, 15–16.

Sackett, P. and Holbrook, A. (1990), "DFA as a Pimary Process Decreases Deficiencies," in Allen, C. W. (Ed.), *Simultaneous Engineering,* pp. 152–155, SME, Detroit.

Shannon, C. E. (1948), "The Mathematical Theory of Communication," *Bell System Technical Journal,* 27, 379–423, 623–656.

Sheldon, D. F., Perks, R., Jackson, M., Miles, B. L., and Holland, J. (1990), "Designing for Whole-life Costs at the Concept Stage," in *Proceedings of ICED,* Heurista, Zurich.

Shewhart, W. A. (1931), *Economic Control of Manufactured Product,*

Simon, H. A. (1981), *The Science of the Artificial,* 2nd ed., MIT Press, Cambridge, MA.

Smith, G. and Browne, G. J. (1993), "Conceptual Foundation of Design Problem Solving," *IEEE Transaction on Systems, Man, and Cybernetics, 23,* 5, September/October.

Snee, R. D., Hare, L. B., Trout, S. B. (Eds.), (1985), "Experimenting with Large Number of Variables," in *Experiments in Industry: Design, Analysis, and Interpretation of Results,* ASQC Press.

Sohlenius, G., Kjellberg, A., and Holmstedt, P. (1999), "Productivity System Design and Competence Management," in *Proceedings of World XIth Productivity Congress,* Edinburgh.

Spotts, M. F. (1973), "Allocation of Tolerance to Minimize Cost of Assembly," *Transactions of the ASME,* 762–764.

Srinivasan, R. S., and Wood, K. L. (1992), "A Computational Investigation into the Structure of Form and Size Errors Based on Machining Mechanics," in *Advances in Design Automation,* Phoenix, AZ, pp. 161–171.

Steinberg, L., Langrana, N. Mitchell, T., Mostow, J. and Tong, C. (1986), "A Domain Independent Model of Knowledge-Based Design," Technical report AI/VLSI Project working paper no. 33, Rutgers University.

Steward, D. V. (1981), *Systems Analysis and Management: Structure, Strategy, and Design,* Petrocelli Books, New York.

Suh, N. P. (1984), "Development of the Science Base for the Manufacturing Field Through the Axiomatic Approach," *Robotics & Computer Integrated Manufacturing, 1.*

Suh, N. P. (1990), *The Principles of Design,* Oxford University Press, New York.

Suh, N. (1995), "Design and Operation of Large Systems," *Journal of Manufacturing Systems, 14,* 3.

Suh, N. P. (1996), "Impact of Axiomatic Design," in *3rd CIRP Workshop on Design and the Implementation of Intelligent Manufacturing Systems,* Tokyo, Japan, pp. 8 17, June, 19–22.

Suh, N. P. (1997), "Design of Systems," *Annals of CIRP, 46,* 1, 75–80.

Suh, N. P. (2001), *Axiomatic Design: Advances and Applications,* Oxford University Press.

Sushkov, V. V. (1994), "Reuse of Physical Knowledge in Creative Design," ECAI-94 Workshop Notes, Amsterdam, August.

Suzue, T. and Kohdate, A. (1988), *Variety Reduction Programs: A Production Strategy for Product Diversification,* Productivity Press, Cambridge, MA.

Swartz, J. B. (1996), *Hunters and the Hunted,*

Swenson, A. and Nordlund, M. of Saab (1996), "*Axiomatic Design of Water Faucet,*" unpublished report, Linkoping, Sweden.

Taguchi, G., (1957), *Design of Experiments,*

Taguchi, G. (1986), *Introduction to Quality Engineering,* UNIPUB/Kraus International Publications, White Plains, NY.

Taguchi, G., and Wu, Y. (1989), *Taguchi Method Case Studies from the U.S. and Europe,* Quality Engineering Series, Vol. 6, ASI.

Taguchi, G., Elsayed, E., and Hsiang, T. (1989), *Quality Engineering in Production Systems,* McGraw-Hill, New York.

Tate, D., and Nordlund, M. (1998), "A Design Process Roadmap As a General Tool for Structuring and Supporting Design Activities," *SDPS Journal of Integrated Design and Process Science, 2,* 3, 11–19, 1998.

Tate, K. and Domb, E. (1997), "40 Inventive Principles with Examples," www.triz-journal.com/archives/1997.

Tayyari, F. (1993), "Design for Human Factors,"in *Concurrent Engineering,* pp. 297–325, edited by H. R. Parsaei and W. G. Sullivan, Chapman & Hall, London.

Thackray, R. (1994), *Correlations of Individual Differences in Non-destructive Inspection Performance, Human Factors in Aviation Maintenance,* Phase 4, Vol. 1, Program Report, DOT/FAA/Am-94/xx, National Technical Information Service, Springfield, VA.

Tsourikov, V. M. (1993), "Inventive Machine: Second Generation," Artificial Intelligence & Society, 7, 62–77.

Tumay, K. http://www.reengineering.com/articles/janfeb96/sptprochr.htm, Business Process Simulation: Matching Processes with Modeling Characteristics, ER Spotlight.

Ullman, D. G. (1992), *The Mechanical Design Process,* McGraw-Hill, New York.

Ulrich, K. T. and Tung, K. (1994), "Fundamentals of Product Modularity," in *ASME Winter Annual Meeting,* DE Vol. 39, Atlanta, pp. 73–80.

Ulrich, K. T. and Eppinger, S. D. (1995), *Product Design and Development,* McGraw-Hill, New York.

Ulrich, K. T. and Seering, W. P. (1987), "Conceptual Design: Synthesis of Systems Components," in *Intelligent and Integrated Manufacturing Analysis and Synthesis,* American Society of Mechanical Engineers, New York, pp. 57–66.

Ulrich, K. T. and Seering, W. P. (1988), "Function Sharing in Mechanical Design," in *7th National Conference on Artificial Intelligence,* AAAI-88, Minneapolis, MN, August 21–26.

Ulrich, K. T. and Seering, W. P. (1989), "Synthesis of Schematic Description in Mechanical Design," *Research in Engineering Design, 1,* 1.

Vasseur, H., Kurfess, T., and Cagan, J. (1993), "Optimal Tolerance Allocation For Improved Productivity," in *Proceedings of the 1993 NSF Design and Manufacturing Systems Conference,* Charlotte, NC, pp. 715–719.

Verno, A., and Salminen (1993), "Systematic Shortening of The Product Development Cycle," presented at the International Conference on Engineering Design, The Hague, Netherlands.

Wagner, T. C. and Papalambros, P. (1993), "*A General Framework for Decomposition Analysis in Optimal Design,*" presented at the 23rd ASME Design Automation Conference, Albuquerque, New Mexico, September.

Wang, J., and Ruxton, T. (1993), "Design for Safety of Make-to-Order Products," in *National Design Engineering Conference of ASME,* 93-DE-1.

Warfield, J. and Hill, J. D. (1972), *A United Systems Engineering Concept,* Battle Monograph, No. 1, June.

Weaver, W. (1948), "Science and Complexity," *American Scientist,* 36, 536–544.

Wheeler, D. (1995). *Advanced Topics in Statistical Process Control: The Power of Shewhart Charts,* SPC Press.

Wheeler, D., and Chambers, D. S. (1992), *Understanding Statistical Process Control,* 2nd ed., SPC Press.

Wei, C., Rosen, D., Allen, J. K., and Mistree, F. (1994), "Modularity and the Independence of Functional Requirements in Designing Complex Systems," DE-Vol. 74, Concurrent Product Design, ASME.

Wirth, N. (1971), "Program Development by Stepwise Refinement," *Communications of the ACM, 14,* 221–227.

Wood, G. W., Srinivasan, R.S., Tumer, I. Y., and Cavin, R. (1993), "Fractal-Based Tolerancing: Theory, Dynamic Process Modeling, Test Bed Development, and Experiment," in *Proceedings of the 1993 NSF Design and Manufacturing Systems Conference,* Charlotte, NC, pp. 731–740.

Woodford, D. (2004), http://www.isixsigma.com/library/content/c020819a.asp.

Yang, K. and El-Haik, B. (2003), *Design for Six Sigma: A Roadmap for Produvct Excellemce,* McGraw-Hill, New York.

Yang, K. and Trewn, J. (1999), "A Treatise on the Mathematical Relationship Between System Reliability and Design Complexity," *Proceedings of IERC.*

Zaccai, G. (1994), "The New DFM: Design for Marketability," *World-Class Manufacture to Design, 1,* 6, 5–11.

Zenger, D. and Dewhurst, P. (1988), *Early Assessment of Tooling Costs in the Design of Sheet Metal Parts,* Report 29, Department of Industrial and Manufacturing Engineering, University of Rhode Island.

Zakarian, A. (2002), Manufacturing Systems Course Material, University of Michigan, Dearborn.

Zhang, C. T. (2003), "40 Inventive Principles with Applications in Service Operations Management," www.triz-journal.com/archives/2003.

Zhang, H. C. and Huq, M. E. (1994), "Tolerancing Techniques: The State-of-the-Art," *International Journal of Production Research, 30,* 9, 2111–2135.

Zwicky, F. (1984), *Morphological Analysis and Construction,* Wiley Interscience, New York.

INDEX